Dear MyCopy Customer,

This Springer book is a monochrome print version of the eBook to which your library gives you access via SpringerLink. It is available to you at a subsidized price since your library subscribes to at least one Springer eBook subject collection.

Please note that MyCopy books are only offered to library patrons with access to at least one Springer eBook subject collection. MyCopy books are strictly for individual use only.

You may cite this book by referencing the bibliographic data and/or the DOI (Digital Object Identifier) found in the front matter. This book is an exact but monochrome copy of the print version of the eBook on SpringerLink.

# Structure–Activity Relationships in Environmental Sciences

# Chapman & Hall Ecotoxicology Series

Series Editors

**Michael H. Depledge**
Director and Professor of Ecotoxicology, Plymouth Environmental Research Centre, University of Plymouth, UK

**Brenda Sanders**
Associate Professor of Physiology, Molecular Ecology Institute, California State University, USA

In the last few years emphasis on the environmental sciences has shifted from direct toxic threats to humans towards more general concerns regarding pollutant impacts on animals and plants, ecosystems and indeed on the whole biosphere. Such studies have led to the development of the scientific discipline of ecotoxicology. Throughout the world socio-political changes have resulted in increased expenditure on environmental matters. Consequently, ecotoxicological science has developed extremely rapidly, yielding new concepts and innovative techniques that have resulted in the identification of an enormous spectrum of potentially toxic agents. No single book or scientific journal has been able to keep pace with these developments.

This series of books provides detailed reviews of selected topics in ecotoxicology. Each book includes both factual information and discussions of the relevance and significance of the topic in the broader context of ecotoxicological science.

*Already Published*

**1. Animal Biomarkers as Pollution Indicators**
D.B. Peakall
Hardback (0 412 40200 9), 292 pages

**2. Ecotoxicology in Theory and Practice**
V.E. Forbes and T.L. Forbes
Hardback (0 412 43530 6), 262 pages

**3. Interconnections Between Human and Ecosystem Health**
Edited by R.T. DiGiulio and E. Monosson
Hardback (0 412 62400 1), 296 pages

**4. ECOtoxicology**
**Ecological dimensions**
Edited by D.J. Baird, P.E.I. Douben, P. Greig-Smith and L. Maltby
Hardback (0 412 75470 3), paperback (0 412 75490 8), 104 pages

**5. Ecological Principles for Risk Assessment of Contaminants in Soils**
Edited by N.M. van Straalen and H. Løkke
Hardback (0 412 75900 4)

# Structure–Activity Relationships in Environmental Sciences

**Monika Nendza**
Analytisches Laboratorium
Luhnstedt, Germany

SPRINGER-SCIENCE+BUSINESS MEDIA, B.V

First edition 1998

© 1998 Springer Science+Business Media Dordrecht
Originally published by Chapman & Hall in 1998
Softcover reprint of the hardcover 1st edition 1998
MyCopy version of the original edition 1998
Typeset in 10/12pt Times by Saxon Graphics Ltd, Derby

DOI 10.1007/978-1-4615-5805-7



∞ Printed on permanent acid-free text paper, manufactured in
accordance with ANSI/NISO Z39.48-1992 and ANSI/NISO Z39.48-1984
(Permanence of Paper).

Springer.com/mycopy

# Contents

# *Series foreword*

Ecotoxicology is a relatively new scientific discipline. Indeed, it might be argued that it is only during the last 5–10 years that it has come to merit being regarded as a true science, rather than a collection of procedures for protecting the environment through management and monitoring of pollutant discharges into the environment. The term 'ecotoxicology' was first coined in the late sixties by Prof. Truhaut, a toxicologist who had the vision to recognize the importance of investigating the fate and effects of chemicals in ecosystems. At that time, ecotoxicology was considered a sub-discipline of medical toxicology. Subsequently, several attempts have been made to portray ecotoxicology in a more realistic light. Notably, both Moriarty (1988) and F. Ramade (1987) emphasized in their books the broad basis of ecotoxicology, encompassing chemical and radiation effects on all components of ecosystems. In doing so, they and others have shifted concern from direct chemical toxicity to humans, to the far more subtle effects that pollutant chemicals exert on natural biota. Such effects potentially threaten the existence of life on earth.

Although I have identified the sixties as the era when ecotoxicology was first conceived as a coherent subject area, it is important to acknowledge that studies that would now be regarded as ecotoxicological are much older. Wherever people's ingenuity has led them to change the face of nature significantly, it has not escaped them that a number of biological consequences, often unfavourable, ensue. Early waste disposal and mining practices must have alerted the practitioners to effects that accumulated wastes have on local natural communities; for example, by rendering water supplies undrinkable or contaminating agricultural land with toxic mine tailings. As activities intensified with the progressive development of human civilizations, effects became even more marked, leading one early environmentalist, G. P. Marsh, to write in 1864: 'The ravages committed by Man subvert the relations and destroy the balance that nature had established'.

But what are the influences that have shaped the ecotoxicological studies of today? Stimulated by the explosion in popular environmentalism in the sixties, there followed in the seventies and eighties a tremendous increase in the creation of legislation directed at protecting the environment. Furthermore, political restructuring, especially in Europe, has led to the widespread implementation of this legislation. This currently involves enormous numbers of environmental managers, protection officers, technical staff and consultants. The ever-increasing use of new chemicals places further demands on government agencies and industries who are required by law to evaluate potential toxicity and likely environmental impacts. The environmental manager's problem is that he needs rapid answers to current questions concerning a very broad range of chemical effects and also information about how to control discharges, so that legislative targets for *in situ* chemical levels can be met. It is not surprising, therefore, that he may well feel frustrated by more research-based ecotoxicological scientists who constantly question the relevance and validity of current test procedures and the data they yield. On the other hand, research-based ecotoxicologists are often at a loss to understand why huge amounts of money and time are expended on conventional toxicity testing and monitoring programmes, which may satisfy legislative requirements, but apparently do little to protect ecosystems from long-term, insidious decline.

It is probably true to say that until recently ecotoxicology has been driven by the managerial and legislative requirements mentioned above. However, growing dissatisfaction with laboratory-based tests for the prediction of ecosystem effects has enlisted support for studying more fundamental aspects of ecotoxicology and the development of conceptual and theoretical frameworks.

Clearly, the best way ahead for ecotoxicological scientists is to make use of the strengths of our field. Few sciences have at their disposal such a well-integrated input of effort for people trained in ecology, biology, toxicology, chemistry, engineering, statistics, etc. Nor have many subjects such overwhelming support from the general public regarding our major goal: environmental protection. Equally important, the practical requirements of ecotoxicological managers are not inconsistent with the aims of more academically-orientated ecotoxicologists. For example, how better to validate and improve current test procedures than by conducting parallel basic research programmes *in situ* to see if controls on chemical discharges really do protect biotic communities?

More broadly, where are the major ecotoxicological challenges likely to occur in the future? The World Commission on Environment and Development estimates that the world population will increase from *c.* 5 billion at present to 8.2 billion by 2025. 90% of this growth will occur in developing countries in subtropical and tropical Africa, Latin America and Asia. The introduction of chemical wastes into the environment in these regions is likely to escalate dramatically, if not due to increased industrial output, then due to the use of pesticides and fertilizers in agriculture and the disposal of damaged, unwanted or obsolete consumer goods supplied from industrialized countries. It may be many years before resources become available to implement effective waste-recycling programmes in countries with poorly developed infra-structures, constantly threatened by natural disasters and poverty. Furthermore, the fate, pathways and effects of chemicals in subtropical and tropical environments have barely begun to be addressed. Whether knowledge gained in temperate ecotoxicological studies is directly applicable in such regions remain to be seen.

The Chapman & Hall Ecotoxicology Series brings together expert opinion on the widest range of subjects within the field of ecotoxicology. The authors of the books have not only presented clear, authoritative accounts of their subject areas, but have also provided the reader with some insight into the relevance of their work in a broader perspective. The books are not intended to be comprehensive reviews, but rather accounts which contain the essential aspects of each topic for readers wanting a reliable introduction to a subject or an update in a specific field. Both conceptual and practical aspects are considered. The Series will be constantly added to and books revised to provide a truly contemporary view of ecotoxicology. I hope that the Series will prove valuable to students, academics, environmental managers, consultants, technicians, and others involved in ecotoxicological science throughout the world.

Michael Depledge
*University of Plymouth, UK*

# *Preface and acknowledgements*

This book is intended to expound the possibilities and limitations of quantitative structure–activity relationships (QSARs) in environmental sciences. The related data and background material are scattered in textbooks and review articles. By assembling the available material and supplementing with my own experience from more than 10 years of practice with QSARs in assessments of environmental hazards and effects, my intention is to give an overview about the method for beginners as well as a compilation of validated models for specialists. For the latter task the chapters had to be self-consistent, without too much reference to other parts of the book; hence some topics are mentioned in different sections verbatim. This should not discourage the newcomer from reading the whole book, because only the important things are repeated.

Although experience shows that most scientists like to stay within the framework of their own training, QSARs require a multidisciplinary approach. The combination of techniques from such varied disciplines as chemistry, toxicology, medicine, pharmacy, ecology, biology, physics and statistics fills the toolbox of QSARs in environmental sciences. Being myself a pharmacist and toxicologist with training in some of the fields mentioned above, I am used to looking for and making use of the tools developed in other faculties. Sharing the expertise of colleagues with a different scientific background is a prerequisite for the understanding and the eventual solution of the problems associated with environmentally relevant chemicals.

I thank all my colleagues for their co-operation and the fruitful discussions that led to this book, especially my teacher Professor Dr J. K. Seydel (Borstel, Germany), Dr G. D. Veith (Duluth, Minnesota, USA) and Professor Dr W. Klein (Schmallenberg, Germany), who provided me with opportunities to work within considerable QSAR projects and to gain experience in this field. Special thanks are due to the unknown reviewer, who did a great deal to improve the manuscript. Any remaining bugs are, however, my own responsibility and I welcome any criticism and suggestions from readers that may help to improve forthcoming editions of this book. Last but not least I thank my husband who, being a chemist, provided technical advice on chemical problems, prepared most of the figures and tables and gave encouragement when I needed it to keep going. The book was written besides the daily work in our laboratory, the Analytical Laboratory Luhnstedt. I thank everybody who took over some of my duties to give me the time to work on this volume.

Luhnstedt, November 1996

# *Abbreviations*

| | |
|---|---|
| Å | Ångstrom ($10^{-10}$ m) |
| AChE | acetylcholine esterase |
| atm | atmosphere ($1.013 \times 10^6$ Pa) |
| ATP | adenosine triphosphate |
| BCF | bioconcentration factor |
| $BOD_n$ | biological oxygen demand within $n$ days |
| $c$ | concentration |
| °C | degree Celsius |
| CMC | critical micelle concentration |
| COD | chemical oxygen demand |
| d | day |
| DA | discriminant analysis |
| DF | degrees of freedom |
| $\Delta G$ | Gibbs free energy |
| $\Delta H$ | free enthalpy |
| DNA | deoxyribonucleic acid |
| DOC | dissolved organic carbon |
| $\Delta S$ | free entropy |
| $EC_{50}$ | median effective concentration |
| $ED_{50}$ | median effective dose |
| EINECS | European Inventory of Existing Chemical Substances |
| $E_S$ | Taft steric substituent constant |
| $\eta$ | viscosity |
| EU | European Union |
| FA | factor analysis |
| $\gamma$ | activity coefficient |
| H | Henry's law constant |
| h | hour |
| HOMO | energy of the highest occupied molecular orbital |
| HPLC | high-pressure liquid chromatography |
| HPVC | high-production volume chemical |
| i.p. | intraperitoneal |

| | |
|---|---|
| IR | infrared spectroscopy |
| $K$ | equilibrium constant |
| $k$ | rate constant |
| $K_d$ | distribution coefficient |
| $K_{oc}$ | soil sorption coefficient, normalized for organic carbon content |
| l | litre |
| $\lambda$ | wavelength |
| $LC_{50}$ | median lethal concentration |
| $LD_{50}$ | median lethal dose |
| LFER | linear free-energy relationship |
| LOEC | lowest observed effect concentration |
| $\log P_{ow}$ | logarithm of the l-octanol/water partition coefficient |
| LSER | linear solvation energy relationship |
| LUMO | energy of the lowest unoccupied molecular orbital |
| MITI | Ministry of International Trade and Industry, Japan |
| MLR | multiple linear regression |
| MR | molar refractivity |
| MS | mass spectroscopy |
| $\mu$ | chemical potential |
| MW | molecular weight |
| $N$ | Avogadro's number ($6.02 \times 10^{23}$ molecules/mol) |
| NMR | nuclear magnetic resonance spectroscopy |
| NOEC | no observed effect concentration |
| OC | organic carbon |
| OECD | Organisation for Economic Co-operation and Development |
| $P$ | partition coefficient |
| Pa | Pascal |
| $\pi$ | lipophilicity substituent constant |
| p.o. | per os (by mouth) |
| PAH | polycyclic aromatic hydrocarbon |
| PC | principal component |
| PCA | principal component analysis |
| pH | negative decadic logarithm of hydrogen ion concentration |
| $pK_a$ | negative decadic logarithm of acid/base dissociation constant |
| PLS | partial least squares |
| $p_v$ | vapour pressure |
| QSAR | quantitative structure–activity relationship |
| $r$ | correlation coefficient |
| $R$ | gas constant |
| rp-HPLC | reversed-phase high-pressure liquid chromatography |
| $r^2$ | explained variance |
| $s$ | standard deviation |
| $\sigma$ | Hammett substituent constant |

| | |
|---|---|
| $S_w$ | solubility in water |
| $T$ | temperature (Kelvin) |
| $t$ | time |
| $t_{1/2}$ | half-life time |
| $T_b$ | boiling point |
| TLC | thin-layer chromatography |
| $T_m$ | melting point |
| US EPA | United States Environmental Protection Agency |
| UV | ultraviolet spectroscopy |
| $V$ | volume, molar volume |
| $x$ | mole fraction |

## Errata

Page 99, Table 4.3. Equation 24 should read:
$$S_w = 55.23x^w/(1 - x^w)$$
and the respective footnote for this equation should read:
$$x^w = x^{oc}(\gamma^{oc}/\gamma^w)$$

Page 101, Table 4.4. Equation 5 should read:
$$\ln p_v \, (\text{atm}) = -(4.4 + \ln T_b) \, [1.803 \, (T_b/T-1) - 0.803 \ln (T_b/T) - 6.8 \, (T_m/T-1)]$$

# *Introduction*

Environmental hazard and risk assessment of chemical substances requires comprehensive information on the exposure, fate and ecotoxicology of the contaminants. Complete data sets are, however, rarely available – not even for high-production volume chemicals (HPVC). One reason for these deficiencies is that testing capacities are limited, which impedes the thorough experimental investigation of all of the more than 100 000 existing and new chemicals. The necessary experimental facts are unavailable for a preliminary screening and ranking of most potentially hazardous compounds. To fill at least some of the data gaps for the time being, various modelling techniques are used to provide sufficiently accurate substitutes until the full measurements become available (e.g. Lipnick, 1985a, 1995; Sheehan *et al.*, 1985; ECETOC, 1986; Klein, Klein and Lange, 1988; Hermens, 1989; Karcher and Devillers, 1990; Hermens and Opperhuizen, 1991; OECD, 1992a, 1993b, 1995; Donkin, 1994; Verhaar *et al.*, 1994; Clements *et al.*, 1995; Feijtel, 1995; Hermens *et al.*, 1995; Lange and Vormann, 1995; Nendza and Hermens, 1995; Zeeman *et al.* 1995; European Commmission, 1996). Based on empirical extrapolations, the models can be used to estimate the parameters relating to the fate and effects of chemicals and hence identify contaminants of special environmental concern. The basic idea is to obtain a ranking of potentially hazardous pollutants; the priority compounds can then be subjected to detailed testing. In this way, the limited resources for experimental investigations can be directed effectively to the chemicals that are most likely to have an environmental impact.

Chemical modelling is based on the premise that a compound's structure, molecular constitution and charge distribution determine all its properties. As an important consequence, it is implied that similar chemical structures have similar properties and behaviour. Furthermore, it is presumed that a compound's properties and reaction rates can be extrapolated from suitable analogues. A prerequisite to describing and interpreting the relevant interactions that result in the noxious effects is the knowledge of the structure of the acting chemical species as well as of the constitution of the receiving environmental targets. The ultimate rationale is the establishment of causal relationships between features of the chemical structures on the one hand

and the observed effects or activities on the other, and the use of these relationships to predict the behaviour of chemicals yet untested. The quantification of the dependence of activity measures on structural descriptors is termed quantitative structure–activity relationships (QSARs). Throughout this volume, the terminology QSAR will be used uniformly with different types of activity, such as for structure–property, structure–reactivity or structure–toxicity relationships.

The history of QSAR research dates back to the early days of modern life science. In the second half of the nineteenth century, during the course of medicinal and pharmacological research on therapeutic agents, relationships between chemical properties and biological effects were recognized. Observations on the extent of toxic effects of organic substances were associated with, for example, the chain length or the water solubility of the compounds. Independent studies by Meyer (1899) and Overton (1897) revealed that an increase in the oil/water partition coefficients parallels an increase in the narcotic effects of various chemicals (Figure I.1).

The physiological explanation for these observations was thought to be the affinity of the narcotics to lipid structures in the organisms, especially in brain and nerve cells. From these observations, inferences were made about the mode of action of narcotic compounds, and the relevance of lipid/water par-

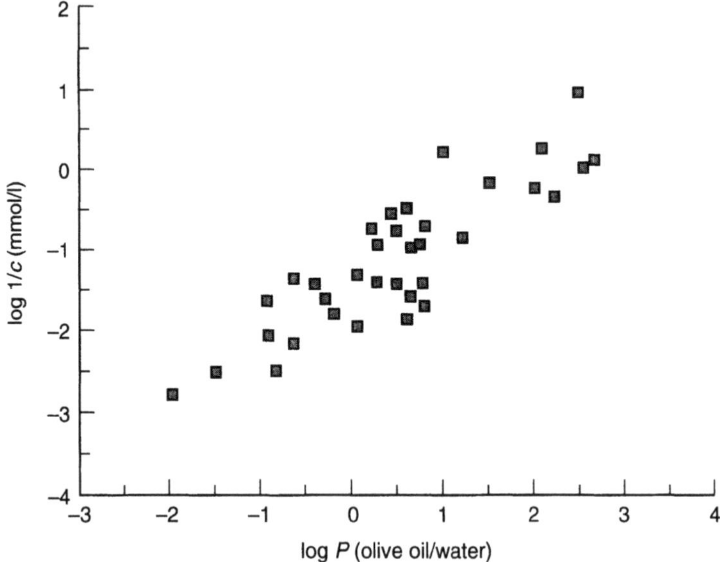


**Figure I.1** Representation of a classic QSAR: relationship between Overton's data on the lowest concentration of test chemicals producing complete narcosis in tadpoles (log l/c in mmol/l) and their olive oil/water partition coefficient (log P); data from Lipnick (1989a).

titioning. In recognition of Meyer's and Overton's pioneering work, correlations between biological activity and partition coefficients are now referred to as Meyer–Overton relationships (Lipnick, 1986, 1989a,b). Subsequent studies questioned the relevance of such relationships, however, because further research revealed that other physico-chemical properties also correlate with biological effects. The receptor concept established by Ehrlich (1910) especially appeared incompatible with the Meyer–Overton theory until it was realized that biological activity may be due to both unspecific distribution processes and receptor interactions. In 1939, Ferguson provided the thermodynamic basis for the principle of Meyer–Overton relationships, such that an equilibrium is established between the chemical concentration, or rather its chemical activity, in the outer water phase and in the inner biophase, where the actual effect is evoked.

Any chemical has to undergo multiple distribution between lipid and water phases, such as membrane passages and adsorption to macromolecules, until it reaches the site of action in an organism. The equilibrium rates of these partitioning processes determine the concentrations of the chemicals in the various compartments of the organism – hence also at the target site – and in this way may be crucial for the extent of the effects. This is principally the same situation as with the processes investigated in physical organic chemistry on the rates of chemical reactions. Consequently, it appeared reasonable to apply the rationales of linear free-energy relationships (LFERs) derived for chemical reactivities to QSAR studies. The analogy between the chemical reactions and the chemical–biosystem interactions becomes evident if both types of processes are considered to be rate-limited, tending to achieve steady states. With regard to this aspect, the Hammett equation (Hammett, 1940) has to be regarded as fundamental for QSAR studies (Hansch, 1993). In physical organic chemistry, the relative reactivities among homologous series of compounds under defined conditions are used to derive quantitative structure–reactivity relationships. Based on the observation of a linear dependency between rate constants and equilibrium constants – for example, the hydrolysis rate (rate constant $k$) of benzoic acid esters and the ionization constant (equilibrium constant $K$) of the corresponding acids – Hammett postulated that substituents contribute a constant effect to the reactivity of compounds ($\sigma$ Hammett):

$$\log k = f(\sigma)$$

The respective rate and equilibrium constants ($k$, $K$) are related to the changes in the free energy of the reactions, $\Delta\Delta G$. The linear Hammett relationships hence imply linear relationships between the free energies of the equilibrium and the rate processes, $\Delta G_{\text{equilibrium}}$ and $\Delta G_{\text{rate}}$. Such linearity should occur only if the changes in the free enthalpy, $\Delta\Delta H$, and in the free entropy, $\Delta\Delta S$, are constant and if a linear relation exists between $\Delta H$ and $\Delta S$. Even in those cases when LFERs have been obtained, these criteria are unlikely to be fulfilled. According to thermodynamic theory, such linearities are highly improbable, and even for the Hammett standard reaction (the ionization of *m*- and *p*-substituted benzoic acids in water at 25°C) it has been doubted that any of these requirements is satisfied (e.g. Sykes, 1982). It therefore has to be realized that interpretations of Hammett equations are based on empirical observations rather than on stringent theory. Although it is not exactly understood why the Hammett equations work, it is one of the most successful concepts in elucidating reaction mechanisms. In spite of their principal shortcomings, LFERs provide a comprehensive basis for structure–activity analyses. The Hammett equation with the electronic descriptor $\sigma$ was extended by a steric parameter, $E_S$ (Taft, 1956) and Hansch and Fujita (1964) provided a hydrophobicity term, $\pi$, such that an expansion of the LFERs became suitable for biological problems:

$$\log k = f(\sigma, \pi, E_S)$$

The application of this principal function to biological and environmental activity data involves searching for model parameters from physical organic systems to account for changes in three major properties among series of chemicals:

$$\text{activity} = f(\text{electronic properties, lipophilicity, steric properties})$$

The variation in effects between the compounds is ascribed to changes in the electronic, lipophilic and steric properties of the substances. The application of statistics, such as multiple regression analysis, to quantify the relationships between descriptors of chemical structures and biological activities was first proposed by Hansch and co-workers (Hansch *et al.*, 1963; Hansch and Fujita, 1964). Since then, QSARs have been widely applied in pharmaceutical, medicinal and agricultural chemistry (e.g. Hansch, 1968a,b; Martin, 1978; Seydel and Schaper 1979; Franke, 1980, 1984; Kubinyi, 1993), but their introduction to the environmental sciences was delayed due to a lack of understanding of the complex systems and feedback mechanisms in ecosystems. Regardless of the endpoint modelled, the QSARs for environmental parameters are based on the same principal methodologies as were developed for drug design.

In general, the interactions between xenobiotics and ecosystem constituents are reversible and tend to establish steady states conditioned by the rate-limiting steps. The reversibility applies, of course, only to the interactions at the molecular level, not to the (long-term) impacts on ecosystems. With respect to hazard and risk assessment, parameters regarding the distribution (e.g. bioconcentration and soil sorption), the reactivity (e.g. degradation) and the toxicity (e.g. aquatic species such as fish, crustaceans, algae) are considered. For the description of the chemicals, mostly structurally derived physico-chemical properties are used, among which the most prominent is $P_{ow}$, the 1-octanol/water partition coefficient. The log $P_{ow}$ parameter quantifies the partitioning of a compound between an aqueous and a lipophilic phase (1-octanol), which is considered to simulate natural hydrophobic phases (e.g. membranes). Computational methods for obtaining log $P_{ow}$ values are available for most classes of organic structures (Leo and Hansch, 1971; Rekker, 1977; Lyman, Reehl and Rosenblatt, 1990; MedChem 1989; ProLogP, 1992; LOGKOW, 1993). Assuming principally analogous processes for the partitioning in different systems (Collander, 1951) and with a readily available descriptor at hand, the first QSARs in environmental sciences concerned effects that can mechanistically be explained from distribution processes.

Bioconcentration is a classic example of the partitioning of organic pollutants between different environmental phases, such as water and aquatic organisms. The accumulation of chemicals in various environmental compartments may be relevant for the magnification of contaminants along foodchains as well as for increases in internal concentration levels that may eventually result in chronic effects. In 1974, Neely, Branson and Blau used partition coefficients to describe the bioconcentration factors (BCF) of organic chemicals in fish. For eight, mostly chlorinated, compounds of intermediate to high lipophilicity (log $P_{ow}$ 2.6–7.6), they obtained a linear correlation between log BCF and log $P_{ow}$, which was suitable for predicting the BCF values of three further compounds within reasonable limits. Linear dependencies between 1-octanol/water partition coefficients, water solubility and bioconcentration were reported by Chiou *et al.* in 1977. The general validity of such linear relationships has, however, been doubted. In analogy to observations in pharmacokinetics and drug design, a maximum uptake has to be assumed for substances with optimum size and lipophilicity. Non-linear relationships are to be expected when considering the entire log $P_{ow}$ range. This fact becomes evident when it is realized that with such linear QSAR models compounds of increasing lipophilicity have a steadily (i.e. infinitely) increasing tendency to bioaccumulate. This holds for certain groups of chemicals in a certain range, but not, for example, for apolar polymers, which have computed log $P_{ow}$ values that are extraproportionally high, yet they are not bioaccumulated (Tulp and Hutzinger, 1978).

Ecotoxicity modelling involves further parallel on-going processes: besides uptake into the organism, transport within the body and interactions with target structures can also be relevant. The early QSAR studies on this topic in the 1970s were limited to small data sets of homologous compounds, to mostly simple endpoints, or used non-interaction related descriptors, whose relevance for the described effects remains unclear from a mechanistic point of view. While investigating the environmental impact of effluents, Kopperman, Carlson and Caple (1974) studied the toxicity of 14 substituted phenols towards the crustacean *Daphnia magna*. The effects were correlated with descriptors of lipophilic and electronic substituent contributions. The resulting QSAR equation was used to interpret phenol toxicity with regard to structural differences (i.e. increased halogen content $\rightarrow$ increased lipophilicity $\rightarrow$ increased toxicity). Furthermore, the fit to the same QSAR model for a series of chemicals was taken to indicate a uniform mode of interaction, and outliers that deviated from a function were attributed to different types of effects. The aquatic environment received further attention in the following years: McLeese, Zitko and Peterson (1979) described different linear relationships between the lethality threshold levels of phenols and anilines towards shrimp and log $P_{ow}$ and substructure indicator variables, but no QSAR model could be derived for clams. In the same year, Kramer, Borns and Böhm (1979) reported the effects of hydrazides on algae, resulting in a non-linear QSAR dependent on log $P_{ow}$ for the aliphatic derivatives, and in linear models for the aromatic hydrazides, clearly distinct for *m*- and *p*-substituted analogues.

The first landmark in ecotoxicity modelling was set by Könemann (1981b), who correlated data on the toxicity of 50 industrial pollutants to fish with log $P_{ow}$. This QSAR was based on a prior model derived with 12 chlorobenzenes, which yielded almost identical results. An extension of the model with a diverse set of mostly chlorinated compounds provided the means for extrapolating to chemicals actually released into the environment. Although Könemann's results were not to be expected, a linear QSAR was derived with log $P_{ow}$ as the only regressor (Figure I.2), which indicated partitioning as the underlying principle of the observed toxicity for a wide variety of substances.

The validity of this QSAR is limited to compounds with log $P_{ow} < 6$, because of the limited water solubility or the hindered uptake into the organisms of highly lipophilic compounds. The function constitutes a baseline model: a compound will be at least as toxic as calculated from this QSAR, unless it is readily metabolized and only a small fraction of the compound in question can reach the site of action. On the other hand, a higher toxicity than calculated by this QSAR has to be expected for all the compounds that produce specific effects. In a multitude of consecutive studies, this QSAR by Könemann has been confirmed with different sets of chemicals or with different endpoints in aquatic toxicology. Since then, the Könemann equation represents a reference system for QSAR modelling in environmental sciences.

A different approach was taken by Enslein and Craig (1978) for the modelling of rat oral $LD_{50}$ data with a large number of indicators of the presence or absence of certain fragments in a structure, log $P_{ow}$ and the molecular weight. Because of some methodological shortcomings, this study in particular provoked the vehement criticism of established QSAR scientists about the transference of their techniques to toxicology and environmental sciences. The argument is that the basics of sound QSAR analyses are neglected (Rekker, 1980): 'The aim of a common QSAR approach is the correlation of biological test data obtained from (a) well-defined interactions between chemical substances belonging to (b) congeneric series of structures and (c) an active site in a biological system.'

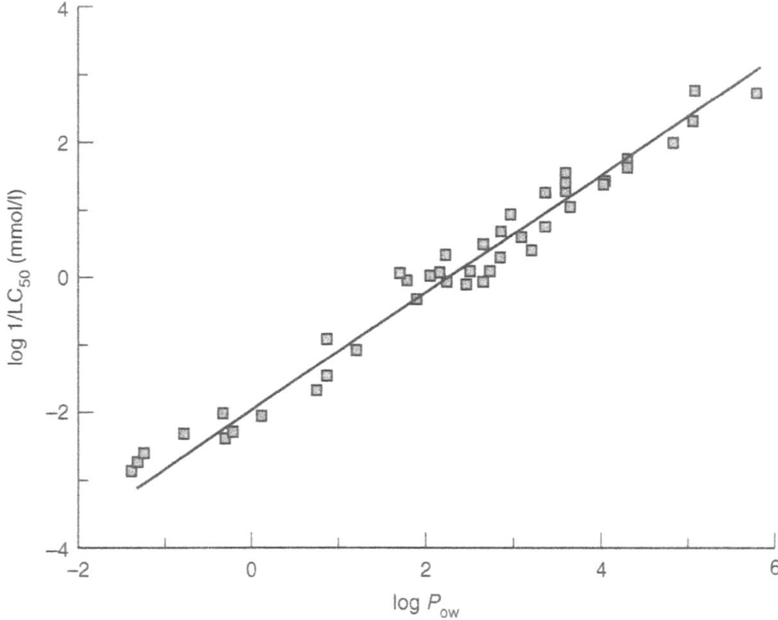

**Figure I.2** Representation of the Könemann equation: the relationship between the toxicity to fish (log $1/LC_{50}$ in mmol/l) of 50 industrial pollutants and their 1–octanol/water partition coefficient (log $P_{OW}$). The regression line is given by log $1/LC_{50}$ = 0.871 log $P_{OW}$ – 1.87 (Könemann, 1981b).

It is evident that all of these requirements cannot be fulfilled with QSAR studies in environmental sciences. The exclusive consideration of defined interactions with a (known) active site may be achievable in studies of drug–receptor (binding site) interactions, but the compliance with this prerequisite is illusory with regard to the endpoints required in environmental hazard assessments. The lethality parameters currently used to indicate toxic potency may reflect a multitude of interactions that in sum evoke the fatal impact. A distinction between these is virtually impossible. Accordingly, the standard endpoint 'death of 50% of the test species ($LC_{50}$)' may be reached due to very different modes of action. Strictly speaking, a comparison between such data for different compounds is likely to be irrelevant. Also, the demand to study solely congeneric series of chemical structures does not meet the reality of environmental hazard assessments, because discharges are rarely of similar compounds but rather of a wide variety of substances of changing composition and concentration. As a consequence, either QSAR studies have to be discarded from the toolbox of environmental sciences due to the inevitable violation of the basic rules, or the methodologies have to be adapted to the given situation, provided that their limitations are obeyed. If the latter choice is taken, the objectives of QSAR studies in environmental sciences are inevitably different from the applications in drug design. Keeping in mind the principal restrictions, the operation of QSARs in environmental sciences can be potentially useful for the following purposes:

*Descriptive QSARs* are studies aimed at describing observed effects of toxicants on environmental components such as organisms, soils or sediments or even on selected subdivisions of these compartments. The objective is to elucidate the mode of interaction between the chemicals and their potential targets in the environment, thus to better understand the principles underlying the toxicity or the environmental impacts. With regard to ecotoxicological research – by, for example, considering effects on the suborganismic, possibly on the cellular or enzymatic level – QSAR techniques can be applied in analogy to medicinal chemistry, bearing in mind that the required knowledge of the mechanism of interaction with a defined target site is generally absent.

*Predictive QSARs* are studies aimed at providing estimates for endpoints and parameters required in environmental hazard and risk assessment. This is a demanding goal, which should be strived for only if the principal shortcomings of the approach are considered, otherwise there is the danger of running into scientifically obscure imponderables.

In any QSAR study, but especially when aiming at predictions, several basic rules and limitations have to be considered, some of which are self-evident: However, it is the disregard of such truisms that led in the past to scientific controversies on the value and applicability of QSARs and environmental models. The rules and limitations refer to two levels: first to the inherent validity of the QSAR itself and, second, to the criteria by which to select the appropriate model for predictions for a given chemical. Of principal importance for the quality of a QSAR model are factors relating to the data set used and to the soundness of the statistics:

The QSAR is derived with a set of chemicals presenting the entire range of compounds concerned (the representative training set).

The training set consists of a sufficient number of chemicals, such that statistics can be applied; for example, at least four to five compounds per parameter (variable) of the model.

The chemicals of the training set span a wide domain of the structure-related parameters, because further applications of the QSAR (e.g. predictions) are limited to extrapolations within this range. The reason for this restriction is that the derived (e.g. linear) relationship cannot be assumed to hold outside the investigated section.

The structure descriptors are selected on the basis of mechanistic considerations to allow meaningful interpretations of the derived models.

The structure descriptors are not intercorrelated to allow the recognition of the significant variables. If, for example, activity data on a set of chlorophenols are investigated with respect to their log $P_{ow}$ and $pK_a$ values, a statistically significant QSAR may be obtained with either one of the descriptors, but an assignment of the relevant descriptor is not feasible, because those are linearly intercorrelated.

The parameters regarding the structure description and the activity measures are quantified with adequate accuracy, preferably on a continuous scale. The data for each variable (calculated as well as experimentally determined), should have been obtained by a sufficiently comparable procedure, because otherwise the variability in the data for one compound may be larger than the variability of consistent data between different compounds. If experimental data from the literature are used, they must be validated.

The activity data are quantified on a molar basis rather than by weight to allow a comparison of compounds of different molecular weight on the same scale. The number of moles (preferably obtained from dose–response relationships) evoking the same effects is an appropriate measure to be related to structural descriptors, which self-evidently are based on molar units.

The QSAR model should be examined graphically with all data points used, including outliers. The visual inspection provides easy access for recognizing statistical artefacts (e.g. chance correlations, two-point correlations, etc.).

Outliers are eliminated from the model only for methodological reasons (erroneous measurements) or mechanistic reasons (different mode of interaction). If such explanations cannot be given, the derived model has to be regarded inappropriate for this data set. The occurrence of outliers may, nevertheless, provide valuable information about the underlying processes.

The principal implications of the selected statistical technique have to be considered. These are evidently different for multiple regression, discriminant analysis or factor analysis.

The statistics of the model should be sufficient with regard to the correlation coefficient, $r$, the standard deviation of the residues, $s$, the $F$-test, etc. At the same time, the QSAR should not be overfitted, because, for example, $r^2 = 0.99$ (i.e. 99% variance explained) is meaningless if the activity data have $\pm$ 20% experimental variability.

The QSAR model has to be validated: for a sufficiently large number of compounds not used to derive the model, predictions have to be calculated and compared to the measured data. Only if there is adequate agreement can the validity of the model be assumed.

While deriving and validating a QSAR model, indications on when and when not to apply the model are obtained (e.g. regarding parameter range(s), substance classes, etc.); such criteria with regard to model limitations need to be given as an integral part of the QSAR.

The application of QSAR techniques requires the recognition that any model is local and that the domain of chemicals of a QSAR model can be characterized in terms of similarity, but that similarity with regard to structure as well as mechanism of interaction may be hard, if not impossible, to define.

In view of the possibilities and the limitations of the method, QSARs in environmental sciences have recently left the realm of academic studies and appeared as a tool in hazard and risk assessment of chemicals. The acceptance of QSARs in the framework of legislation and chemicals regulation differs among countries. In the EU, the initial testing of new chemicals is mandatory but in the US the Environmental Protection Agency (EPA) has to evaluate new compounds within 90 days, before their manufacture, often with little or no experimental data available. For these evaluations, a multitude of tools is used, among them QSARs; QSARs deliver data usable for decision-making about risk, necessary testing and controls. In contrast to this approach, the EU requires that a predefined set of tests must normally be conducted at the pre-marketing stage, with the exception of a few parameters. For the basic tier of assessment of chemicals, in most countries (except Japan), experimental determinations of bioconcentration data are not required; instead, extrapolations are made from the partition coefficient $\log P_{ow}$ of the compounds: no substantial bioconcentration is assumed for substances with $\log P_{ow} < 3$ (BCF $< 100$). Chemicals with $\log P_{ow}$ between 3 and 6 are classified as highly accumulating, eventually resulting in the demand for testing. Superlipophilic compounds characterized by $\log P_{ow} > 6$ and a molecular weight $> 500$ are classed as modestly accumulating. Basically, this assessment scheme represents the application of QSARs for legislative purposes.

EINECS, the European Inventory of Existing Commercial Substances, is listing $> 100\,000$ chemicals marketed before 18 September 1981 (BMU, 1992). Only a few of the necessary data have been measured, however, so a broad field is open for the application of QSARs. One objective is the filling of gaps in the data to identify the compounds that need to be investigated most urgently. QSARs cannot substitute for chemicals' testing, but they can indicate the compounds that need testing as a priority.

Provided the limitations of QSARs in environmental sciences are considered and the respective application criteria are complied with, structure–activity relationships can be a powerful tool for elucidating modes of interaction, screening the validity and the plausibility of experimental data, detecting of outliers and predicting activity parameters for product development, identification of priority pollutants and range finding for further testing.

# PART ONE

## *Theoretical Background*

Environmental impact may occur upon interactions of chemicals with their environmental counterparts due to intermolecular forces; the resulting effects depend on the structure and conformation of both reactants. Quantitative structure–activity relationships (QSARs) are used to recognize and utilize the systematic relationships between the principal properties of the chemicals and their biological, ecotoxicological and pharmacological activity. The principle of QSARs consists of relating the activities observed for a series of chemicals to a set of theoretical parameters, which are assumed to describe the relevant properties of their structures quantitatively. Derivation and application of QSARs hence requires three essential prerequisites:

descriptors of the chemicals' *structure*
measures of the chemicals' *activity*
statistical techniques to quantify the *relationship*.

By describing the biological or physico-chemical activity of compounds from their chemical structures, QSARs represent a generalization of observations obtained with few compounds, which are assumed to hold for an entire class of chemicals. The change in activity among this series of chemicals is described as a function of the differences in their structures:

$$\text{activity} = f \, (\text{structure})$$

The activity measures in environmental sciences concern ecotoxicological effects as well as fate- and exposure-related parameters such as partitioning between the water, air and soil phases, or degradation/transformation. Because of the different nature of these endpoints, different techniques may be appropriate for the respective structure–activity analyses. Nevertheless, whichever sophisticated method is applied, it has to be stressed repeatedly that the basis of any QSAR modelling is the experimental data, and that their quality (i.e. variability) determines the reliability and soundness of the derived models.

The two principal approaches to QSARs comprise the substructure-based methods and the linear free-energy relationship technique (LFER), which relates the activity to various kinds of physico-chemical parameters (Hansch, 1968a,b). The principle of LFERs assumes that incremental changes in the structure of organic chemicals introduce incremental changes in the free energy of their reactions. The individual processes underlying a particular activity will occur with different probabilities for different structures. Presuming that a constant number of molecular interactions is required at the target site ($C_t$: internal concentration) to produce a definite response (e.g. $LC_{50}$), the corresponding exposure concentration ($C$: external concentration) is indirectly proportional to the probabilities that the chemicals reach the site of action ($Pr_1$) and interact with the target molecules ($Pr_2$):

$$C_t = C \, Pr_1 \, Pr_2 \text{ const.}$$

Logarithmic transformation yields the general expression for Hansch-type QSARs:

$$\log 1/C = \log Pr_1 + \log Pr_2 + \text{const.}$$

The interaction rates (i.e. probabilities) can be led back to the basic properties of the chemicals and the respective relationship can be formalized by a multidimensional function:

$$\log 1/C = a \, D_1 + b \, D_2 + c \, D_3 + \dots + \text{const.}$$

with $D_1$, $D_2$ and $D_3$ denoting the descriptors of different physico-chemical/structural properties and $a$, $b$ and $c$ the coefficients quantifying the contributions of the respective properties to the activity. The logarithm of the activity measure can thus be described as a linear combination of functions characterizing, for example, the lipophilic, electronic or steric properties of the chemicals. The Hansch approach directly points out what physico-chemical forces may be acting, and how they influence the biological activity/property under study. The resulting equation is of a clear, immediately understandable form and allows a lateral comparison with other Hansch equations (i.e. with other mechanisms of action) to reveal more general phenomena. The selection of the descriptors to be considered, their appropriate transformations and the order of the respective dependencies (linear/non-linear) rest on the underlying model hypothesis (Figure 1.1).

Either a steady-state or a kinetic model may be assumed, with the basic processes being:

transport and partitioning
formation of the chemical/target complex
reactions of the chemical/target complex.

The overall probability of effects depends on the rates of the individual processes, and hence on the fraction of the applied dose of the compounds to undergo the relevant interactions within the observation time. Transport and partitioning are governed predominantly by the distribution of the chemicals

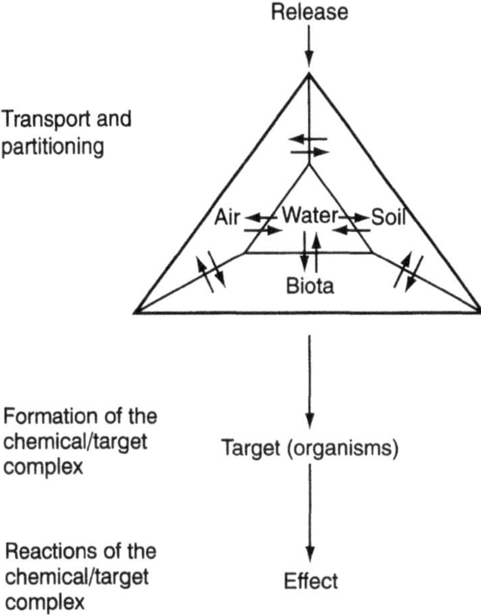



**Figure 1.1**  Three basic processes are involved in producing effects by chemicals released into the environment: 1, transport and partitioning; 2, formation of the chemical/target complex; 3, reactions of the chemical/target complex.

between lipophilic and aqueous phases (lipophilicity). During migration to the target site, repeated passive transfer from aqueous to lipid to aqueous phases occurs; hence exclusive affinity for either the aqueous or the lipophilic phases results in negligible transport rates. The formation of the chemical/ target complex depends on the forces of interaction between the two components, mostly hydrophobic bonding, their charge, their shape and their size. The reactions of the chemical/target complex are, with regard to the specific site, often controlled by the compounds' electron distribution. Thus, the effects of xenobiotics on environmental systems may be caused by three principal properties: the lipophilic, electronic and steric features of the compounds. Because the different properties of the chemicals may determine each of the relevant interactions to a different extent, QSAR models for the entire process aggregate a summation of the individual contributions. A discrimination may be achieved only if the activity data are analysed on the distinct interactions separately.

Generating QSARs for application in environmental sciences has to consist of several principal elements (Figure 1.2):

Assumption of a model hypothesis with regard to the relevant processes for the activity (definition of the endpoint).
Definition of the chemical class to be investigated.

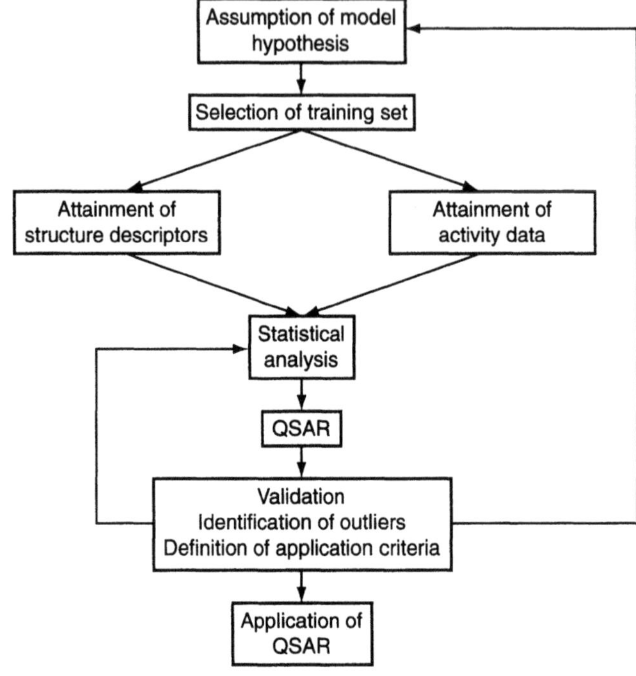


**Figure 1.2**   Principal elements of developing a QSAR model.

Selection of process-related descriptors.

Selection of a representative training set of compounds covering the entire domain of relevant properties.

Generation of the activity data for a well-defined endpoint.

Generation of the descriptor data.

Derivation of the QSAR.

Validation of the QSAR.

Identification of outliers.

Definition of application criteria for the QSAR.

Application of the QSAR for screening purposes, testing the validity of test results, substitution of missing information, and (partial) replacement of experiments in chemicals' testing.

# 1

# Descriptors of the chemical structures

Descriptors of the chemical structures are used to characterize and quantify properties of the entire molecules or relevant substructural fragments or the physical or chemical properties of the substances. The quantitative descriptors are based on the molecular structures as represented by constitutional formulas, hence an unambiguous definition of the structures is required, also with regard to the geometry, conformation, isomerism, tautomerism, stereochemistry, etc. The structural descriptors in a QSAR model have to represent the entire spectrum of the molecular or submolecular interactions leading to a particular property (i.e. hydrophobic, polar, electrostatic and steric interactions). Preferably, they should reflect the rate-limiting processes that are involved in producing the activity investigated. More than 200 different parameters have been reported (Dearden, 1990). These may be classified (Franke, 1985) with regard to their

coverage of relevant interactions between contaminants and ecosystem components;
general applicability to non-homologous sets of chemicals;
interpretability with regard to the interactions modelled;
unambiguous definition;
accessibility/availability.

The available structural descriptors comply with these criteria to varying extents. Their use in QSAR modelling in environmental sciences depends primarily on the type of endpoint to be described. The activity data may be quantified with various degrees of precision either on a continuous or a categorical scale. With regard to the objective of either investigating underlying processes or obtaining a tool for predictive purposes, different sets of descriptors may be appropriate.

The invention of ever new parameters in the past decades led to only limited progress towards QSAR modelling, because most of them have no general significance and they are applicable only to the specific data set for which they were derived. Despite this flood of descriptors, some important interactions still cannot be readily parameterized (e.g. the strength of hydrogen bonding). The array of descriptors included in a QSAR analysis should in any case be selected explicitly on considerations of the underlying processes and interactions; the alternative approach of throwing the book of parameters at the problem and then letting statistics decide invites the generation of meaningless chance correlations inherent in any large data set. The larger the dimensionality of the initial data matrix, the higher is the risk for chance correlations (Topliss and Edwards, 1979). Often, then, the derived QSARs merely reflect the perseverance of the investigators to exploit many different variables on the activity data of mostly a small number of compounds, until some combination of these descriptors delivers a statistically acceptable result.

The most commonly used descriptors cannot reflect enantiomerism, due to their determination in achiral systems. Predictions based on these descriptors must yield the same activity for both enantiomers of a chiral compound. But in natural (chiral) systems their impacts may be dramatically different, as, for example, the teratogenicity of thalidomide. Especially methods for calculating descriptors do not generally discriminate between enantiomers, not even between diastereomers, which leads to identical predictions for different compounds. The (optical) isomerism of chemical structures must, therefore, be deliberately considered in QSAR studies.

An appropriate set of descriptors that serve as the independent variables in the consecutive statistical analysis should have several principal characteristics:

Coverage of a broad property/descriptor range.
No intercorrelations between descriptors used simultaneously.
Even distribution over the property/descriptor range.
The ratio of the number of the independent parameters to the number of test compounds is not to exceed 1/5.
Tabulated or calculated descriptor values should have been validated experimentally for the series of compounds under investigation.

As a fundamental principle, even the best selection of the descriptor set cannot lead to a meaningful and representative QSAR if the set of chemicals used for the model derivation is not carefully designed.


## 1.1 LIPOPHILICITY

Lipophilicity has emerged as the key parameter for assessing the potential environmental impact of contaminants. This property constitutes a measure for the preference of a substance for either aqueous or non-aqueous phases. The partitioning between compartments of different polarity determines the rate and the direction of the transport of chemicals in the environment and thus their accumulation in some of its components. Therefore, lipophilicity appears in QSARs in environmental studies in two ways – as an endpoint by itself as well as a chemical descriptor to model further distribution-related parameters.

The partitioning between two non-miscible phases may be regarded as a continuous exchange of the solute ($x$) between two solvents with a zero net transfer at equilibrium ($K_P = K_{-P}$) (Figure 1.3):

$$x_{\text{aqueous phase}} \underset{K_{-p}}{\overset{K_{p}}{\rightleftharpoons}} x_{\text{non-aqueous phase}}$$

The two concurrent processes, the transfer from the water to the non-aqueous phase and the transfer from the non-aqueous to the water phase, are related by an equilibrium constant, which is expressed as the unitless partition coefficient ($P$):

$$P = \frac{\text{equilibrium concentration of chemical in non-aqueous phase (mol/l)}}{\text{equilibrium concentration of chemical in water (mol/l)}}$$

After release into a multiphase system, a chemical distributes between the phases according to the respective equilibrium rate constants. A large partition coefficient relates to lipophilic substances, of which a larger proportion transfers into non-aqueous phases, and a smaller partition coefficient relates to more hydrophilic substances, of which a larger proportion tends to remain in the water phase; that is, the higher the partition coefficient for the given two-phase system, the higher is the concentration of the chemical in the non-aqueous phase relative to the water phase when a steady state is achieved. Equal molar fractions in both phases yield a distribution coefficient of one. Substances of low lipophilicity are usually small molecules with polar, possibly charged, moieties. If $P$ approaches zero, the chemical will be so insoluble in non-polar (fatty) phases that it is unlikely to permeate lipid membranes and it will remain localized in the first aqueous phase it contacts. Conversely, as $P$ approaches infinity, the chemical will be so insoluble in water that it tends to become trapped in non-polar phases, where it accumulates. Exceptions occur with those compounds that are actively transported through membranes

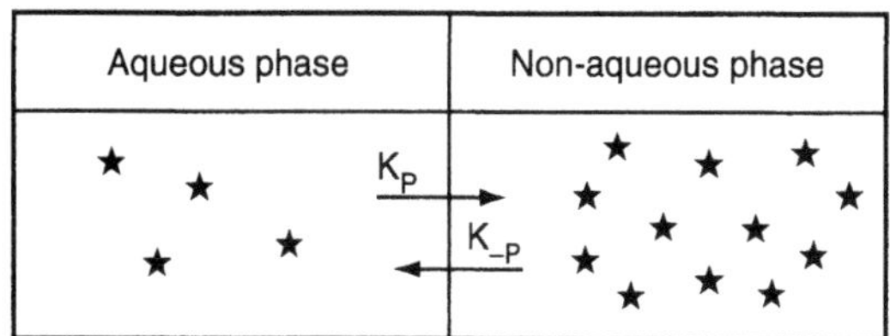


**Figure 1.3**  A partition coefficient ($P$) is defined as the ratio of the steady-state concentrations of a chemical in the non-aqueous and the aqueous phases. The figure shows the partitioning of a hypothetical compound (asterisks) with a partition coefficient $P = 3$ between an aqueous phase (left) and a non-aqueous phase (right).

or with amphiphiles that do not always follow log $P_{ow}$ in their accumulation and distribution.

The partitioning between aqueous and non-aqueous phases in the environment concerns biotic and abiotic phases. The lipophilic phases comprise a wide diversity of component types. In organisms, fatty tissue, cell membranes and serum proteins constitute non-aqueous compartments into which chemicals may partition. In the abiotic ecosphere, soil and sediments as well as dissolved particulate matter in water bodies may be temporary hydrophobic sinks. All such lipid compartments are surrounded by water that serves as a route for mutual interactions between the different partitioning processes (Figure 4.2). With regard to distribution in environmental systems, a steady-state equilibrium is rarely fulfilled, but within a dynamic system local pseudo-equilibria may occur. The respective partitioning processes are hence influenced by influx and efflux of material from and into the surroundings. Partition coefficients have been determined for a large number of chemicals, with $P$ values ranging from $10^{-4}$ to $10^8$, hence covering at least 12 orders of magnitude. Most environmentally relevant substances reveal log $P_{ow}$ values between 0 and 5.

Because of their definition as equilibrium constants, partition coefficients do not refer to the total amount of chemicals in the system but rather indicate how large a fraction of the release tends to transfer into the different phases. Partition coefficients may be used as an extrathermodynamic reference scale for characterizing lipophilicity as one of the most important properties of small molecules acting on macromolecular systems in aqueous solutions; for a comprehensive review of hydrophobicity see, for example, Taylor (1990). In QSAR analyses, the log $P_{ow}$ term is frequently used to account for two major processes related to partitioning:

transport (e.g. membrane passages);
hydrophobic bonding (e.g. to target sites, which can again be membranes).

From thermodynamic considerations, the equilibrium partitioning of an organic compound in a solvent/water system is determined by the difference between the partial molar excess free energy ($\Delta G^e_{ow}$) of the compound in the organic solvent (o) and in the water phase (w) with molar volumes $V_o$ and $V_w$, respectively (Schwarzenbach, Gschwend and Imboden, 1993):

$$\Delta G^e_{ow} = \Delta G_{ow} + RT \ln \frac{V_w}{V_o}$$

The free energy of transfer ($\Delta G_{ow}$) drives the net distribution until equilibrium is achieved and the chemical potential ($\mu$) of the solute

$$\mu = \mu^0 + RT \ln \gamma x$$

(where $\gamma$ is the activity coefficient and $x$ the mole fraction) is equal in both phases:

$$\mu_o = \mu_w$$

and hence

$$RT \ln \gamma_o \, x_o = RT \ln \gamma_w \, x_w$$

This equation can be rewritten as

$$RT \ln \frac{x_o}{x_w} = RT \ln \gamma_o - RT \ln \gamma_w$$

The term $x_o/x_w$ is a measure of the relative abundance of the solute in the two phases at equilibrium (i.e. a partition constant on a mole fraction basis). According to

$$c = \frac{x}{V}$$

the partition coefficient $P$ based on molar concentrations ($c$) can be rephrased as

$$\ln P = \ln \frac{c_o}{c_w} = \ln \frac{x_o / V_o}{x_w / V_w} = \ln \frac{\gamma_w / V_o}{\gamma_o / V_w}$$

and hence $P$ provides an expression of the difference between the non-ideality of the solution of the compound in the organic solvent and in the water phase:

$$\ln P = \ln \gamma_w - \ln \gamma_o + \ln \frac{V_w}{V_o}$$

$$= -\frac{\Delta G_{ow}^e}{RT} + \text{const.}$$

It is assumed that $\gamma$ at dilute concentrations is independent of both the compounds' concentration and the presence of organic solvent molecules in the water phase, and hence the logarithm of the partition coefficient ($\ln P$ resp. $\log P_{ow}$) is directly proportional to the change in the free energy $\Delta G$ at the transfer of a molecule (e.g. from the organic to the aqueous phase).

These derivations imply that $\log P_{ow}$ always concerns only one single defined species of a substance and due regard has to be paid to the stereochemistry, the occurrence of conformation changes, tautomerism, hydrogen bond formation and/or ionization to different extents in the two phases under the given conditions. (The consequences of varying degrees of ionization at ambient pH on the apparent partitioning are demonstrated in section 1.2.3.)

Although log $P_{ow}$ is widely used as a descriptor of chemical structures, it is a measure of a multitude of properties of the respective compounds, related to their size and polarity (e.g. Taft *et al.*, 1985; Kamlet *et al.*, 1988a,c; ElTayar, Testa and Carrupt, 1992). The log $P_{ow}$ values principally contain information on all kinds of interactions in the 1-octanol/water system. However, the manifestation of a property such as polarity is reflected by log $P_{ow}$ to different extents for different compound classes. Chemicals with approximately identical log $P_{ow}$, such as acetophenone (log $P_{ow}$: 1.58) and phenol (log $P_{ow}$: 1.46), both of which are assumed to be non-specific toxicants, differ considerably in their effects on fish (LC$_{50}$ acetophenone: 162 mg/l; LC$_{50}$ phenol: 28 mg/l). The difference in their toxicity, which for both is basically the result of partitioning into non-polar biophases, has been ascribed to different contributions of polar interactions with the target sites, resulting in a discrimination of non-polar and polar non-specific toxicants. Within each of these two compound classes, log $P_{ow}$ suffices to describe the variation in toxicity between class members, but apparently systematic divergences occur between these two classes that cannot be compensated by the log $P_{ow}$ parameter. Accordingly, log $P_{ow}$ may not represent a unique property for all types of compounds.

### 1.1.1 1-OCTANOL AS A REFERENCE MODEL FOR NON-AQUEOUS PHASES

The selection of the appropriate solvent system to model actual membranes and other hydrophobic compartments has long been debated, but the 1-octanol/water system is now being accepted as the general reference system. For special study objectives, however, different water/solvent (hydrophobic phase) systems may be more suitable to investigate distribution processes (e.g. Leahy *et al.*, 1992b). Because of the different composition of the various non-aqueous environmental phases including membranes, a search for one universal model solvent would be futile (Donkin, 1994). Elaborated for applications in medicinal chemistry and pharmacology, the 1-octanol/water partition coefficient is a reasonable measure for the interaction of organic compounds with biologically important molecules (Hansch, 1968b) and also generally serves well for the assessment of partitioning in environmental systems.

The 1-octanol/water system provides empirical descriptors of the lipophilicity of compounds that work in QSAR analyses (Table 1.1), even though the principal nature of the underlying processes is not yet thoroughly understood.

The water-saturated 1-octanol phase resembles natural non-polar phases in such that its polarity is similar to that of biological membranes and the OH-function of 1-octanol is assumed to model the hydrogen-bonding donor/acceptor and dipolar effects of real macromolecular systems.

**Table 1.1** Advantages and disadvantages of using the 1-octanol/water partitioning system as a reference scale.

| Advantages | Disadvantages |
| --- | --- |
| Membrane analogous polarity | Bulk phase without structure |
| H-bond donor/acceptor capacity | Not adopted for inorganic pollutants |
| Because of water content in 1-octanol phase no desolvatization of polar moieties | Determination in the range < 0 and > 5 subject to large uncertainties |
| Very low vapour pressure and UV-transparency facilitate experimental determinations | Strongly dependent on ionized fraction at ambient pH values |
| Large database available | |

Differences occur with regard to the three-dimensional arrangement of the non-aqueous phases. Most biological lipids consist of membranes, in which the molecules are predominantly arranged in bilayers. The lipid phase thus has a distinct structure and restricted spatial dimensions. Because the 1-octanol phase is a bulk phase, presumably with little or no structure, organic solutes may display different activity coefficients and partitioning behaviour in 1-octanol than in membranes. The more ordered a lipid phase is – in, for example, micelles and membranes – the more important becomes the entropy, resulting in non-linearity of the relationship between the respective partition coefficients and log $P_{ow}$.

1-Octanol is quite insoluble in water and vice versa; hence the partition coefficients are not strongly temperature dependent, being mostly in the order of 0.001 to 0.01 log $P_{ow}$ units per $K$ (Lyman, 1990b). Even the low mutual solubility – at equilibrium the aqueous phase contains $4.5 \times 10^{-3}$ mol/l 1-octanol and the organic phase contains 2.3 mol/l water (Lyman, 1990b) – reveals so that the 1-octanol/water partition coefficient does not equal the ratio of the compounds' solubilities in the individual solvents. This miscibility of the solvent phases, as well as ionization and association phenomena, limits the stringent validity of the presumed Nernst distribution, which, moreover, requires infinite dilution. Hence, at concentrations > 0.01 mol/l, $P_{ow}$ is frequently observed to be dependent on the solute concentration.

Early studies with olive oil as the non-aqueous phase suffered from the inhomogeneity of the organic solvent, because it is a complex mixture with different composition depending on its origin and processing; hence reproducibility of the measurements is limited. Various organic solvents were subsequently used for determining partition coefficients, which differ greatly depending on the polarity and the molar volume of the solute as well as the solvent (Table 1.2).

The presumed relationships between partitioning in different solvent/water systems (organic phases A and B) according to the Collander equation (1951)

$$\log P_{AW} = a \log P_{BW} + \text{const.}$$

holds only for analogous processes while the ratio of the respective activity coefficients remains constant. From practical experience, the 1-octanol/water system reveals sufficient similarity to the respective environmental media to allow the derivation of such Collander-type correlations. Also the extensive experimental data, gathered especially at Pomona college by the group of Hansch and Leo, indicated the 1-octanol/water system to be better as a general model for partitioning processes between lipids and water than other solvent/water combinations (Hansch and Leo, 1979). The prominence of the 1-octanol/water partitioning system stems from its practical productiveness and, despite its principal shortcomings (Table 1.1), it is widely accepted, particularly because of the large database available and the easy access to estimates via computerized calculations.

### *1.1.2 EXPERIMENTAL DETERMINATION OF PARTITION COEFFICIENTS*

The experimental determination of partition coefficients always involves exposing the test compound to a two-phase system of water and a non-aqueous phase. The classical shake-flask method has been standardized in the form of an OECD guideline (OECD, 1981b); the mutually presaturated (for 24 h) 1-octanol and water phases, the latter containing the dissolved test compound, are thoroughly mixed by shaking until equilibrium is attained; the phases are separated consecutively by centrifugation, the water phase is filtered to remove emulsified 1-octanol droplets, and the test compound is quantified in both phases. The mass-balance is established and then P is calculated. Preferably, the experiments are replicated at different solute concentrations. Extreme $P_{ow}$ values may be accessible by modifying the ratio of the phase volumes. The shake-flask technique usually provides reliable results in the log $P_{ow}$ range $0 \pm 4$ if conducted by well-experienced personnel. Its major limitation is the

**Table 1.2** Partition coefficients obtained in different solvent/water systems (Dunn, Grigoras and Johansson, 1986).

| | Solvent | | | | | |
|---|---|---|---|---|---|---|
| **Solute** | 1-Octanol | Diethylether | Trichloro-methane | Benzene | Tetrachloro-methane | Hexane |
| **Benzamide** | 0.64 | –0.22 | 0.11 | –0.71 | –1.54 | –2.30 |
| **Pyridine** | 0.65 | 0.08 | 1.43 | 0.41 | 0.23 | –0.21 |
| **Aniline** | 0.90 | 0.85 | 1.42 | 1.00 | 0.60 | –0.30 |
| **Phenol** | 1.46 | 1.64 | 0.37 | 0.36 | –0.36 | –0.70 |
| **Benzoic acid** | 1.87 | 1.89 | 0.50 | 0.21 | –0.22 | –0.72 |
| **2-Naphthol** | 2.70 | 1.77 | 1.74 | 1.74 | 0.99 | 0.30 |

non-applicability for highly lipophilic compounds ($\log P_{ow} > 5$). Problems may also arise from:

compound impurities
solvent impurities
glass/surface adsorption effects
formation of stable emulsions
concentration dependency
non-attainment of equilibrium
ion-pair formation
extractable ions in buffered aqueous phases.

Alternative experimental methods comprise several modifications of the shake-flask procedure, such as the slow-stirring method, which is especially suitable for highly lipophilic compounds (Bruijn *et al.*, 1989; Brooke *et al.*, 1990), and various chromatographic techniques.

The chromatographic techniques make use of the partitioning between the stationary lipophilic and the mobile hydrophilic phases. They differ from the equilibrium partitioning in flask experiments principally by using a dynamic process, which may be more similar to the distributions in environmental systems. The TLC-derived $R_m$ parameter (Biagi *et al.*, 1975) provides easy access to lipophilicity measures, but the data suffer from their specificity among chemical classes; that is, different interactions with the stationary phase require different coatings of the stationary phase and/or different solvent mixtures for the mobile phase, which make a standardization almost impossible. An analogous approach uses rp-HPLC capacity factors ($k'$) to quantify lipophilicity

$$k' = \frac{t_r - t_0}{t_0}$$

calculated from $t_r$, the retention time of the test compound, and $t_0$, the elution time of an unretained reference compound, such as urea or thiourea (Fini, Brusa and Chiesa, 1981; Harnisch, Möckel and Schulze, 1983). The solute is eluted with organic/aqueous solvent mixtures (acetonitrile, methanol/water, buffer) of different organic fractions (mostly $\geq 30\%$) from octyl- or octadecylmodified silicagel and $k'$ is extrapolated to 0% organic phase. In doing so, mostly a linear relationship between $k'$ and the eluent composition is presumed, but because of variations in physical properties with composition of, for example, aqueous acetonitrile or aqueous methanol, curvature, in either direction, has been encountered (Taylor, 1990). Also, the use of a hydrocarbon stationary phase restricts the estimation of $\log P_{ow}$ values due to phase dissimilarities with 1-octanol. The general use of this HPLC method, as for the above-mentioned TLC method, is limited because they yield different partitioning scales for compounds of different parent structures and hence they are not comparable among heterogeneous sets of chemicals.

This restriction does not apply to a modified rp-HPLC technique for the determination of partition coefficients (Unger, Cook and Hollenberg, 1978; Unger and Feuerman, 1979; Unger and Chiang, 1981; Taylor, 1990), which utilizes the distribution between 1-octanol, coated on, for example, octa-decylmodified silicagel as the stationary phase, and 1-octanol saturated water or buffer as the eluent. The RP-18 columns are conditioned with a mixture of 1-octanol, water and methanol (20:10:30 volume parts), then purged with 1-octanol saturated water to remove excess 1-octanol and equilibrated with 1-octanol saturated buffer of an appropriate pH value. This same buffer is used to elute the test compounds, which offers an important advantage: the effect of the dissociation of ionizable compounds on the apparent lipophilicity under environmental conditions can be accounted for, which thus provides partition coefficients that correspond to the state of the substances actually evoking the observed effects. The results generally agree well with those obtained by the shake-flask procedure and also tiny amounts of even impure samples can be analysed within a short time. The measurement of lipophilic amines of considerable basicity may require the masking of silanol functions of the solid phase with dimethyloctylamin (DMOA) in the eluent to avoid interactions (Unger and Chiang, 1981). The concentration of the injected sample should be high enough to allow easy detection and small enough to avoid overloading the 1-octanol coating, eventually resulting in accelerated elution. The flow rates have to be adjusted to provide a complete distribution of the test compounds; elution that is too rapid involves the risk of insufficient contact between the aqueous phase and the 1-octanol layer, which results in partition coefficients that are too low.

Especially when calibrated with standard compounds relative to the classical shake-flask method, rp-HPLC is the method of choice for the experimental determination of partition coefficients. The liability to large variation in measured $\log P_{ow}$ for the highly lipophilic compounds ($\log P_{ow} \geq 5$) subjects the experimental data to major uncertainties and for such molecules calculation is generally preferable (Taylor, 1990). Reported $\log P_{ow} > 7$ (e.g. for some chlorinated dioxins up to 11), corresponding to concentration differences between the aqueous and the 1-octanol phase of about $10^7$ and more are not exact values, but reflect the accuracy of the experimental analytical techniques used; they may be subject to major uncertainties of $\pm 2$ log units. The respective data have to be treated with great caution in QSAR studies, because at best they indicate the order of magnitude of the compounds' lipophilicity.


## *1.1.3 COMPUTATION OF PARTITION COEFFICIENTS*

The universal relevance of partitioning in the assessment of chemicals provoked early attempts to estimate the respective values directly from the chemical structures, especially for preliminary screening purposes. Based on the assumption of an additive, constitutive nature of partition coefficients, Hansch and co-workers derived a lipophilicity substituent constant analogous to the $\sigma$ Hammett approach (Hansch *et al.*, 1963; Hansch and Fujita,1964):

$$\log \frac{P_X}{P_H} = \log P_X - \log P_H = \pi_X$$

According to this equation, the lipophilicity constant $\pi$ is a measure for the contribution of the substituent X to the lipophilicity of the substituted compound as compared to the non-substituted parent compound (hydrogen atom instead of X). The $\pi$ value for hydrogen is by definition set to zero due to the substitution of H by H resulting in no net difference; for example, the $\pi$ values are for $C(CH_3)_3$: 1.98, $SCF_3$: 1.44, $CH_3$: 0.56, $CO_2CH_3$: –0.01, $NO_2$: –0.28, OH: –0.67 and $NH_2$: –1.23 (Hansch, 1975). A positive $\pi$ value indicates an increase in lipophilicity by this substituent and a negative $\pi$ value a decrease. The contribution of a substituent to a compound's lipophilicity can then be expressed by a linear free-energy relationship, as $\pi$ is proportional to the difference in change of the free energy occurring upon transfer of the non-substituted and the substituted compound from one phase into another phase. The system-specific coefficient $\rho$ was cancelled from the equation due to being defined as unity for the 1-octanol/water reference system. The $\pi$ scale is applicable only within congeneric series of aromatic compounds – it fails for aliphatics. The presumed additivity of the $\pi$ values does not hold when the following occur:

intramolecular hydrogen bonding
ortho effects
conjugation effects
heteroaromatics.

A more general approach, also based on the assumption of an additive, constitutive nature of partition coefficients, was proposed by Rekker (1977). The $\log P_{ow}$ of a compound corresponds to the sum of lipophilicity contributions of its constituents. The structures are virtually decomposed into substructural fragments $(f_i)$ with individual contributions $(a_i)$ to the lipophilicity of the entire molecules. Additionally, correction terms $(b_i)$ account for interactions between different fragments in the molecules $(F_i)$:

$$\log P_{ow} = \Sigma\, a_i\, f_i + \Sigma\, b_i\, F_i$$

Principally the same fragment $(f_i)$ and factor $(F_i)$ approach was used by Hansch and Leo (1979), proceeding in a methodologically different way. Rekker used a large database with heterogeneous chemicals whereas Leo and Hansch used a rather small set of especially selected compounds to obtain the fragment values by statistical analysis (multiple regression) of representative sets of experimental log $P_{ow}$ data. By diligent elaborations, the contributions to log $P_{ow}$ of a wide variety of substructures/fragments, covering most organic compounds, were quantified. However, the mere summation of the respective fragment values to estimate log $P_{ow}$ values assumes that the compounds consist of isolated pieces and ignores the interactions within the molecules because of, for example:

type of bond between fragments
branching
attachment to aromatic or aliphatic moieties
multiple halogenation
proximity of polar fragments
intramolecular hydrogen bonding.

The respective contributions to the molecules' overall lipophilicity also had to be quantified in the calculations of log $P_{ow}$. This task proved demanding due to the mutual, partly counteracting effects. A complete coverage has not yet been achieved. The two log $P_{ow}$ estimation techniques, differing in the definition of the particular fragments and factors, have been continuously refined and are both available fully computerized (MedChem, 1989; ProLogP, 1992). The problem of lacking fragment values may be avoided by defining fragment values for atoms in relation to neighbouring atoms or fragments; the respective atom/fragment contribution method is also available as commercial software (LOGKOW, 1993). All these group contribution methods are limited to the calculation of log $P_{ow}$ values of inert organic compounds, and they cannot be reliably applied for:

inorganic compounds
surface-active compounds
chelating compounds
organometallic compounds
partly/fully dissociated compounds
compounds of extremely high or low lipophilicity
mixtures
impure compounds.

The log $P_{ow}$ values calculated with either program for organic chemicals are generally reliable, even though discrepancies occur among the estimates and also in comparison to experimental data (Figure 1.4); that is, the programs fail for a variety of compounds to varying extents, mostly if the very high (log $P_{ow}$ > 5) or very low (log $P_{ow}$ < 0) lipophilicity range is concerned and/or if intramolecular interactions are not adequately accounted for by the respective algorithms. Neither procedure can hence be ranked as the best, and some experience is needed to rate the reliability of the estimated data. With regard to so-called superlipophilic substances especially, where the calculated

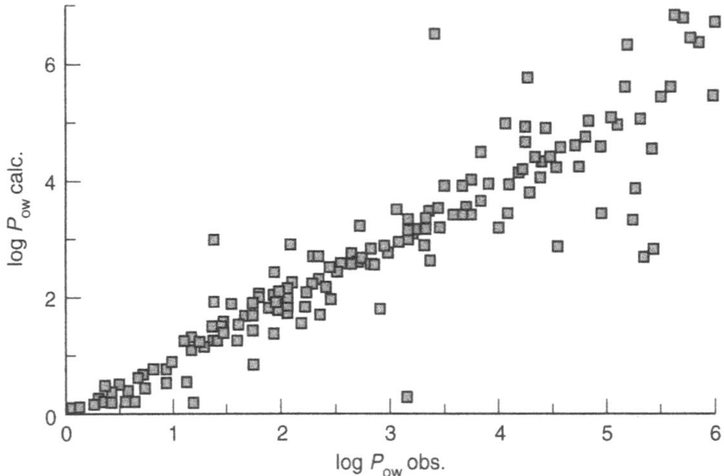

**Figure 1.4** Comparison of measured (obs.) and calculated (calc.) log $P_{ow}$ values (MedChem, 1989). The agreement is generally sufficient for compounds with log $P_{ow} < 5$, whereas there are major deviations for superlipophilic chemicals. The discrepancies may be due to shortcomings in the calculation methods as well as in the reported experimental data. The major uncertainties in the measured log $P_{ow}$ values $> 5$ are the reason why the experimental data are not necessarily true; in several cases, where the respective compounds were re-analysed, the calculated data were even better.

data frequently deviate from the experimental values, the latter are also subject to major uncertainties (often $\pm$ 2 log $P_{ow}$ units and more). The apparently too high calculations were in several cases later confirmed by refined experimental procedures. Accordingly, for practical purposes, a computed log $P_{ow} > 5$ should be read: this compound is lipophilic with a log $P_{ow}$ of $\pm$ 20% about the estimated value, so a calculated log $P_{ow}$ of 7.5 can be taken as evidence that the actual value lies between 6 and 9. This margin of error corresponds to that encountered with experimental determinations in this lipophilicity range. Depending on the further use of this datum (most environmentally relevant threshold values correspond to log $P_{ow}$ values of about 3), only in a few cases will an experimental explication of such estimates be necessary.

The calculation of log $P_{ow}$ values by hand using a fragment method requires considerable experience to consider the necessary correction factors. The principal procedure is illustrated for two simple compounds in Figures 1.5 and 1.6.

Detailed tabulations of fragment and factor values are given by Hansch and Leo (1979), Rekker (1977) and Lyman (1990b) together with elaborated instructions and examples.

Alternative approaches to the calculation of log $P_{ow}$ values from fragment contributions comprise solvent regression equations – extrapolations from

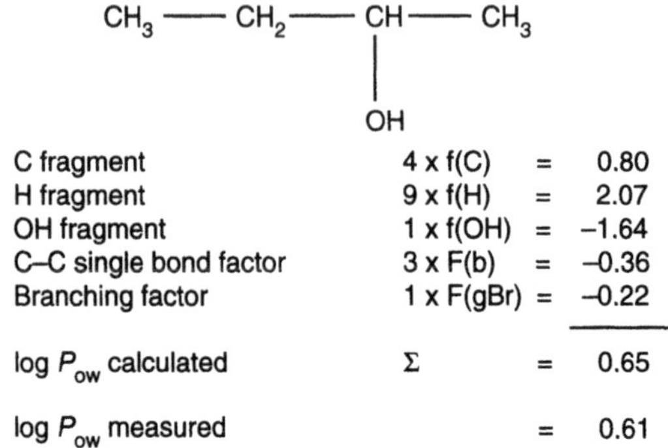

| C fragment | 4 x f(C) | = | 0.80 |
| H fragment | 9 x f(H) | = | 2.07 |
| OH fragment | 1 x f(OH) | = | −1.64 |
| C–C single bond factor | 3 x F(b) | = | −0.36 |
| Branching factor | 1 x F(gBr) | = | −0.22 |

| log $P_{ow}$ calculated | Σ | = | 0.65 |

| log $P_{ow}$ measured | | = | 0.61 |

**Figure 1.5**   Calculation of log $P_{ow}$ for butanol according to the fragment method. The fragment values for the substructures of the compound are summed and modified by factor values for bond types and branching (Lyman, Reehl and Rosenblatt, 1990). For this example, the difference between the calculated (0.65) and the measured (0.61) log $P_{ow}$ is negligible.

other water/solvent systems – and estimations from activity coefficients, $\gamma$ (Lyman, 1990b).

$$P_{ow} = 0.151 \frac{\gamma_i^{w,\infty}}{\gamma_i^{o,\infty}}$$

If experimental activity coefficients are unavailable, they can be estimated by, for example, the UNIFAC approach (Fredenslund, Gmehling and Rasmussen, 1977). The UNIFAC model is based on the group contribution concept, calculating the activity coefficients from two parts. The combinato-

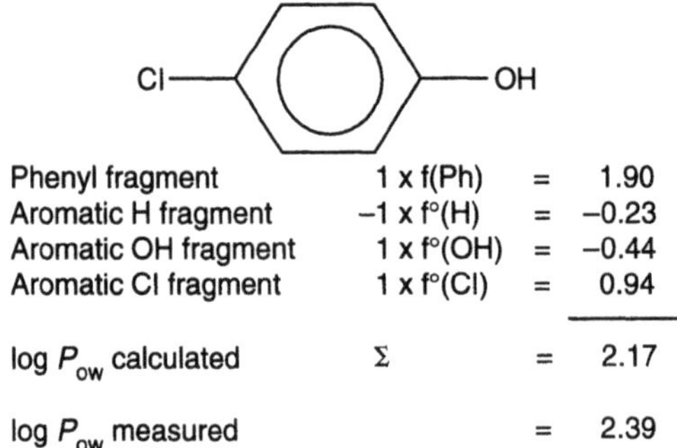

| Phenyl fragment | 1 x f(Ph) | = | 1.90 |
| Aromatic H fragment | −1 x f°(H) | = | −0.23 |
| Aromatic OH fragment | 1 x f°(OH) | = | −0.44 |
| Aromatic Cl fragment | 1 x f°(Cl) | = | 0.94 |

| log $P_{ow}$ calculated | Σ | = | 2.17 |

| log $P_{ow}$ measured | | = | 2.39 |

**Figure 1.6**   Calculation of log $P_{ow}$ for chlorophenol according to the fragment method. The fragment values for the substructures of the compound are summed (Lyman, Reehl and Rosenblatt, 1990). For this example, the agreement between the calculated (2.17) and the measured (2.39) log $P_{ow}$ is less than with the previous example (Figure 1.5), but still within the experimental variance.

rial part takes entropy effects into account and depends on the size and the shape of the molecules (i.e. the group surface volume and area). The residual part is governed mainly by the energetic interactions of different groups in the molecules (i.e. group interactions). When $\log P_{ow}$ values are estimated from activity coefficients they reveal an average error of $\pm 1$ log unit with major variability for lipophilic compounds with $\log P_{ow} > 6$. But also with extended parameter tables (Chen, Holten-Andersen and Tyle, 1993), missing group contributions and the neglect of proximity effects of substituents in environmentally relevant chemicals still limit the applicability of this method. The recently presented method of estimating $\log P_{ow}$ from the free energy of solvation and the molecular contact surface, calculated by quantum-chemical methods (Schüürmann, 1995), has not been validated for diverse data sets. For practical purposes in environmental hazard assessment, $\log P_{ow}$ values calculated by the classic fragment method, preferably computerized – so as not to overlook necessary correction factors – revealed generally sufficiently accurate input for consecutive evaluations.

## 1.2 ELECTRONIC PARAMETERS

Parameters aiming at a description of the electron density distribution of molecules should be used in QSAR studies only if specific, well-defined, interactions are modelled and the respective features of both reaction partners are known. The mutual interactions due to electrostatic interactions (ion–ion interactions, ion–dipole interactions, dipole–dipole interactions) and hydrogen bonding may control the extent and strength of intermolecular chemical/target (receptor) interactions. The necessary information is rarely available for QSARs in environmental sciences; the few exceptions possibly being pesticide/target interactions and abiotic transformations. For the majority of environmentally relevant chemicals with regard to most endpoints, the inclusion of electronic parameters in QSARs is justified merely on a statistical basis, and as such can be unmasked as chance correlations. Nevertheless, besides the investigation of potentially common modes of interaction between a target site and a series of contaminants, electronic parameters have two important applications in the context of QSARs in environmental sciences, and also when they do not serve as descriptors in the models: in accounting for ionization and dissociation; and in the rational design of test series (section 3.1).

Obviously, compound properties such as partitioning between water and lipid phases will be different if they are (partly) dissociated. Accounting for ionization and dissociation of chemicals at ambient pH values by direct use of an electronic descriptor such as $pK_a$ is inadequate; rather the relevant parameters (e.g. $\log P_{ow}$) have to be adopted for the fractions of ionized/unionized species (Taylor, 1990; Kubinyi, 1993); for details, see section 1.2.3.

## 1.2.1 ELECTRONIC DESCRIPTORS

The use of $\sigma$ Hammett substituent constants in QSAR analyses has been prominent since the early studies by Hansch and Fujita (1964) in medicinal chemistry. The electron-withdrawing/releasing potency of substituents on an aromatic system has been formalized by the Hammett equation (Hammett, 1940). The relative reactivities among homologous series of compounds under defined conditions – for example, the logarithms of the rate and equilibrium constants for the hydrolysis of substituted benzoic acid esters (rate constant $k$) and the ionization constants ($K$) of the corresponding acids, were found to be linearly related:

$$\log k_x = \rho \log K_x + c$$

The index x denotes the respective substituent in the $m$- or $p$-position of the benzene ring; $\rho$ and c represent the slope and the intercept of the function, respectively. Subtracting the analogous equation for the non-substituted ester and the corresponding acid with x = hydrogen (H) yields

$$\log k_X - \log k_H = \rho \, (\log K_X - \log K_H)$$

which can be rewritten as

$$\log \frac{k_X}{k_H} = \rho \log \frac{K_X}{K_H}$$

If the ionization constant $K_X$ is available, a constant $\sigma_X$

$$\sigma_X = \log \frac{K_X}{K_H} = pK_{(H)} - pK_{(X)}$$

can be derived for each substituent investigated ($\log K = - pK$). Substituting the logarithm of the ratio $K_X/K_H$ in the previous equation with $\sigma_X$ yields the so-called Hammett equation:

$$\log \frac{k_X}{k_H} = \rho \sigma_X$$

The dimensionless $\sigma$ values encode the substituent constants that express the polar effect of the substituents relative to hydrogen. The stronger the electron withdrawal from the reaction centre, the larger is the numeric value of $\sigma$. The $\sigma$ value for hydrogen is by definition set to zero because of the substitution of H by H resulting in no net difference. Representative $\sigma$ values are for $p$-$SO_2Cl$: 1.04, $p$-$NO_2$: 0.78, $m$-Cl: 0.37, $p$-Cl: 0.24, $p$-F: 0.06, $p$-$CH_3$: –0.14, $p$-$C_2H_5$: –0.15 and $p$-$O^-$: –0.81, respectively (Perrin, Dempsey and Serjeant, 1981). The $\sigma$ values are generally assumed constant for a given substituent, but evidently they vary (e.g. with the substitution pattern on the parent compound). The influence of a substituent such as chlorine on a reaction centre must differ in the *o-*, *m-* or *p-* position on an aromatic ring due to variant inductive, mesomeric and steric effects. Subsequently, numerous $\sigma$ scales were derived to account for inductive and resonance contributions of the substituents, such as $\sigma_I$ and $\sigma_R$, or $\sigma^+$ and $\sigma^-$ for + M– and –M–substituents (Taft, 1956).

For the purpose of characterizing chemicals for designing and selecting training sets for QSAR analyses, $\sigma$ values can be retrieved from tabulations (e.g. Perrin, Dempsey and Serjeant, 1981; Hansch, Leo and Taft, 1991). If several substituents are present on a parent structure, the sum of the corresponding $\sigma$ values is used. Especially for *o*-substituents, specific $\sigma$ values for the given chemical class have to be used.

The second application of $\sigma$ values for QSARs in environmental sciences is for the estimation of $pK_a$ values. In contrast to the substituent constants, the latter represent properties of the entire molecules and are used to quantify the degree of dissociation at a given (e.g. ambient) pH. In this way, like log $P_{ow}$, $pK_a$ values may appear in two ways: as an endpoint by itself in hazard-assessment schemes and as a chemical descriptor to account for environmental conditions affecting chemical properties. For a large number of chemicals, $pK_a$ values are available in the literature and may also be retrieved from databases. If such are not available and experimental determinations are not feasible, estimates may be obtained from a compound class-specific Hammett equation for aromatic substances or a Taft equation for aliphatic acids:

$$pK_a = pK_a{}' - \rho \, (\Sigma \, \sigma_x)$$

The $pK_a$ value of the substituted derivative can be calculated from that of the parent compound ($pK_a{}'$), the substance class-specific constant $\rho$, which is a measure of the sensitivity of the reaction to substituent effects with respect to the ionization of benzoic acids ($\rho = 1$), and tabulated (e.g. Perrin, Dempsey and Serjeant, 1981; Hansch, Leo and Taft, 1991) Hammett $\sigma$ or Taft $\sigma^*$ values (Table 1.3).

**Table 1.3** Examples of Hammett ($\sigma$) and Taft($\sigma^*$) equations for estimating $pK_a$ values (Perrin, Dempsey and Serjeant, 1981)

| Chemical class | $pK_a$ |
| --- | --- |
| Phenols | $9.92 - 2.23 \, \Sigma \, \sigma$ |
| Benzoic acids | $4.20 - 1.00 \, \Sigma \, \sigma$ |
| Anilines | $4.58 - 2.88 \, \Sigma \, \sigma$ |
| Substituted pyridines | $5.25 - 5.90 \, \Sigma \, \sigma$ |
| Substituted pyrimidines | $1.23 - 5.90 \, \Sigma \, \sigma$ |
| Aliphatic alcohols | $15.9 - 1.42 \, \Sigma \, \sigma^*$ |
| Aliphatic thiols | $10.2 - 3.50 \, \Sigma \, \sigma^*$ |
| Aliphatic amides | $22.0 - 3.10 \, \Sigma \, \sigma^*$ |

Further ways of describing electronic features of chemicals comprise spectroscopic data (e.g. IR, UV or NMR measurements), which correlate with other electronic descriptors within chemical classes (Seydel and Schaper, 1979; Kubinyi, 1993).

Hydrogen bonding, intra- as well as intermolecular, even though recognized as an important interaction, is hardly encountered in QSARs due to the insufficient availability of comprehensive descriptor scales. If used at all, it is often restricted to an indicator parameter characterizing either the presence or absence of hydrogen bond donor or acceptor functions (e.g. Ou, Ouyang and Lien, 1986; Rubino and Yalkowsky, 1987). The respective solvatochromic (LSER) parameters $\alpha$, hydrogen bond donor acidity, and $\beta$, hydrogen bond acceptor basicity (e.g. Taft *et al.*, 1985; Kamlet *et al.*, 1988a,b; Abraham *et al.*, 1989; Hickey and Passino-Reader, 1991), have to be determined empirically and hence are not universally applicable for making predictions (Leahy *et al.*, 1992a). Replacement of the LSER hydrogen bonding descriptors by quantum-chemically derived quantities was not very successful (Wilson and Famini, 1991). Approaches such as the evaluation of differences in partition coefficients ($\Delta\log P$) in water/solvent systems of variant hydrogen bonding capacity (Altomare *et al.* 1991; ElTayar *et al.*, 1991, 1992; Waterbeemd and Kansy, 1992; Abraham *et al.*, 1994) or the water-dragging effect (Fan *et al.*, 1990) require experimental data that are frequently not available for most environmental chemicals. The description of hydrogen bonding on the basis of thermodynamics (i.e. enthalpy factors, $E$, and free-energy factors, $G$, for proton donors and proton acceptors, respectively) yields a set of 12 descriptors for use simultaneously (Raevsky *et al.*, 1992a,b, 1994), but requires large data sets for modelling purposes. This method has not yet been thoroughly validated for environmental endpoints, where the hydrogen bonding partners of xenobiotics are neither known nor may be assumed to be of constant constitution. The enormous problems of unambiguous hydrogen bonding parameterization stem from the insufficient recognition of the underlying principles on a quantitative basis. Further complications arise from the fact that a universal scale of hydrogen bonding ability is unlikely to exist. Compounds vary not only in strength but also in ranking order of their hydrogen bonding as a function of the respective solvent properties (Leahy *et al.*, 1992a).

Quantum-chemical methods (Dewar and Thiel, 1977; Dewar *et al.*, 1985; Stewart, 1989) may be applied to obtain a large variety of stereo-electronic descriptors such as ionization potential, polarizability, electron affinity, dipole moment, charge densities, HOMO- and LUMO-energies or atomic charges using semi-empirical or *ab initio* calculations (Table 1.4).

The descriptors obtained generally refer to the ground state of the molecules under gas-phase conditions, which may not correspond to the solvatized compounds or their transition states involved in interactions in environmental phases. Additionally, the currently used quantum-chemical descriptors do not contain any entropy terms. The different semi-empirical calculation schemes, such

**Table 1.4** Examples of quantum-chemical parameters

| HOMO | Energy of highest occupied molecular orbital | $= -IP$ | eV |
|---|---|---|---|
| LUMO | Energy of lowest unoccupied molecular orbital | $= -EA$ | eV |
| DIFF | Difference between HOMO and LUMO | $= LUMO - HOMO$<br>$= 2 \times HARDNESS$ | eV |
| HARDNESS | Lewis acid strength | $= \frac{1}{2}(IP - EA)$<br>$= \frac{1}{2}(-HOMO + LUMO)$<br>$= \frac{1}{2} DIFF$ | eV |
| EN | Electronegativity | $= \frac{1}{2}(IP + EA)$<br>$= \frac{1}{2}(-HOMO - LUMO)$ | eV |
| IP | First ionization potential | $= -HOMO$ | eV |
| EA | Electron affinity | $= -LUMO$ | eV |
| HEAT | Heat of formation | | kcal/mol |
| DIPOL | Dipole moment | | debye |

as MOPAC, AM1 and PM3 (QCPE, 1986, 1990) feature different model assumptions. Because of these differences in the model algorithms the resultant output may differ considerably. For example, the use of $E_{LUMO}$ from any one of these different schemes requires profound knowledge about the differences in the model algorithms – respectively, the relevance of these differences with regard to the endpoint or activity modelled. The consequences of these facts are twofold. First, data produced by the inappropriate algorithm for the given problem are likely to render meaningless chance correlations and, second, data on a given parameter (e.g. $E_{LUMO}$) must be calculated consistently according to the same scheme for QSAR model derivations as well as for consecutive predictions. The use, for example, of a CNDO-derived quantum-chemical parameter to predict the activity of a compound with a QSAR model established with AM1 descriptors is doomed to failure. Quantum-chemical parameters can now be readily generated because of the rapidly increasing computer power. This obvious advantage is obscured by the liability of producing chance correlations by the uncritical statistical treatment of such large data arrays. Another shortcoming in the use of quantum-chemical descriptors for QSAR analyses is posed by the crucial impetus of the selected geometry of the molecules on the calculated parameters. Generally, an energetic minimum conformation is used as a starting point, but this may differ significantly among different computer programs. Further bias in the relevance of the calculated values results from the inability to decide on the actual conformation of the molecules, which is responsible for the activity to be described. Hence, if a quantum-chemical descriptor does not work in a QSAR model (e.g. $E_{LUMO}$), it may not be the wrong parameter, but it may be calculated for the wrong (i.e. not the active) conformer.

## 1.2.2 POLARIZABILITY DESCRIPTORS

Polarizability-related parameters such as molar refractivity are mostly described as steric descriptors with the acknowledgement of some extra electronic information. Their dual character combines elements that account for the general steric bulk as well as for the polarizability, which refers to the displacement of charges within molecules influenced by an electric field (i.e. induced dipoles). Evidently, molar polarizability influences various global properties of compounds, such as their hydrophobicity, by contributing to the solvation energies in different solvents (Lewis, 1989; Schüürmann, 1990b). The polarizability factor $\alpha$ corresponds to the dipole $\mu_d$ induced by the electric field $E$:

$$\alpha = \frac{\mu_d}{E}$$

The molar polarization $(P_E)$

$$P_E = \frac{4}{3}\pi N \alpha$$

is a measure for the separation of charges per mole in a non-polar substance in an electric field of unity ($N$ is Avogadro's number: $6.02 \times 10^{23}$ molecules/mol). $P_E$ is accessible from the dielectric constant $\varepsilon$ according to the relationship:

$$P_E = \frac{\varepsilon - 1}{\varepsilon + 2} \frac{MW}{d}$$

with MW denoting the molecular weight and $d$ the density. Substitution of $\varepsilon$ by $n^2$ ($n$ = refractive index) according to the Maxwell relation $\varepsilon_{\lambda = \infty} = n^2_{\lambda = \infty}$ yields the Lorentz–Lorenz equation for the molar refractivity MR:

$$MR = \frac{n^2 - 1}{n^2 + 2} \frac{MW}{d}$$

This formula represents the two components in this parameter: the term $(n^2 - 1)/(n^2 + 2)$ indicates the delocalizability of the electrons in the molecule (i.e. the aptness of the electron system to be polarized). The relation MW/$d$ reveals MR to depend on the molar volume. The larger the polar part of a molecule, the larger is its MR value. Atomic molar refractivity contributions are assumed to be additive and can be retrieved from tabulations (e.g. Hansch, 1975; Seydel and Schaper, 1979). The closely related parachor, the product of the molar volume and a surface tension term, is of negligible relevance for QSAR studies and cannot be recommended for use, except for the empirical estimation of boiling points (section 4.5).

The dual nature of these polarizability descriptors poses severe problems with regard to their use in QSAR analyses. Only when the test set is designed so that intercorrelations with, for example, steric or lipophilic parameters can be excluded will the descriptors provide distinctive information. The same restrictions apply to the application of polarizability descriptors for QSARs in environmental sciences, as already stated for electronic descriptors. Only if the exact constitution of the reactants is known will it be possible to confirm that, for example, MR indicates interactions with a polar binding site, and that a significant lipophilicity contribution indicates interactions with a hydrophobic binding site.

### 1.2.3 IMPACT OF IONIZATION AND DISSOCIATION ON PARTITIONING

Chemicals released into the environment are generally evaluated with respect to their non-dissociated form, but this neglects the effects of the dissociation on further properties, such as partitioning and distribution, solubility in various media, adsorption/desorption, accumulation and degradability. Hence there can be a significant misjudgement of the chemicals' potential hazard. The proportions of the species of a compound (neutral, anionic, cationic or zwitterionic) are dependent on the pH value of the surrounding phase. Evidently, the deprotonated (anionic) form of an acid has a different polarity than the non-dissociated form, which results in reduced lipophilicity and increased water solubility. As a result, the apparent adsorption to soil and sediment and the bioaccumulation of this compound may decrease as compared to the neutral form. Only in few cases does the opposite effect occur; for example, when ionic compounds interact with charged membrane components acting as an ion trap.

Generally, the partitioning of ionized species into other than aqueous phases is highly unlikely; the apparent distribution is related to the log $P_{ow}$ values of the non-dissociated form by the following equations for acids and bases, respectively:

$$\log P_{app.} = \log P_{ow} - \log (1 + 10^{pH-pK_a}) \quad \text{(acids)}$$
$$\log P_{app.} = \log P_{ow} - \log (1 + 10^{pK_a-pH}) \quad \text{(bases)}$$

The above terms are applicable when the concentrations of the ionized species in the organic phase are negligible and the ambient pH is relatively close to the respective $pK_a$ values. The pH partition profiles take the form of a sigmoidal curve (Figure 1.7).

The apparent log $P_{ow}$ corresponds to the log $P_{ow}$ of the unionized compound at pH values $< pK_a$ for acids or $> pK_a$ for bases (log $P_{ow} = \log P_{unionized}$). Increased ionization due to pH changes results in a linear decrease in the apparent log $P_{ow}$ until another plateau is achieved where only the ionized form determines the partitioning (log $P_{ow} = \log P_{ionized}$). The situation becomes much more complex if the molecule contains multiple dissociable groups or if ion pair formation has to be accounted for.

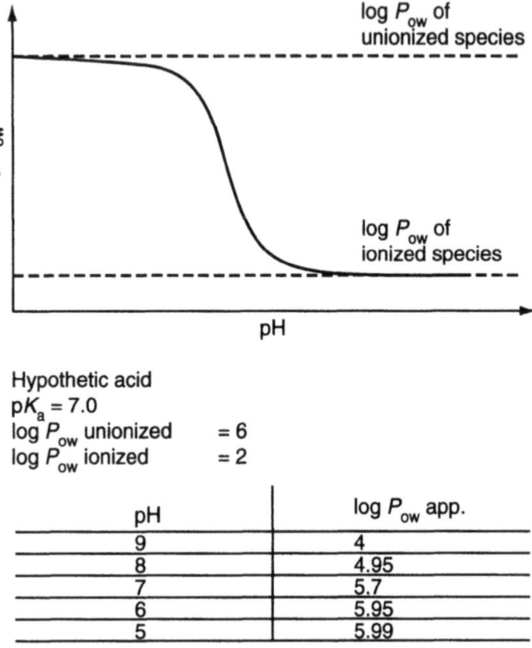



| pH | log $P_{ow}$ app. |
|---|---|
| 9 | 4 |
| 8 | 4.95 |
| 7 | 5.7 |
| 6 | 5.95 |
| 5 | 5.99 |

**Figure 1.7**  pH dependence of the apparent log $P_{ow}$ value of a hypothetical weak acid with p$K_a$ 7.

If a compound has more than one functional group possibly involved in protonation/deprotonation, it is recommended to calculate p$K_a$ values for each of these and then decide on their relevance for the investigation. In ampholytes such as an aminophenol (or hydroxyaniline depending on the class of compounds under study), both groups may be ionized under environmental conditions and hence both may contribute to the actual speciation of the compound.

The consideration of the pH effects on log $P_{ow}$ applies to those QSARs where the lipophilicity descriptor accounts for the rate-limiting, mostly transport, processes. The apparent log $P_{ow}$ values at ambient pH ranges for some toxicants are given in Figure 1.8. In toxicity modelling, when only one species is assumed responsible for the specific effects, it is more appropriate to correct the effect concentrations for the respective fraction (Taylor, 1990; Kubinyi, 1993).

## 1.3 GEOMETRIC DESCRIPTORS

The shape and the size of molecules rule a variety of processes related to the fate and effects of environmental contaminants, such as:

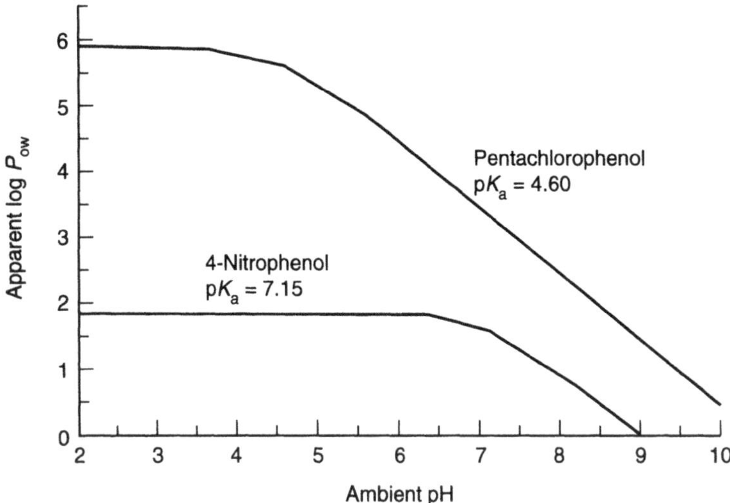



**Figure 1.8**  pH dependence of the apparent log $P_{OW}$ values of two acids with different p$K_a$ values. At ambient pH values about 7, the apparent log $P_{OW}$ of pentachlorophenol (p$K_a$ 4.60, log $P_{OW}$ 5.86) is significantly reduced as compared to the non-dissociated form. The effect of ionization at ambient pH is much lesser for a weaker acid such as 4-nitrophenol (p$K_a$ 7.15, log $P_{OW}$ 1.91). The pH profiles are calculated according to: log $P_{app.}$ = log $P_{OW}$–log $(1+10^{pH-pK_a})$. Reproduced from Nendza and Hermens (1995) with kind permission from Kluwer Academic Publishers, Dordrecht.

size-limited diffusion
fit to target sites
shielding of reactive moieties.

Membranes and other organic barriers are permeable only to compounds of an effective diameter $\leq$ 9.5 Å (Opperhuizen *et al.*, 1985). Larger molecules are hindered to diffuse through membranes and, unless active transport mechanisms act, they cannot reach intracellular targets. To be able to react with specific binding sites (e.g. receptors) the xenobiotic has to have a complementary constitution; the lesser the degree of fit the lesser the intrinsic activity. Bulky substituents near reactive centres in a molecule may shield this moiety and thus reduce its reactivity; for example, *o*-alkyl substituted esters are less likely to hydrolyse than analogous compounds with a different substitution pattern.

A principal complication in the quantification of geometric descriptors in QSAR analyses is the determination of the relevant three-dimensional structure of a compound on which to base the calculations. Because most molecules are flexible, they may have different geometries in different surroundings. The use of different plausible conformations for calculated stereo-electronic descriptors may result in differences of up to 0.5 orders of magnitude.

Three-dimensional QSAR approaches (Kubinyi, 1993) attempt to model active site interactions. However, although molecular modelling, comparative molecular field analysis (CoMFA) and conformational similarity analysis are interesting tools in drug design, their applicability for modelling environmental processes is unrealistic, because complex composite effects have to be addressed rather than pure ligand/target interactions and the exact conformation of neither the toxicants nor the interaction sites is generally known. A principal problem is the fact that conformations *in vacuo* (i.e. all conformations calculated by force field, semi-empirical or *ab initio* methods) may be significantly different from the conformations in aqueous solutions and at the binding site of a target.

Size-related parameters such as steric bulk may be represented by the molar volume $V$, which is accessible from the molecular weight MW and the density $d$ of the substance:

$$V = \frac{MW}{d}$$

The van der Waals volume $V_{vdW}$ may be estimated by fragment additive procedures using empirically determined van der Waals volumes for atoms or fragments (Bondi, 1964) or atomic van der Waals radii and optimized structures:

$$V_{vdW} = \Sigma\, V_{vdW(atoms,\ fragments)}$$

The respective atom and fragment values have been tabulated (e.g. Seydel and Schaper, 1979); the major disadvantage of these methods, however, is the neglect of the fact that volume and surface depend on the conformation of the molecule. The solvent-accessible volume and surface area depend on the size of the solvent molecule, generally water, and this is considered additionally. The solvent-accessible surface is defined as the locus of the centre of a spherical solvent rolled over the van der Waals surface of the solute.

The classical steric parameters are the Taft (1956) $E_S$ constants, which were developed in analogy to the $\sigma$ Hammett constants based on rate constants for the acidic hydrolysis of aliphatic esters $R\text{-}COOC_2H_5$, where polar substituent effects are presumed negligible. The $E_s$ constants reflect the shielding of a reactive centre by the substituent $R$; tabulations are given by, for example, Seydel and Schaper (1979).

The STERIMOL parameters (Verloop, Hoogenstraaten and Tipker, 1976) are based on van der Waals radii, standard bond length and angles for so-called reasonable conformations of the molecules; the derived constants describe the length of a substituent along the binding axis between the substituent and the parent molecule (L) and the substituent width in the four directions perpendicular to the L axis as well as to each other ($B_1$–$B_4$).

Based on the three-dimensional structures, preferably with their geometry optimized by force field methods, diameters of molecules along various axes can be calculated (Opperhuizen *et al.*, 1985; Schüürmann, 1990a). The molecules are orientated such that a vector along the largest diameter corresponds to the *x*-axis ($D_{max.}$), the second largest diameter orthogonal to the latter to the *y*-axis ($D_{eff.}$), and the minimal diameter ($D_{min.}$) perpendicular to both others to the *z*-axis (Figure 1.9).

If the principal direction of transport of a molecule is assumed to be along the *x*-axis, the effective diameter $D_{eff.}$ then conforms to the minimal pore width required for the passage through a barrier (e.g. a membrane). Because the calculation of the respective dimensions is not always practical, molecular weight has been used as a substitute in simplified assessments for regulatory purposes; for example, size-limited uptake into organisms resulting in reduced bioconcentration is assumed for compounds with MW > 500 (Umweltbundesamt, 1990).

## 1.4 TOPOLOGICAL DESCRIPTORS

Topological descriptors reflect the connectivity between atoms in a molecule based on graph theory (Randic, 1975; Kier and Hall, 1976; Basak, 1990; Sabljic, 1990; Balasubramanian, 1994). This approach is intended for planar structures; for three-dimensional bodies it is limited to strictly homologous series with identical topology and stereochemistry. Hence, the connectivity indices ($\chi$) refer only to the two-dimensional (hydrogen-suppressed) formulas, and steric (i.e. three-dimensional) features are not considered. The various $\chi$ parameters differ by the number i of adjacent bindings ($^{i}\chi$: ith order), by the type and degree of branching (path $\chi_{p}$, cluster $\chi_{c}$) and by the consideration of atom-specific valence corrections ($\chi^{v}$). The particular strength of this type of descriptor concerns the parameterization of branching among series of closely related analogues. One shortcoming of these descriptors is that they do not necessarily allow a unique parameterization of chemicals, because

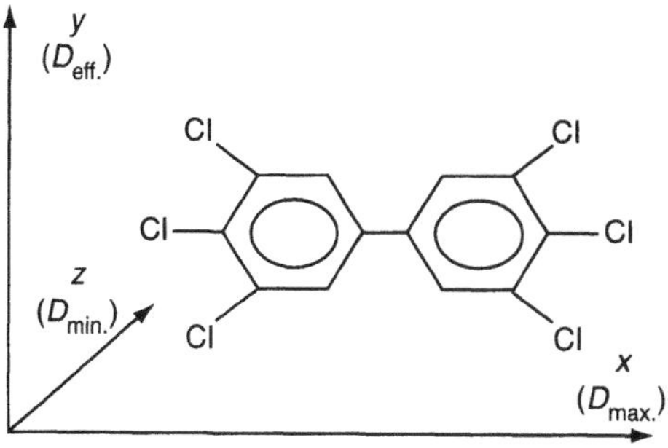

**Figure 1.9**  Orientation of 3, 3', 4, 4', 5, 5'-hexachlorobiphenyl (PCB 169) for the determination of its maximum, minimum and effective diameter.

identical topological indices may result for different structures (Balasubramanian, 1994).

The calculation of connectivity indices from a molecular formula is extremely simple (Figure 1.10). However, QSARs based on special sets of topological descriptors, like other unconventional parameters, have to be checked thoroughly before application, because the descriptor definition may be ambiguous and hence not reproducible. Examples are models of chemical toxicity to bacteria based on autocorrelation vectors (Devillers *et al.*, 1986). These yield a zero value for path-length five in phenol, which implies that no path over five atoms occurs in this molecule, or modified connectivity indices, whose generation is not explained (Gombar and Enslein, 1989). The arbitrary adaptation of parameter values may yield statistically improved models, but these are unsuitable for reliable application. The use of connectivity indices as molecular descriptors was reduced to absurdity by statistically excellent correlations with the length of the Beilstein citations (in cm) for diverse alkanes and even the telephone numbers of employees of a fish cannery in Sweden (Edward, Johanson and Wold, 1988). Nonetheless, even when connectivity indices are generated correctly, and are reproducible, the scientific value of using connectivity indices in QSAR analyses is limited (Taylor, 1990; Kubinyi, 1993) because of the lack of interpretability in mechanistic terms, especially when different connectivity indices are used to describe the same activity for different subsets of compounds (e.g. section 1.6); and the high probability for chance correlations from the frequently uncritical combination of a large number of interrelated connectivity terms.

Connectivity indices, either used individually or in various combinations, have been demonstrated to correlate with the major physico-chemical

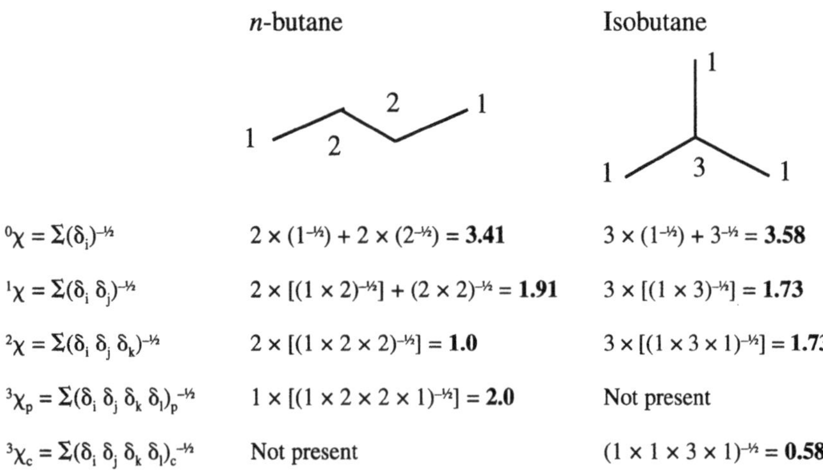


$^0\chi = \Sigma(\delta_i)^{-\frac{1}{2}}$     $2 \times (1^{-\frac{1}{2}}) + 2 \times (2^{-\frac{1}{2}}) = \mathbf{3.41}$     $3 \times (1^{-\frac{1}{2}}) + 3^{-\frac{1}{2}} = \mathbf{3.58}$

$^1\chi = \Sigma(\delta_i\,\delta_j)^{-\frac{1}{2}}$     $2 \times [(1 \times 2)^{-\frac{1}{2}}] + (2 \times 2)^{-\frac{1}{2}} = \mathbf{1.91}$     $3 \times [(1 \times 3)^{-\frac{1}{2}}] = \mathbf{1.73}$

$^2\chi = \Sigma(\delta_i\,\delta_j\,\delta_k)^{-\frac{1}{2}}$     $2 \times [(1 \times 2 \times 2)^{-\frac{1}{2}}] = \mathbf{1.0}$     $3 \times [(1 \times 3 \times 1)^{-\frac{1}{2}}] = \mathbf{1.73}$

$^3\chi_p = \Sigma(\delta_i\,\delta_j\,\delta_k\,\delta_l)_p^{-\frac{1}{2}}$     $1 \times [(1 \times 2 \times 2 \times 1)^{-\frac{1}{2}}] = \mathbf{2.0}$     Not present

$^3\chi_c = \Sigma(\delta_i\,\delta_j\,\delta_k\,\delta_l)_c^{-\frac{1}{2}}$     Not present     $(1 \times 1 \times 3 \times 1)^{-\frac{1}{2}} = \mathbf{0.58}$

**Figure 1.10**  Examples of the calculation of connectivity indices for butane isomers.

descriptors (e.g. Boecklen and Niemi, 1994); the latter, also accessible from computations, provide more meaningful information. Because the various connectivity indices do not supply any additional interpretable information, there is no essential need for their use.

## 1.5 INDICATOR VARIABLES

Besides the scaled (continuous) descriptors, dichotomous indicators of either the presence or absence of molecular features have often been used in QSAR analyses. Indicator variables are assigned values of '1' or '0' corresponding to 'yes' or 'no' if certain substructures, substituents, substitution patterns occur (e.g. di-ortho substitution, isomerism, etc.). Their use in multiple regression analysis yields the group contribution, given by the respective regression coefficient, to the total activity.

The application of indicator variables is meaningful only if aspects of the observed activity (e.g. elevated toxicity) are attributable to the respective features and cannot be accounted for by the physico-chemical descriptors; for example, the systematic deviation of polynitroaromatics from baseline toxicity QSARs may be accommodated by an indicator of multiple nitro-substitution. A prerequisite for the use of such dummy parameters is that the respective structural features convey an all-or-none contribution to the observed activity, and hence there is no partial grading in their effects. The Free–Wilson method (section 3.2.1) is a QSAR approach that relies solely on indicator variables accounting for the effects on the overall activity of the absence or presence of substituents or substructures in a certain position of the parent molecule. The substructure approach may well yield comprehensive, interpretable models if applied to endpoints such as degradation parameters, where the effects are due to interactions with specific structural moieties in the molecules. The absence or presence of readily metabolizable substructures determines the rate of biodegradation of xenobiotics. The microbial community in, for example, a soil is more likely to attack functional groups such as aldehydes, alcohols, esters or amines than tertiary butylgroups or halogen on branched molecules. Meaningful substructure-based QSARs for the respective endpoints require that the selected fragments are evenly represented by a sufficient number of compounds in the data set used to derive the model, to avoid an indicator variable merely describing the peculiarities of a single substance (e.g. the experimental error in this datum).

## 1.6 RELATIONSHIPS BETWEEN DESCRIPTORS OF CHEMICAL STRUCTURES

For the derivation and application of QSARs, it is a prerequisite to realize which descriptors are mutually related (i.e. contain similar information). For a sound QSAR model the descriptor variables must reflect the relevant types

**Table 1.5** Intercorrelations between descriptors of chemical structures for a set of diverse organic compounds, characterized by the correlation coefficient; the second figure (in parentheses) gives the number of compounds available for each pair of parameters (modified from Nendza and Russom, 1991).

| | MW | $T_m$ | $T_b$ | HV | $\log p_v$ | $\log S_w$ | $\log P_{ow}$ | $pK_a$ | H1 | H2 | ME | TE | BE | DI | MR | $V$ | SA | TT |
|---|---|---|---|---|---|---|---|---|---|---|---|---|---|---|---|---|---|---|
| MW | 1.00 (741) | | | | | | | | | | | | | | | | | |
| $T_m$ | 0.53 (470) | 1.00 (470) | | | | | | | | | | | | | | | | |
| $T_b$ | 0.85 (732) | 0.73 (463) | 1.00 (732) | | | | | | | | | | | | | | | |
| HV | 0.76 (732) | 0.69 (463) | 0.94 (732) | 1.00 (732) | | | | | | | | | | | | | | |
| $\log p_v$ | 0.64 (438) | 0.63 (295) | 0.98 (438) | 0.85 (438) | 1.00 (438) | | | | | | | | | | | | | |
| $\log S_w$ | 0.68 (703) | 0.47 (452) | 0.59 (695) | 0.40 (695) | 0.44 (431) | 1.00 (703) | | | | | | | | | | | | |
| $\log P_{ow}$ | 0.70 (720) | 0.24 (461) | 0.57 (712) | 0.40 (712) | 0.39 (432) | 0.98 (703) | 1.00 (720) | | | | | | | | | | | |
| $pK_a$ | 0.24 (263) | 0.29 (177) | 0.29 (262) | 0.26 (262) | 0.32 (137) | 0.04 (255) | 0.001 (260) | 1.00 (263) | | | | | | | | | | |
| H1 | 0.25 (716) | 0.45 (469) | 0.48 (707) | 0.66 (707) | 0.32 (432) | 0.17 (688) | 0.22 (703) | 0.29 (261) | 1.00 (716) | | | | | | | | | |
| H2 | 0.34 (726) | 0.44 (468) | 0.54 (719) | 0.74 (719) | 0.34 (431) | 0.10 (698) | 0.13 (714) | 0.23 (261) | 0.96 (714) | 1.00 (726) | | | | | | | | |
| ME | 0.92 (622) | 0.47 (412) | 0.80 (617) | 0.73 (617) | 0.66 (408) | 0.58 (607) | 0.62 (617) | 0.18 (240) | 0.41 (612) | 0.47 (621) | 1.00 (622) | | | | | | | |
| TE | 0.93 (680) | 0.52 (443) | 0.84 (675) | 0.81 (675) | 0.56 (416) | 0.55 (654) | 0.59 (668) | 0.29 (263) | 0.50 (671) | 0.58 (680) | 0.95 (622) | 1.00 (680) | | | | | | |
| BE | 0.78 (680) | 0.36 (443) | 0.72 (675) | 0.64 (675) | 0.64 (416) | 0.60 (654) | 0.66 (668) | 0.02 (246) | 0.30 (671) | 0.33 (680) | 0.90 (622) | 0.79 (680) | 1.00 (680) | | | | | |

| | | | | | | | | | | | | | | | | | | |
|---|---|---|---|---|---|---|---|---|---|---|---|---|---|---|---|---|---|---|
| DI | 0.17 (680) | 0.19 (443) | 0.25 (675) | 0.29 (675) | 0.10 (416) | 0.04 (654) | 0.07 (668) | 0.30 (246) | 0.30 (671) | 0.30 (680) | 0.17 (622) | 0.25 (680) | 0.06 (680) | 1.00 (680) | | | | |
| MR | 0.90 (735) | 0.47 (470) | 0.86 (726) | 0.75 (726) | 0.73 (434) | 0.70 (703) | 0.75 (720) | 0.13 (262) | 0.24 (716) | 0.32 (726) | 0.93 (622) | 0.86 (680) | 0.93 (680) | 0.11 (680) | 1.00 (735) | | | |
| $V$ | 0.75 (722) | 0.25 (464) | 0.64 (713) | 0.55 (713) | 0.50 (438) | 0.59 (687) | 0.68 (701) | 0.002 (261) | 0.19 (700) | 0.24 (709) | 0.87 (622) | 0.75 (667) | 0.90 (667) | 0.06 (667) | 0.90 (716) | 1.00 (722) | | |
| SA | 0.78 (722) | 0.28 (467) | 0.70 (715) | 0.64 (715) | 0.58 (432) | 0.61 (694) | 0.70 (710) | 0.04 (260) | 0.30 (710) | 0.34 (722) | 0.89 (622) | 0.82 (680) | 0.93 (680) | 0.09 (680) | 0.93 (722) | 0.97 (707) | 1.00 (722) | |
| TT | 0.01 (725) | 0.04 (470) | 0.02 (716) | 0.03 (716) | 0.05 (437) | 0.08 (689) | 0.10 (704) | 0.02 (262) | 0.10 (716) | 0.12 (714) | 0.12 (612) | 0.09 (671) | 0.08 (671) | 0.13 (671) | 0.05 (719) | 0.21 (709) | 0.14 (710) | 1.00 (725) |

Abbreviations: MW=molecular weight; $T_m$=melting point; $T_b$=boiling point; HV=heat of vaporization; log $p_v$=log vapour pressure; log $S_w$=log water solubility; H1, H2=total (H1) and weighted (H2) number of hydrogen bonding fragments; ME=minimized electronic energy; TE=total energy; BE=binding energy; DI=dipole moment; MR=molar refractivity; $V$=molar volume; SA=surface area; TT=number of tautomerizable substructures.

of interaction and hence need to be selected so that each represents one distinct property of the compounds. The assemblage of (interrelated) parameters in a model to satisfy statistical ambitions will not yield a meaningful relationship but will obscure its significance and predictive power. The understanding of the inherent descriptor collinearities (Craig, 1971) is also important for a comparative evaluation of different QSARs on the same endpoint.

The principal relationships between descriptors of chemical structures in a large array of compounds can be demonstrated. The correlation matrix (Table 1.5) obtained for a variety of parameters for several hundred compounds in more than 20 chemical classes (Nendza and Russom, 1991) reveals that hardly any descriptor is truly independent and that several highly significant correlations result.

There is a correlation ($r \geq 0.9$) between, for example:

calculated log $P_{ow}$ and log water solubility;
boiling point and heat of vaporization;
boiling point and log vapour pressure;
surface area and molar refractivity;
molar volume and molar refractivity;
molar volume and surface area.

Virtually no correlations with other descriptors occur for the $pK_a$ values, dipole moments and melting points. The ambiguous nature of some parameters is disclosed by their considerable intercorrelations ($r \geq 0.7$) with a large number of other variables. Molecular weight is influential for many descriptors (although derived on a molar basis), either lipophilic ($r = 0.70$), electronic ($r = 0.78–0.93$) or steric ($r = 0.75–0.78$). Molar refractivity encodes various aspects of chemical properties as illustrated by its relations with molecular weight ($r = 0.90$), boiling point ($r = 0.86$), heat of vaporization ($r = 0.75$), log vapour pressure ($r = 0.73$), log water solubility ($r = 0.70$), log $P_{ow}$ ($r = 0.75$), molecular energy ($r = 0.86–0.93$) and steric bulk parameters ($r = 0.90–0.93$), whereas those are not necessarily interrelated ($r = 0.39–0.98$).

Multiple intercorrelations between descriptors of chemical structures are illustrated best using multivariate statistics (section 3.2.2). A principal component analysis of the data set of 18 descriptors (Table 1.6, Figure 1.11) revealed that > 80% of the information content of these descriptors is expressed by four factors that explain 54.7%, 15.8%, 8.1% and 5.6% of the total variance, respectively.

The first factor is related to steric bulk, the second to hydrogen bonding, the third may be interpreted in terms of volatility-related properties and the fourth in terms of solubility-related properties. By coincidence with simple regression, the $pK_a$ values, the dipole moments and the melting points separate into individual factors, indicating that their information is different from that of the other parameters. The possible mutual intercorrelations between

**Table 1.6** Principal component (PC) analysis of descriptors of chemical structures for a set of diverse organic compounds: (a) > 80% of the explained variance is expressed in the first four PCs; (b) the loadings of the original descriptor variables in the VARIMAX rotated factor matrix reflect the grouping of the parameters (i.e. high loadings in the same PC indicate high intercorrelations between the descriptors).

| (a) | PC | Eigenvalue | (%) | Sum (%) |
|---|---|---|---|---|
| | 1 | 9.91 | 54.7 | 54.7 |
| | 2 | 2.89 | 15.8 | 70.5 |
| | 3 | 1.47 | 8.1 | 78.6 |
| | 4 | 1.02 | 5.6 | 84.2 |
| | 5 | 0.81 | 4.5 | 88.7 |
| | 6 | 0.68 | 3.7 | 92.4 |
| | 7 | 0.57 | 3.1 | 95.6 |
| | 8 | 0.29 | 1.6 | 97.1 |
| | 9 | 0.28 | 1.6 | 98.7 |
| | 10 | 0.11 | 0.6 | 99.3 |
| | 11 | 0.04 | 0.2 | 99.6 |
| | 12 | 0.03 | 0.2 | 99.7 |
| | 13 | 0.02 | 0.1 | 99.8 |
| | 14 | 0.02 | 0.1 | 99.9 |
| | 15 | 0.01 | 0.1 | 100 |
| | 16 | 0.001 | 0.0 | 100 |
| | 17 | −0.02 | 0.0 | 100 |
| | 18 | −0.10 | 0.0 | 100 |

| (b) | PC 1 | PC 2 | PC 3 | PC 4 | PC 5 | PC 6 |
|---|---|---|---|---|---|---|
| MW | 0.74 | 0.13 | 0.28 | 0.28 | −0.13 | 0.20 |
| $T_m$ | 0.16 | 0.30 | 0.37 | 0.19 | −0.12 | 0.83 |
| $T_b$ | 0.54 | 0.31 | 0.65 | 0.24 | −0.12 | 0.28 |
| HV | 0.45 | 0.55 | 0.58 | 0.17 | −0.07 | 0.21 |
| $\log p_v$ | −0.41 | −0.15 | −0.89 | −0.08 | 0.15 | −0.19 |
| $\log S_w$ | −0.50 | 0.18 | −0.15 | −0.79 | 0.01 | −0.28 |
| $\log P_{ow}$ | 0.60 | −0.20 | 0.14 | 0.76 | 0.01 | 0.02 |
| $pK_a$ | −0.01 | −0.14 | −0.14 | −0.01 | 0.97 | −0.08 |
| H1 | 0.15 | 0.94 | 0.11 | −0.16 | −0.13 | 0.13 |
| H2 | 0.18 | 0.95 | 0.15 | −0.09 | −0.05 | 0.09 |
| ME | −0.87 | −0.25 | −0.21 | −0.13 | 0.09 | −0.14 |
| TE | −0.74 | −0.40 | −0.17 | −0.24 | 0.16 | −0.15 |
| BE | −0.90 | −0.15 | −0.24 | −0.15 | −0.05 | −0.07 |
| DI | 0.05 | 0.17 | 0.04 | −0.03 | −0.13 | 0.05 |
| MR | 0.87 | 0.10 | 0.35 | 0.24 | −0.04 | 0.13 |
| V | 0.95 | 0.07 | 0.12 | 0.15 | 0.04 | 0.03 |
| SA | 0.94 | 0.17 | 0.17 | 0.20 | 0.01 | −0.02 |
| TT | 0.10 | 0.05 | −0.04 | −0.07 | −0.01 | −0.02 |

For an explanation of the abbreviations, see Table 1.5.

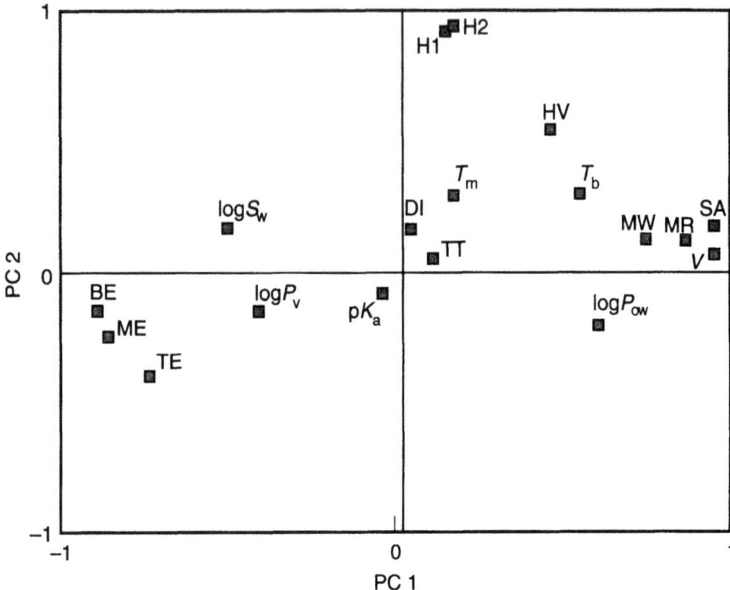


**Figure 1.11** Principal component plot (PC 1 versus PC 2) demonstrating collinearities among chemical descriptors and physico-chemical properties (data from Tables 1.5 and 1.6).

descriptors of chemical structures thus make it essential that each data set used for the derivation or the validation of a QSAR is selected with regard to the avoidance of chance correlations, which are almost inevitable if not deliberately prevented.

# 2

# *Activity and effects parameters*

Two types of parameters have to be considered in assessments of chemical contaminants in the environment: exposure-related factors that determine the spatial and temporal abundance of the contaminants, and effects-related endpoints that indicate the potential toxicity. The observed interactions and effects are assumed to be related directly to the concentration of the compounds at the site of action. By appropriate combination of the respective information, a probability of negative impacts can be obtained – risk assessment (Chapter 9). The most toxic agents are not necessarily the most dangerous ones. The statement by Paracelsus that 'sola dosis facit venenum' still holds true (Ariens, Mutschler and Simonis, 1978); any substance, xenobiotic or biogenetic, will be toxic if applied in a sufficiently high dose.

## 2.1 PROCESSES AND INTERACTIONS

The principal processes resulting in an environmental impact can be assumed to be intermolecular interactions between the xenobiotics and the target systems. The components involved are considered chemically defined structures, which vary with regard to the effects encountered. By analogy to fundamentals in pharmacology, the effects are mostly evoked via a chain of consecutive processes, which may mutually affect each other (Figure 2.1).

The underlying processes and interactions concern:

diffusion
interphase partitioning
adsorption
covalent binding
enzymatically catalysed reactions (metabolism)
chemically/physically induced degradative reactions.

Passive diffusion is the major mechanism of transport for xenobiotics in abiotic environments and inside organisms. Diffusion is caused by a gradient in concentration of the chemicals. The driving force is the thermal movement of molecules and not flux of the respective solvent. Fick's first law of diffusion

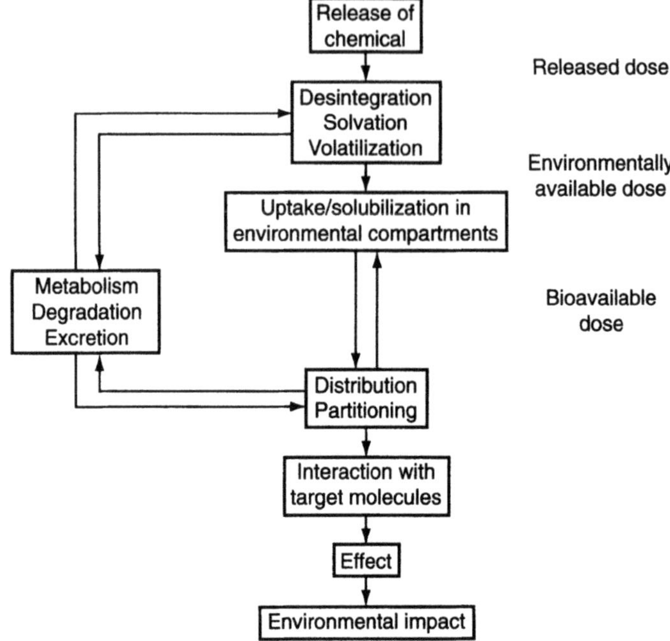



**Figure 2.1**   Scheme of processes involved in producing environmental impacts.

$$\frac{\mathrm{d}Q}{\mathrm{d}t} = -DA\frac{\mathrm{d}c}{\mathrm{d}x}$$

relates the amount of chemical transported per unit time ($\mathrm{d}Q/\mathrm{d}t$) to the effective concentration gradient ($\mathrm{d}c/\mathrm{d}x$), the area of diffusion interface ($A$) and the compound specific diffusion coefficient ($D$). The last quantity depends on the size of the molecules ($r$: radius of molecule/particle), the viscosity of the solvent ($\eta$) and the temperature ($T$) according to:

$$D = \frac{RT}{6\pi\eta rN}$$

with $N$ denoting Avogadro's number ($N = 6.02 \times 10^{23}$ molecules/mol). Ultimately, diffusion yields a homogeneous distribution of the compounds within a given compartment. Depending on the phase properties of, for example, different tissues, different maximum concentrations may result in different compartments. Lipid diffusion involves passive diffusion as well as interphase partitioning (section 1.1). The transport of chemicals between two aqueous compartments, separated by a lipid membrane, depends on their partitioning between the first aqueous compartment and the membrane, their diffusion across the membrane and the consecutive constitution of another partitioning equilibrium between the membrane and the second aqueous compartment (Figure 2.2).

The respective concentrations are established according to Nernst distributions:

$$c_{m1} = P_1\, c_{w1} \quad \text{and} \quad c_{m2} = P_2\, c_{w2}$$

with $c_{m1}$ and $c_{m2}$ denoting the equilibrium concentrations at the membrane surface in contact with the aqueous compartments 1 and 2, $c_{w1}$ and $c_{w2}$ the concentrations in the water phases, and $P_1$ and $P_2$ the membrane/water partition coefficients for the partitioning between the membrane and the aqueous compartments 1 resp. 2. The effective concentration gradient ($- \mathrm{d}c_m/\mathrm{d}x$) across the membrane of width $x$

$$-\frac{\mathrm{d}c_m}{\mathrm{d}x} = \frac{(P_1\, c_{w1}) - (P_2\, c_{w2})}{x}$$

results in an efflux of the compound from the first aqueous compartment according to Fick's first law of diffusion

$$-\frac{\mathrm{d}Q_1}{\mathrm{d}t} = -DF\,\frac{(P_1\, c_{w1}) - (P_2\, c_{w2})}{x}$$

This relationship indicates that permeation across membranes is significantly determined by the respective lipid/water partition coefficients of the compounds. If significant interactions between the chemical and the membrane occur (e.g. hydrogen bonding), the transfer across the membrane may be slower than indicated from pure diffusion. The differences in properties of the lipid phases give rise to unequal distribution among, for example, body tissues and liquids. The resultant equilibrium concentrations are a function of both the biotic/abiotic system and the chemicals investigated. Hence, the

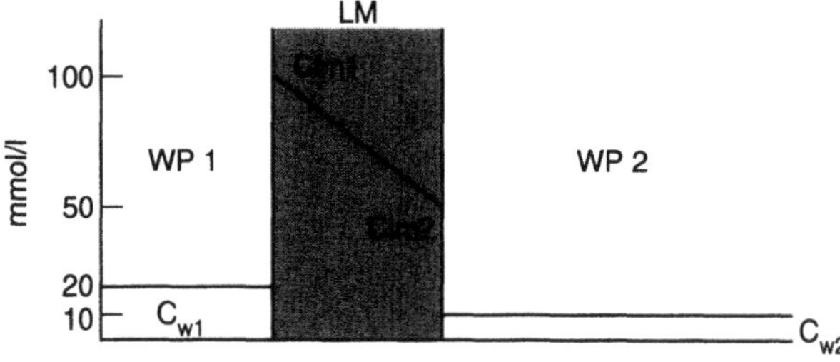


**Figure 2.2**  Concentration gradient resulting from lipid diffusion over a membrane for a compound with $P = 5$. WP 1 = water phase 1; WP 2 = water phase 2; LM = lipid membrane; Cw1 = concentration in WP 1 (20 mmol/l); Cw2 = concentration in WP 2 (10 mmol/l); Cl ml = concentration at the membrane surface of WP 1 (100 mmol/l); Cl m2 = concentration at the membrane surface of WP 2 (50 mmol/l). Reproduced from Pfeifer, Pflegel and Borchert (1984) with permission from Ullstein Mosby, Wiesbaden.

chemical must be of adequate water solubility to achieve sufficient concentrations in the aqueous media adjacent to the lipid barrier and to utilize the available permeation surfaces. The $pK_a$ value of the contaminant and the pH conditions on both sides of the membrane play an essential role, because lipid diffusion occurs predominantly with the neutral species – its ionized form is unlikely to be transported this way. The ionized species may instead bind to charged head groups of the membranes. Differences in pH on both sides of the membrane frequently cause asymmetric distributions of chemicals within an organism and the steady-state concentrations (i.e. accumulation) may not be equal in all compartments. The example (Figure 2.3) illustrates the relevance of differential ambient pH conditions, which have to be taken into account when assessing the distribution of dissociating chemicals between different compartments.

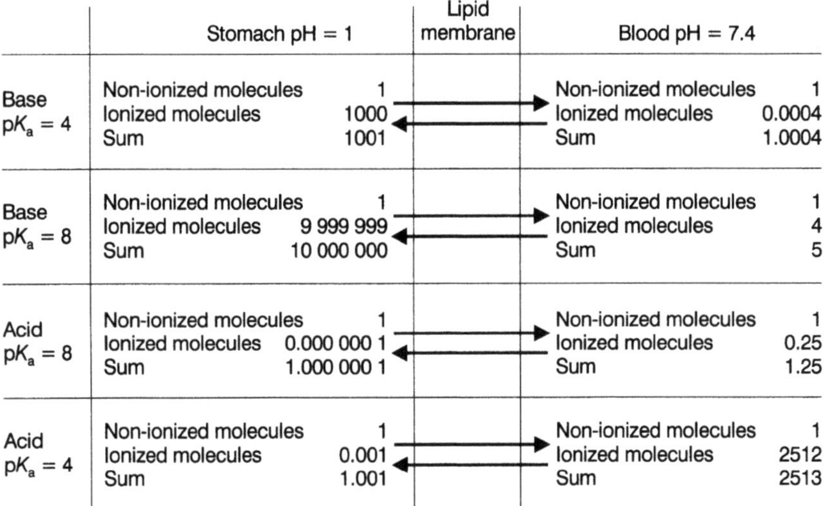



**Figure 2.3** Examples of the pH-dependent asymmetric distribution of ionizable compounds between compartments of different pH – here the acid medium in a stomach (pH 1) and the neutral medium in adjacent blood vessels (pH 7.4). the relative abundance of the non-ionized compounds in each compartment is normalized to 1. Weak bases (e.g. $pK_a$ 4) are predominantly non-ionized in the blood and ionized in the stomach. These compounds are hardly absorbed from the stomach, but may be eliminated from the blood into the stomach. Bases with relatively high $pK_a$ (> 7.4) are mostly ionized in both compartments and thus their uptake from the stomach into the blood is severely hindered. Weak acids (e.g. $pK_a$ 8) are predominantly non-ionized in both compartments and thus can readily pass the lipid barrier, depending on their concentration gradient. Stronger acids (e.g. $pK_a$ 4) are mostly non-ionized in the low-pH stomach and ionized in the blood, resulting in a preferential diffusion into the compartment with higher pH (i.e. the blood). Reproduced from Pfeifer, Pflegel and Borchert (1984) with permission from Ullstein Mosby, Wiesbaden.

The distribution does not take place instantaneously upon the chemicals' emission, but requires some time that depends both on the xenobiotic and on the constitution of the multiphase system concerned. Until the establishment of a steady state, the parallel processes – uptake, distribution and elimination into and from the individual compartments – yield changes in the concentrations of the chemicals in each of the respective phases. Although the increase or decrease in the concentrations of the compounds due to the uptake, distribution, transformation and elimination may result from a variety of processes, the change in concentration over time can generally be described by a few differential equations according to zero-, first- or second-order kinetics (Figure 2.4).

In most cases, they follow first-order kinetics:

$$\frac{dc_p}{dt} = -k\, c_p$$

Therefore, the rate of change in the concentration of a compound ($c$) by the process (p) is at any time ($t$) directly proportional to its concentration in the given compartment and determined by the first-order rate constant ($k$). Integration and taking the antilogarithm yield the respective exponential expression:

$$c_p(t) = c_p^{\,0} e^{-k\,t}$$

Substituting $c_p(t)$ by $c_p^{\,0}/2$ supplies the half-life time for the first-order process:

$$t_{1/2} = \frac{\ln 2}{k}$$

For zero-order processes, the rate of change in the concentration of the compound is constant, independent of its concentration in this given compartment, and determined by the zero-order rate constant ($^0k$).

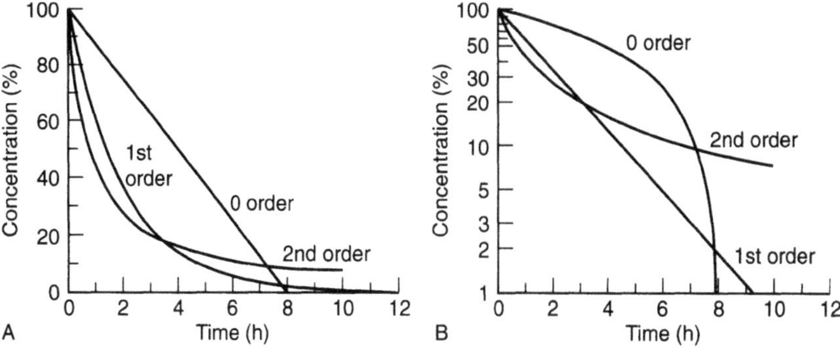


**Figure 2.4** Time dependence of chemical concentrations (%) according to zero-, first- and second-order kinetics: A: linear presentation; B: semilogarithmic presentation.

$$\frac{dc_{\mathrm{p}}}{dt} = -^0k$$

The zero-order rate constant corresponds to the slope of the non-logarithmic presentation of the kinetics. The respective half-life, however, depends on the concentration. Second-order kinetics are relevant for bimolecular reactions when the concentrations of both reactants are rate-determining. When the concentration of the environmental reaction partner is sufficiently large to be assumed to be constant as compared to the concentration of the contaminant, the kinetics are simplified to pseudo first order.

Assuming an open one-compartment model with uptake ($k_{\mathrm{u}}$) and elimination ($k_{\mathrm{e}}$) obeying first-order kinetics, the time course of the compounds' concentration ($c_{\mathrm{p}}(t)$) can be described by:

$$c_{\mathrm{p}}(t) = \frac{Dfk_{\mathrm{u}}}{V_{\mathrm{d}}(k_{\mathrm{u}} - k_{\mathrm{e}})}\,(e^{-k_{\mathrm{e}}t} - e^{-k_{\mathrm{u}}t})$$

depending on the applied dose ($D$), the fraction adsorbed ($f$) and the volume of distribution ($V_{\mathrm{d}}$). This kind of function was first described by Bateman in 1910 for the decay of a radioactive substance yielding radioactive daughter products that are also decaying (Pfeifer, Pflegel and Borchert, 1984). The so-called Bateman function is generally applicable, when two first-order processes occur concurrently (Figure 2.5).

The repeated application of a chemical eventually results in the establishment of a steady state, where constant relationships between the concentrations in the different compartments are achieved and uptake and elimination are at equilibrium. The ratio of the steady-state maximum and minimum concentrations to the respective values after the first dosage corresponds to the

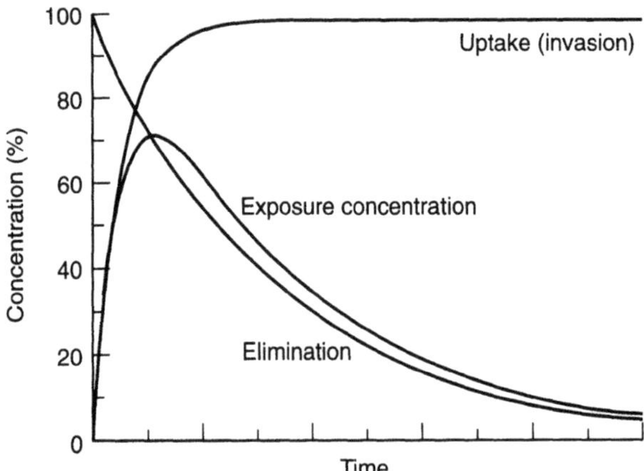


**Figure 2.5** Time dependence of the exposure concentration on the uptake (invasion) and the elimination of a chemical substance according to the Bateman function.

accumulation factor $R$, which is dependent on the dosage interval and the elimination half-life time. Upon continuous exposure for an extended period of time (i.e. the dosage interval is zero), the concentrations in the different compartments depend solely on the elimination rate. The input then corresponds to zero-order kinetics and the time course of the concentration, asymptotically approaching steady state (Figure 2.6), can be described by:

$$c_\mathrm{p}(t) = \frac{{}^0k_\mathrm{u}}{V_\mathrm{d}k_\mathrm{e}}(1 - e^{-k_\mathrm{e}t})$$

These process kinetics can be assumed, for example, for the distribution between and the accumulation in the different phases (compartments) in the environment for chemicals continuously released from (non-)point sources.

The adsorption to abiotic and biotic surfaces is governed principally by the same processes as lipid diffusion, but with the partition coefficients replaced by the respective adsorption coefficients for the materials. Two principal interactions contribute to adsorption, which may be regarded as like partitioning between an aqueous and a solid phase:

Physisorption:
    van der Waals interactions
    hydrophobic interactions
    charge-transfer interactions
    dipole–dipole interactions
    hydrogen bonding.
Chemisorption:
    ligand exchange and ionic binding
    direct or induced ion-dipole interactions
    covalent binding.

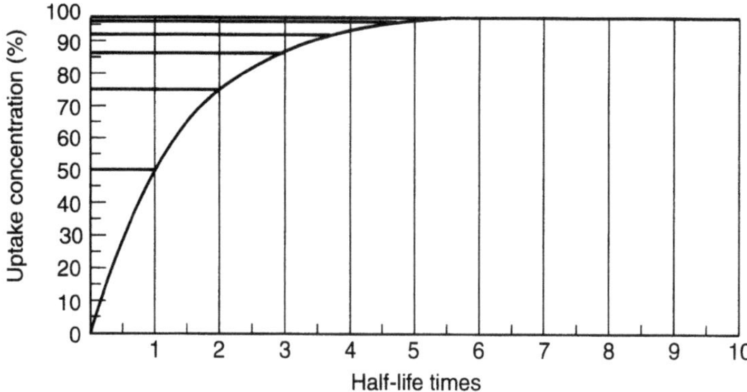


**Figure 2.6** Time dependence of the concentration of a compound in an organism during uptake from water as a percentage of the steady-state level.

Adsorption is rarely a simple single process; rather the various interactions may operate simultaneously and the predominating contributions depend on the structure and composition of the particular chemical and the sorbent as well (Schwarzenbach, Gschwend and Imboden, 1993).

Transformation reactions concern chemicals in the environment by abiotic (chemical and photochemical) and biotic (microbial) pathways.

Chemical reactions:
    nucleophilic substitution
    elimination
    ester hydrolysis
    oxidation
    reduction.
Photochemical reactions:
    fragmentation
    intramolecular rearrangements
    hydrogen atom abstraction
    dimerization
    electron transfer.
Enzymatic reactions:
    oxidation
    reduction
    hydrolysis
    conjugation.

The various degradation reactions, which may differ significantly in reaction products and rates, depending on the environmental conditions, scarcely result in complete mineralization of the compounds (formation of $CO_2$, $H_2O$ and inorganic salts), but mostly yield organic derivatives of the contaminant (Figure 2.7).

From this perspective, it has to be considered that degradation may in some cases produce even more toxic metabolites (toxification) and the fate and effects of these need to be accounted for in a meaningful hazard assessment for the parent compound.


## 2.2 EFFECTS-RELATED PARAMETERS

Toxicity is not a specific property of any particular contaminant, but the result of interactions between the substance and the species exposed to it. The toxic potential of chemicals to affect cells, organs and organisms adversely is determined by the affinity for and the effectivity of the interactions between the chemical agent and the biological components (e.g. the receptors) in the organism, as well as the dose or concentration of the contaminant. The potential toxic effects may partly be compensated by the defence mechanisms of the affected biota (repair, regeneration and regulation at different levels of the

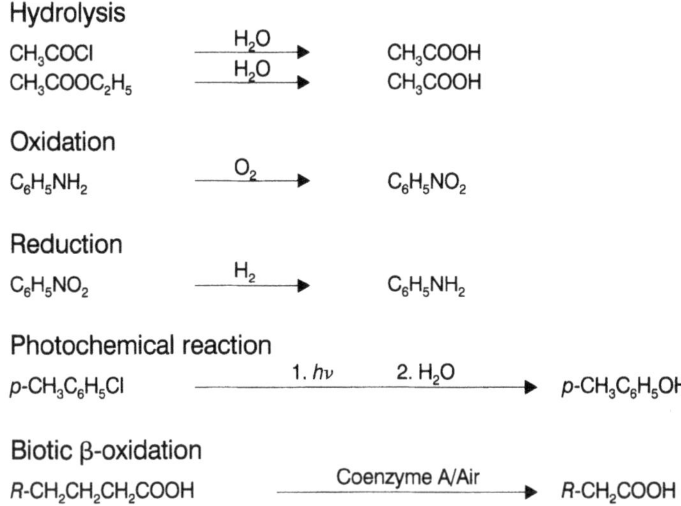


**Figure 2.7**  Examples of some environmentally relevant degradation processes.

biosystem). Consequently, toxic concentrations derived from any biotest system always relate to this specific situation and the conventionally defined experimental conditions (Nusch, 1991) and extrapolations to different species or environments may lead to grossly false results.

Ecotoxicity is supposed to be concerned with any damaging effects of chemicals on species, populations and their environment, through which the structure, function, stability or repair mechanisms of the ecosystem may be impaired. The actual impact may occur at different trophic levels of ecosystem organization with the targets including:

Individuals:
  interactions with biomolecular entities (e.g. enzymes, receptors and membranes)
  interference with metabolic pathways
  interference with feedback regulations
  neurotoxicity
  mutagenicity and carcinogenicity
  teratogenicity
  effects on behaviour.
Populations:
  changes in species density
  changes in the predominant species
  changes in species abundance and biomass
  changes in the spatial distribution of organisms
  changes in the fluctuation of species

changes in the reproduction rate
changes in the foodweb
changes in the occurrence of competitors.

The task for ecotoxicity assessment is to provide qualitative and quantitative indicators of the potential environmental impacts of chemicals, such as changes in the abundance of individual species or the diversity of the species community, taking into account the concentration and the time of exposure (dose–response relationships). Toxic action at the level of the organism may be classified according to Ariens (1984):

Reversible (selective) interactions with specific sites of action:
clear concentration–effect relationships
acute and subacute effects
additivity of effects
effects disappear after elimination of the toxicants.
Irreversible interactions with target molecules:
occurrence on exposure to reactive toxicants
chemical lesions caused by the attack of critical biopolymers (e.g. DNA) and functional proteins (e.g. enzymes and receptors)
high probability of synergism
long latency until effects are expressed after exposure to small concentrations over an extended period of time
persistent damage.
Accumulation in lipid-rich tissues:
occurrence upon exposure to persistent, metabolically stable, compounds
mobilization upon 'melting' of fat reservoirs
acute and subacute effects
additivity of effects
effects disappear after elimination of the toxicants.

Effects data applicable for QSAR analyses are reported in terms of a constant equivalent response – that is, isoeffective concentrations that correspond to the number of moles of the compounds evoking the same effect. Activity measures such as toxicity parameters should be based on a dose–response curve, not only on a single point determination. Various factors act as determinants in dose–response relationships, such as route of exposure, duration of exposure and species sensitivity (Figure 2.8).

No single test system can serve as an indicator of the entire spectrum of the toxic potential of xenobiotics. Indeed, it will provide only one aspect of the possible impairments. The classical laboratory toxicity tests always relate to the standardized conditions under which they are conducted. Their generally good reproducibility can be achieved only by a substantial abstraction from the (variable) conditions in the real environment. The extrapolation of laboratory test data (obtained in a monofunctional system) to numeric threshold

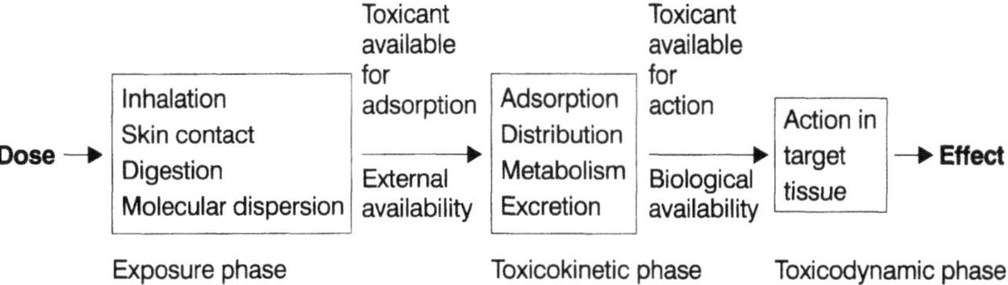

**Figure 2.8** The main phases of toxic action. Reproduced from Ariens (1984, p. 433) by courtesy of Marcel Dekker, Inc., New York.

values for natural ecosystems (representing multifunctional systems) is accordingly limited. Not even the most complex model systems will predict all conceivable risk for ecosystems. However, simple toxicity tests will indicate the potential toxicity of chemicals at the ecosystem level.

Toxicants will affect different species by different modes, due to differences in the receiving targets, and to different extents, due to differences in their sensitivity. Thus a chemical may have not only one mode of action but may also act by different mechanisms on different species. The classic examples are the photosystem inhibitors that have high toxicity towards algae and plants, but only non-specific effects on fish and *Daphnia*. This difference in toxicity occurs because the respective targets are lacking in fish and daphnids. Acetylcholinesterase inhibitors are highly toxic in all species with the respective enzymes, such as daphnids, fish and mammals, whereas plants, algae and bacteria are affected to a lesser extent by a different mechanism. Accordingly, toxicity is not a compound-specific property; its mode and extent are also determined by the organisms affected. Knowledge of a compound's mode of action towards different species, preferably at the biomolecular level, is hence essential when selecting an appropriate QSAR for predicting toxic effects.

The quantification of toxicity does not generally refer to a compound's concentration at the site of action. Instead, it is extrapolated from the administered dose (LD/ED) or the concentration in the surrounding medium (LC/EC). Assuming an identical mode of action, constant concentrations of different contaminants at the target site (or rather constant activities accounting for the non-ideal behaviour of the toxicant in the target phase; Ferguson, 1939) were postulated to be equipotent (McCarty, 1987). Compounds with the same mode of action are thus assumed to have identical intrinsic activity, and differences in their effects are solely due to their pharmacokinetic behaviour. The same argument is used for the critical-volume hypothesis (Abernethy, Mackay and McCarty, 1988; Jaworska and Schultz, 1993), which relates, for example, non-specific effects to the volume occupied in membranes by the toxic agents. A particular type of effect is then evoked when the abundance of any chemical, or the sum of the chemicals present, reaches a critical threshold (CBB: critical body burden). Different modes of action may be discerned with different concentration levels at the respective targets. Because of difficulties in identifying the exact target sites and in measuring the local amounts of toxicants, surrogate measurements (e.g. concentrations in the exposure medium) are generally used. The extrapolation of critical body residues from ambient concentrations does not overcome these short-comings (McCarty *et al.*, 1991, McCarty and Mackay, 1993); lethal body bur-den are not necessarily constant and may be influenced by exposure levels and exposure times in relation to the survival time of the tested animals as well as their lipid content (Wolf *et al.*, 1992a; Wezel *et al.*, 1995). Furthermore, a comparison of measured and calculated critical body burden ($CBB_{calc.} = LC/EC_{50} \times BCF$) suggests that the estimates are higher than the actual concentrations in the organisms (Figure 2.9); that is, toxicity may occur at levels much lower than $CBB_{calc.}$.

These discrepancies can be attributed only partly to the non-attainment of steady state in toxicity measurements as compared to BCF determinations (McKim and Schmieder, 1991). The use of calculated CBB data is hence of limited reliability and may not protect the organisms exposed.

When any species is exposed to xenobiotics, the minimum effects observed are evoked by partitioning into lipid phases. This so-called non-specific toxi-city (baseline toxicity) is caused by membrane perturbation, predominant with many environmentally relevant chemicals. The membrane lipid and pro-

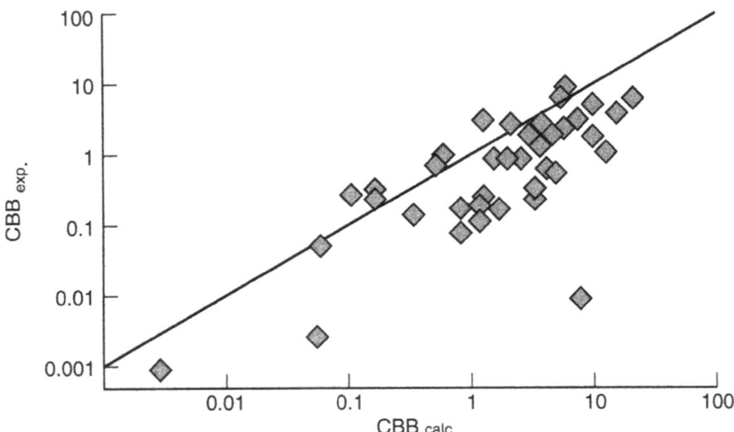


**Figure 2.9** Comparison of measured and calculated ($LC_{50} \times BCF$) critical body burden (CBB in mmol/kg). Data taken from McKim and Schmieder (1991) and Sijm, Schipper and Opperhuizen (1993). The calculated threshold residues in the organisms appear sys-tematically higher than the actual ones, thus resulting in an underestimation of toxicity. This has been attributed partly to the non-attainment of steady state in toxicity measure-ments as compared to BCF determinations.

tein components hence have to be regarded as the most relevant targets for environmental toxicants, regardless of the species. Alterations in the lipid matrix may adversely influence the functioning of membrane proteins. However, although baseline effects caused by water/membrane partitioning can be observed in all kinds of organisms, there are considerable differences in their sensitivity: tests with plants, microorganisms and *in vitro* systems tend to be less sensitive than those with animals with a nervous system. Nerve cells may be more responsive because of a mode of action common to them and the so-called non-specific effects may not be caused by a single toxicological phenomenon (Donkin, 1994). Furthermore, the varying sensitivity may be due to different pharmacokinetics in the different test set-ups. Deviations from baseline toxicity may occur in either direction. If further specific targets for a given chemical are present in an organism, then the toxicity may be drastically enhanced. A toxicity lower than baseline is encountered only with contaminants that are unstable over the observation time (e.g. due to metabolic effects); in other words, the observed effects do not refer to the chemical itself but to its degradation products. The essential consequence is that a compound that is sufficiently stable is always at least as toxic as the baseline, (non-)linearly related, for example, to log $P_{ow}$. However, it may be much more noxious for some species because it attacks specific targets.

The biological parameters used in environmental hazard assessments are usually obtained from several species regarded as representative of selected groups of organisms in the aquatic, terrestrial and soil ecospheres:

Aquatic ecosphere:
    fish (secondary consumers)
    *Daphnia* (primary consumers)
    algae (primary producers).
Terrestrial ecosphere:
    mammals (secondary consumers)
    plants (primary producers).
Soil ecosphere:
    earthworm (secondary decomposer)
    microorganisms (primary decomposer).

The respective standard endpoints ($LD_{50}/LC_{50}$) concern impacts that are much more severe than are tolerable in the environment – usually the death of 50% of the exposed individuals after an acute exposure (short span of the individuals' life time) or a chronic exposure (one to several consecutive generations of a species). This high lethality is tested for practical reasons: the dose–response curves are steepest for the 50% observed effects and such concentrations can be quantified with high precision (Stephan, 1977; Molinengo, 1979). The alternative of registering the concentrations with no observable effects (NOEC) may be biased towards the effects monitored, which may not coincide with those really occurring.

## 2.3 QUALITY REQUIREMENTS FOR ACTIVITY AND EFFECTS DATA

A comprehensive set of substance-specific data for environmental hazard assessment requires information on well-defined endpoints concerning

transport
bioavailability
ecotoxicity
transformation.

The data-quality requirements for QSAR models relate to several aspects of the experimental procedure, data transformation and the selection of the appropriate test compounds. Only if the input data of a QSAR meet the highest quality standards may a sound model be derived. Because the accuracy of predictions can never be better than the variability of the respective measurements (usually ± 20% and more), validity assessment of the activity and effects data is crucial in QSAR derivations. The data should be generated by tests that are methodologically and mechanistically defined. The latter is not trivial for parameters such as biodegradability, soil sorption and ecotoxicity. With regard to the considerable variability of measurements, inter- and also intra-laboratory, the test results, especially when collected from different literature sources, should be critically evaluated with respect to:

a standardized protocol
reproducibility of results
interpretability of data with regard to the relevance of effects.

The experimental data underlying any QSAR model should preferably be determined by standardized protocols, but the absolute values will still vary in reliability and precision, which must affect the accuracy of the predictions based on the acquired QSARs. Besides sufficient precision and reliability, the activity and effects data for QSAR analyses should principally meet these additional requirements:

coverage of a broad activity range;
even distribution over the activity range;
identical mode of interaction (parallel dose–response curves);
activity expressed as a function of applied concentration (e.g. $EC_{50}$);
activity expressed in molar concentrations because, for example, applied doses
    on a weight/weight basis are not comparable for different compounds;
effects data (%) to be transformed accordingly;
variability in the test data has to be accounted for;
the test system should not be more complex than can be mechanistically
    understood;
considerations have to be given to the time course of the test (e.g. if steady
    state has been achieved).

The compounds subject to a QSAR analysis must differ sufficiently in activity to provide the basis for a representative model and to avoid the description of local phenomena. If the activity range is too narrow it may over-value the inherent variance in the experimental data, which may significantly exceed ± 20% for parameters such as soil sorption, bioconcentration and ecotoxicity. The difference in activity between compounds must therefore clearly exceed that among the replicated measurements for each of the test substances. Otherwise, the derived QSAR may show an artificial relationship between the compounds' structure and the experimental scatter, but not the effects under investigation.

The basic paradigm that QSARs are always specific to the mode of interaction demands that activity and effects data reflect analogous processes among the series of test compounds with regard to this endpoint. The necessary information may be taken from dose–response curves and observations during the experiments, such as the onset and the time course of the effects. Ideally, the assignment of chemicals to a mode of action class is based on the comprehensive understanding of the underlying processes. If chemicals interact by the same mechanism with the target site – that is, their effects differ quantitatively but not qualitatively – they must do so by exerting the same principal properties; hence they must obey the same QSAR. However, the reverse conclusion does not hold true. Those compounds that conform to a given QSAR do not necessarily act by a uniform mode; an apparent coincidence may occur by chance within the studied set of compounds. If an appropriate selection of substances is taken from the different classes, different QSARs for the same endpoint are likely to result. From the fit or misfit (outlying) with a QSAR, only two conclusions can hence be drawn (these are not valid in the reverse direction):

Identical mode of action must yield identical QSARs.
Different QSARs imply different modes of action.

Conversely, identical QSARs do not prove identical modes of action and different modes of action do not always result in different QSARs. It thus follows that no QSAR can be applied for all environmentally relevant compounds successfully, but it is valid only for those that interact according to the presumed underlying mode of interaction. To obtain reliable predictions, the chemicals have to be classified by mode of interaction with the target, generally from structural analogies. Once the class has been determined, the corresponding QSARs can be identified. For the selection of the appropriate model further limitations have to be considered, such as statistical quality, parameter range, detected outliers and validation status.

# 3

# *Statistical methods*

For the formal description of relationships between activity measures and structural descriptors of compounds various statistical techniques can be used, such as:

(Non-)linear (multiple) regression:
    independent variables: physico-chemical descriptors (Hansch approach)
    independent variables: substructure indicators (Free-Wilson method)
Multivariate statistics:
    principal component analysis (PCA)
    factor analysis (FA)
    partial least-squares analysis (PLS)
Classification methods:
    pattern recognition
    cluster analysis
    discriminant analysis.

The method of choice depends on the type of activity to be modelled and on the quality and quantity of the data. The derivation of meaningful QSARs always consists of three crucial steps:

rational design of test series
statistical modelling
validation.

Sound QSARs for predictive purposes have to satisfy two criteria:

The models provide an accurate description of the activity of all compounds it is based on.
The predictions are made with as few parameters as possible in relation to the degrees of freedom in the data.

The objective of any statistical modelling is not to obtain correlations but to account for mechanistic relationships. A good model describes the observed activity or effects with only few, generally $\leq 3$, process-related properties. In contrast, a model with a large number of arbitrary variables is highly likely to represent a chance correlation and will reflect a lack of expertise in the application of statistical tools. It is a trivial fact that increasing the number of descriptor variables ultimately leads to statistically perfect correlations, especially if the number of variables equals the number of activity data points. Such a relationship may exactly reproduce the given data set (i.e. $r^2 = 1.00$) but it lacks scientific credibility and has definitively no predictive power. Proficient QSAR modelling thus requires at least a basic understanding of the constraints of statistics.

## 3.1 RATIONAL DESIGN OF TEST SERIES

The amount of information obtainable from any QSAR analysis, and hence ultimately its predictive power, largely depends on the initial design of the underlying experimental study. Only if the compounds included in the test data set are true representatives of the chemical class(es) concerned, can the maximum information be extracted from the data. The members of the training set, which are used to derive the model, should be as diverse as possible with respect to their structural features. In statistical terms this means that they should reveal minimum intercorrelation and maximum variance in the properties regarded relevant for the effects studied.

When selecting compounds for investigations the underlying issue is the problem of defining the structural and functional similarity among the chemicals. Two compounds are considered similar if they resemble each other in one or more critical aspects regarding, for example, their structural, stereo-electronic or physico-chemical features (Basak and Grunwald, 1994, 1995; Basak, Bertelsen and Grunwald, 1995). Similarity measures rely on the distance between descriptor coordinates of the chemicals in the $n$-dimensional property space. The closer the respective data points, the more similar are the compounds. Various statistical techniques provide an assessment of similarity distance functions, where an increased number of descriptor variables does not necessarily improve the effectiveness of similarity recognition (Hunter, 1994). Depending on the descriptor set evaluated, different aspects of similarity among substances can be represented. Congruence of steric and geometric properties of compounds does not imply any similarity with regard to their reactivity or their mode of toxic action.

The rational selection of a test series for QSAR analysis (Craig, 1971; Schaper, 1983) is thus an essential prerequisite for avoiding chance interdependences between descriptor variables. From the pool of imaginable derivatives, the training-set compounds need to be selected based on data for a variety of descriptor variables. Clearly, only theoretical and computable or tabulated parameters characterizing the range of properties among this series of chemicals can be considered at this stage. In the case of two or three potentially relevant variables – generally the case for environmentally relevant endpoints – the selection of the 'most diverse' substances for the training set can be done by visual inspection of the data distribution (Figure 3.1).

If an unsuitable selection is made (Figure 3.1A) it will be impossible with the subsequent QSAR analysis to decide merely on a statistical basis which

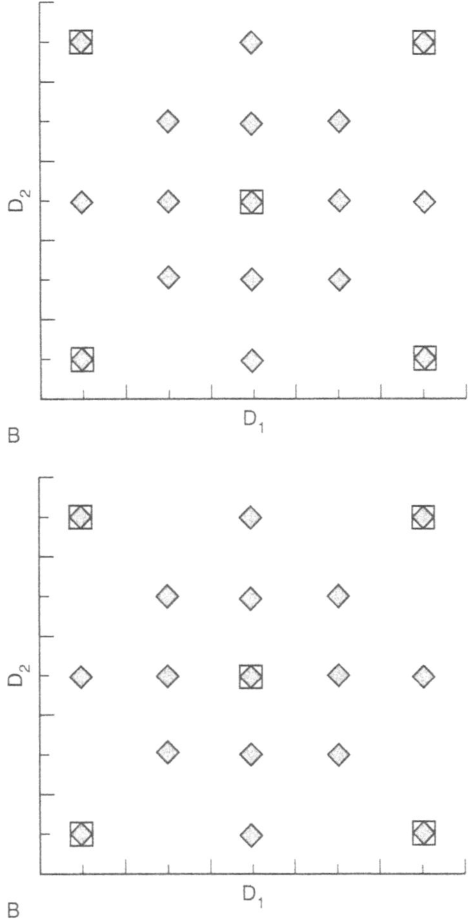

**Figure 3.1**  Selection of compounds for a training set for a QSAR derivation based on their properties $D_1$ and $D_2$. The compounds selected for the QSAR analysis are indicated by the frames. A: inappropriate selection of the test compounds due to the high intercorrelation between their descriptors $D_1$ and $D_2$; B: selection of the test compounds with regard to maximum variations and minimum intercorrelation between their descriptors $D_1$ and $D_2$.

property, either $D_1$ or $D_2$, determines the observed effects and to what extent. If only one parameter is relevant, no decision can be taken for either of the variables because they will describe the activity data equally well. A practical example is the frequent preference for series of chlorinated phenols, where the intercorrelation between $\log P_{ow}$ and $pK_a$ values is high ($r > 0.9$), which guarantees statistically good QSARs but does not provide any basis for the recognition of the relevant properties. In contrast, the selection shown in Figure 3.1B represents a training set that spans the entire range of both variables $D_1$ and $D_2$, but avoids any correlation between them. Based on such a set of compounds, a meaningful and unbiased interpretation of the resulting QSAR becomes feasible and the extent of observed effects can be definitively attributed to structural features of the compounds.

The selection of the training set becomes more sophisticated if more than three descriptors are assumed to be relevant. A manual selection is not feasible with a parameter space that is more than three-dimensional and computational techniques have to be applied. These mostly use multivariate statistics to account for underlying multiple intercorrelations among variables, such as PCA and PLS (Hellberg, 1986; Tosato *et al.*, 1991; Lindgren *et al.*, 1995), to find an acceptable compromise with respect to collinearity, variance and synthetic/commercial accessibility of the chemicals.

There is no excuse for not making an effort to choose the appropriate compounds to form the basis of the intended QSAR. Even if little information is available on the relevant descriptors, there is at least always the possibility of using a $\pi/\sigma$-substituent constants diagram of the candidate chemicals to assess the intercorrelations between lipophilic and electronic properties.

## 3.2 STATISTICAL MODELLING

Statistical modelling of data sets is one of the three essentials in QSAR analysis. There are powerful tools for unravelling the principles and rules hidden in the pool of experimental data. Before a successful analysis, however, some basic truisms have to be realized:

Qualitative understanding of the activity under study is essential for its quantitative description by a QSAR.
Statistics are a powerful tool in QSAR analysis, but nothing more than a tool.
There are no causal relationships between the statistical and the scientific significance of a QSAR model.

### B3.2.1 REGRESSION ANALYSIS

Regression analysis has been the favoured statistical method for QSAR analyses because it is both simple and illustrative. Regression analysis is an exact mathematical procedure (Draper and Smith, 1981; Weisberg, 1985) to correlate independent $X$ variables (the descriptors $X_1$, $X_2$,...) with dependent $Y$ variables (the activity measures $Y$) in the form of

$$Y = a\,X_1 + b\,X_2 + ... + c$$

where $a$ and $b$ represent the corresponding regression coefficients for the $X$ variables (i.e. the slopes), and $c$ the intercept. By regression analysis an expression is derived where the sum of the squared residuals of the underlying data from the function is minimal. The $X$ variables may be used either directly or after scaling, which compensates for large differences in the absolute values of different descriptors and allows a better comparison of the weight of the individual parameters in explaining $Y$. Principally, only the $Y$ variable is assumed to be subject to errors (i.e. data scatter), but this evidently also applies to $X$ variables such as physico-chemical data. The effects on the QSAR model of dealing with data with a potentially substantial error always has to be taken into account.

Regression models with several $X$ variables – multiple linear regression (MLR) – may result from either a manual selection (interaction-based) or from an automated selection (statistics-based) of the compound descriptors. For automated stepwise variable selection, variables are either added (forward selection) or eliminated (backward selection), which contribute the most or the least, respectively, to the significance of the model. Because intercorrelations between the variables are neglected by this technique, there is the danger that it will produce a local optimum instead of the global optimum of the model. This pitfall may occur if, for example, one variable reveals a much larger correlation with the activity data than the other descriptors, and hence is always included in the model for statistical reasons. It may mask the fact that a combination of two other parameters could provide a much better, and possibly more meaningful, description of the observed activity data.

To comply with good statistical practice, several criteria have to be met for a significant regression model based on either the physico-chemical descriptors (Hansch approach) or the indicator variables (Free–Wilson method):

The $Y$ data must have been obtained on a continuous scale, such as that for $LC_{50}$ data (Figure 3.2).

The $X$ and $Y$ data should cover several orders of magnitude such that differences in the data for the set of compounds definitively exceed the variability among replicated data for each individual compound (Figure 3.3).

The $X$ and $Y$ data must be evenly distributed over the entire range of the model to avoid two-point correlations (Figure 3.4).

The minimum of data points per independent $X$ variable is 4–5; the desirable range is > 10–15.

Regression analysis is not applicable for modelling discrete response data such as biodegradability (classifying compounds either 'readily degradable' or 'non-readily degradable'), because two-point correlations are the inevitable result (Figure 3.2).

The significance of a derived QSAR model is indicated by the statistical terms:

correlation coefficient $r$
standard deviation $s$
$F$ value.

The correlation coefficient $r$ is a relative measure of the goodness of fit of the data points to the model, and takes the value of 1.0 for the perfect relationship.

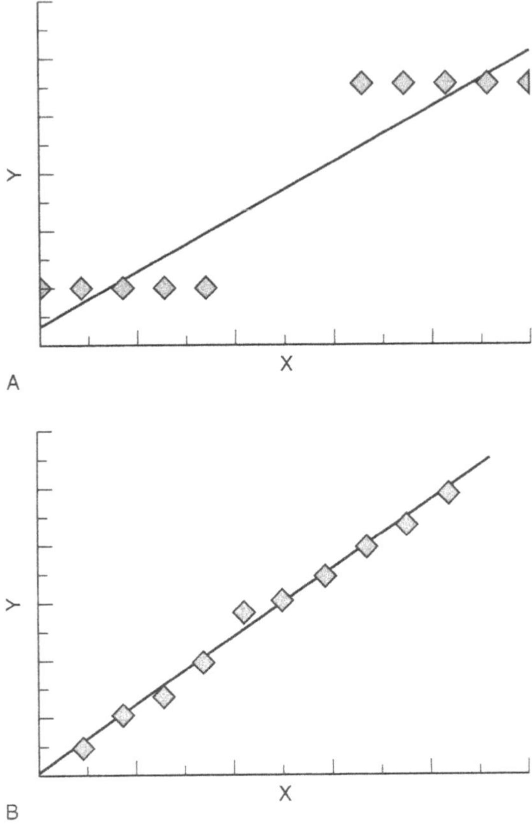

**Figure 3.2**   Example of the inappropriate application of regression analysis to dichotomous activity data (A) as compared to continuous activity data (B). Data distributed as in A are unsuitable for regression analysis because it results in a two-point correlation.

$$r = \frac{\sum \Delta X \Delta Y}{\sqrt{(\sum \Delta X^2 \sum \Delta Y^2)}}$$

$\Delta X$ and $\Delta Y$ represent the deviation of the actual $X$ and $Y$ data from the calculated function. Because the $r$ value depends on the overall variance of the dependent variable, and therefore on the number of data points and their range, it relates only to the specific data set analysed and gives no indication of the predictive power of the model. The squared correlation coefficient $r^2$ represents the explained variance (i.e. the variance in the $Y$ values accounted for by the $X$ variables), mostly expressed as a percentage. Ideally, the model covers all the variance of the $Y$ data except the experimental scatter. It is essential, therefore, to be aware of the uncertainty in the underlying activity data; for example, *in vivo* toxicity data generally vary by > 20% among

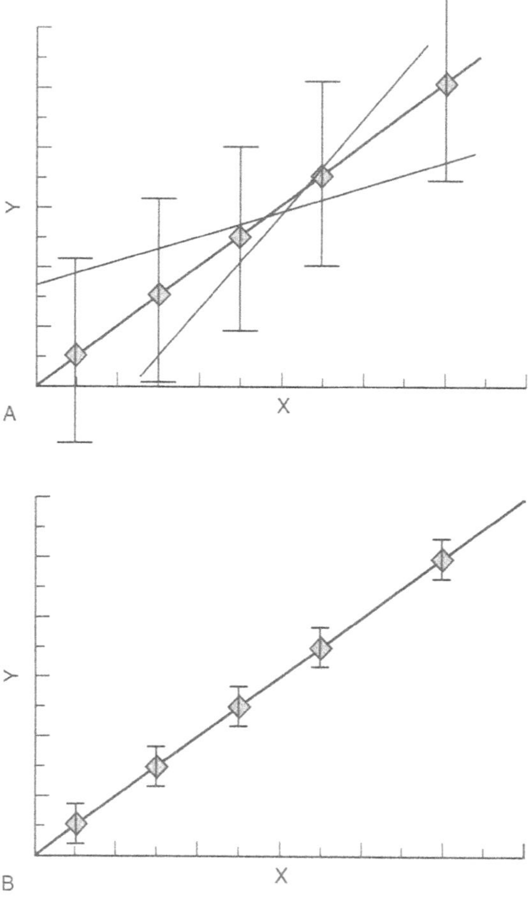

**Figure 3.3**  Example of the inappropriate application of regression analysis to highly variable activity data (A) as compared to activity data with low variability as compared to the activity range covered (B). The high uncertainty in data set A does not allow the definition of the true regression model; rather a bundle of different lines can be fitted to these data.

replicates. If the input data for $Y$ indeed scatter by $\geq 20\%$, a $r^2$ value of 0.8 (i.e. 80% variance explained and $r = 0.9$) thus indicates sufficient fit of the model. Consideration of further $X$ variables is inadvisable so as not to produce an overfitted model. With such an uncertainty in the $Y$ data, any discussion about whether a QSAR with $r = 0.989$ or $r = 0.992$ is better becomes superficial. The QSAR with a high $r$ value is hence not necessarily 'the best', because it can also be a statistical artefact describing the experimental scatter in this particular data set.

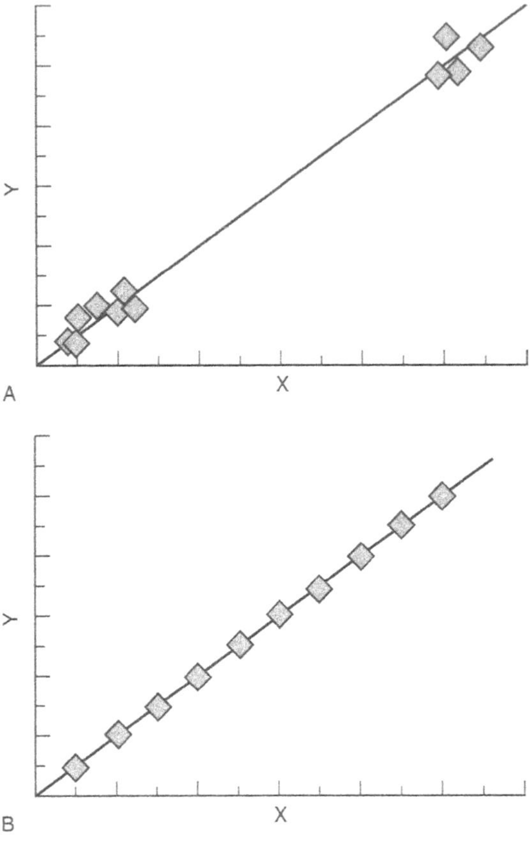

**Figure 3.4**  Example of the inappropriate application of regression analysis to clustered activity data (A) as compared to evenly distributed activity data (B). Data distributed as in A are unsuitable for regression analysis because it results in a two-point correlation.

The standard deviation $s$ is a measure of the goodness of fit of a correlation, which is required for the determination of confidence intervals and which approaches 0.0 for the perfect relationship.

$$s = \sqrt{\frac{\Sigma(\Delta Y)^2}{\mathrm{DF}}}$$

The $s$ value depends on the distance of the observed $Y$ values from the derived function and on the degrees of freedom (DF)

$$\mathrm{DF} = n - k - 1$$

which are determined by the number of compounds in the data set ($n$) and the number of $X$ variables ($k$) in the model. As a consequence, the $s$ value will be smaller the more compounds are described by a QSAR with lesser descriptors. The standard deviation of a model should not exceed the standard deviation of the activity data, which ranges about 0.3 log units for most biological data.

The $F$ value is a measure of the statistical significance of a correlation, which is determined mostly by the number of $X$ variables used in the model in relation to the number of data points included.

$$F = \frac{r^2 \mathrm{DF}}{k(1 - r^2)}$$

The $F$ values, which are tabulated for different degrees of freedom at different significance levels (e.g. Seydel and Schaper, 1979), must correspond to the $\geq 95\%$ significance level to indicate the significance of a regression model.

The most effective check of the soundness of a regression model is always the graphical representation of the derived QSAR in relation to all data points including outliers, because then any mistakes become clearly evident (for comprehensive validation of statistical models see section 3.4). The easy access to statistical methods tempts researchers to play around with the data. Even though this approach may sometimes yield surprising findings, it has to be made clear that the results have to conform with statistical and scientific quality standards and that the objective remains to derive as simple as possible models to describe the data of interest in a meaningful way.

### *(a) Linear free-energy relationships (LFERs): the Hansch approach*

The Hansch approach uses multiple regression statistics to describe changes in the observed effects among chemicals from differences in their physico-chemical properties. The basic presumption is the linear dependence of the activity on the free energy (linear free-energy relationships LFERs). The regression coefficients for the descriptors in the QSAR model then indicate the relative contributions of the respective properties to the magnitude of the effect. A large positive coefficient characterizes an enhancing effect on the activity, whereas a negative coefficient flags those properties of compounds that decrease an activity, such as toxicity.

The range of the descriptors covered by the compounds included in the model instantaneously sets the limits for predictive applications of the QSAR, because there is no certainty that the relationship holds beyond this domain. With $\log P_{ow}$-dependent models, for example, it is frequently found that they are linear between $\log P_{ow}$ 0 and $\log P_{ow}$ 5, but reveal curvature outside this range. Accordingly, predictions based on improper extrapolations outside the linear domain covered by the QSAR may be wrong by several orders of magnitude.

The major advantage of the Hansch approach is the utilization of principal physico-chemical properties of the chemicals for descriptors. Hence the derived models can allow interpretations on a mechanistic basis, if the technique is applied correctly:

The data are evenly distributed over the range of the activity and the descriptor(s).
The descriptors used simultaneously in a QSAR are not intercorrelated.
The range covered by each parameter exceeds significantly the variability in the parameter for the individual compounds.
The activity is due to a uniform mode of interaction.
The ratio of the number of compounds in the data set to the number of parameters in the model is at least 5:1.
The QSAR has been validated with a set of compounds not used for the derivation of the model.

In addition to the QSAR equation and the respective statistics, further information is necessary for predictive applications:

compound class(es) covered
mode of action accounted for
parameter(s) range covered
recognized outliers (i.e. (classes of) compounds for which the model is not valid).

Sometimes, the derivation of a 'simple' Hansch equation is the result of a complex series of analyses and hypotheses on the mechanisms underlying the biological activity/property under study. The final equation thus summarizes such preliminary work. The correct application of Hansch-type QSARs provides a powerful tool for estimating parameters in environmental sciences and its prevailing efficiency is illustrated throughout Part 2.

### (b) Substructure models: the Free–Wilson method

Like the continuous physico-chemical descriptor $X$ variables, indicators of the presence or absence of certain substructures have also been treated by multiple regression analysis. As modified by Fujita and Ban (Seydel and Schaper, 1979), this group contribution method can be a useful alternative to the LFER approach, if only limited knowledge is available about the relevant molecular properties or no uniform physico-chemical descriptors for the various compounds in the data set are accessible. For activities and properties of compounds that may be attributed to the occurrence of certain substructures in the molecules (e.g. biodegradation; section 4.8), Free–Wilson-type substructure models have their major application in environmental sciences.

The activity of the parent compound is taken as the reference and the contributions of the substituents of the derivatives are accounted for by the indicator variables; thus the Free–Wilson model can take an analogous form of the Hammett equation:

$$\log 1/C = \Sigma\, a_i I + A_p$$

$C$ is the activity measure, $I$ represents the indicator variables that can take values of either 0 or 1, $a_i$ is the group contribution of substituent $X_i$ relative to $a_H = 0$, and $A_p$ is the activity (expressed as log $1/C$) of the non-substituted parent compound. The group increments are a reflection of the sum of all structural effects influencing the activity of the derivatives. Except for the molecular structure of the compounds and the respective activity measures, no further information is required, hence the analysis can be conducted fast and simply. The Free–Wilson method is more efficient when the structures are more complex (i.e. those with the greatest variety of substitution sites and substituents). A ranking of the substituents by their effects on a compound's activity can reveal which substituent in which position is most effective at promoting the observed type of activity. Although substructure-based QSAR modelling appears to be quite a simple approach, the basics of statistics have to be respected, and large sets of compounds are often needed:

Each relevant substructure has to occur in a sufficient number of compounds. Each substructure has to occur in combination with different other substructures.
For comprehensive predictive power of the model, a wide variety of substructures has to be considered.

The width of the confidence interval for the activity increments and the calculated activity values allow the identification of conspicuous substructures, which deviate with respect to the compounds' overall activity. Also it reveals in which position the activity contributions obey the underlying additivity assumption (independent contributions) or not. This obvious advantage of the Free–Wilson method is also its major weakness: its predictive use is strictly limited to substances with substructures in the data set used to derive the model. The occurrence of other structural moieties may affect the activity strongly. The neglect of this basic principle is likely to result in erroneous estimates.

### (c) Non-linear regression models

The linear models used most often represent only a restricted section of structure–activity relationships. There is no reason to assume, for example, that the enhancing effect of the increase in lipophilicity on a compound's toxicity to fish will perpetuate for chemicals representing the extremes of the lipophilicity scale (log $P_{ow} \gg 6$ or $\ll 0$). Rather, non-linearity in these relationships is to be expected because of the establishment of a highly complex set of equilibria between the aqueous and non-polar phases at different rates from the release of the chemical into the system until the interaction with the target site. The linearly increasing section of the QSAR may progress into a plateau or into a more or less rapid decrease in activity after a further increase in the respective compound property. A variety of factors may contribute to a non-linear dependence of activity on structural features:

The activity is measured before a steady state has been reached.
The solubility limits are exceeded for some compounds.
Micelles formation occurs for highly lipophilic compounds.
Diffusion across membranes is hindered for very large molecules ($\geq 9.5$ Å).
Transport to the target site involves partitioning between phases of different lipophilicity, hence the rate constants for the partitioning processes are related to log $P_{ow}$ by different LFERs that may counteract to different extents (differences in kinetics of transport and distribution).
The model assumptions become invalid if a certain parameter range is surpassed (e.g. changing mode of interaction).

The resulting relationships may be described with polynominal functions (simplest case: parabolic) or bilinear functions, depending on the data set and the computing efforts (Figure 3.5).

Based on investigations of transport phenomena involved in the pharmacokinetics of various drugs, Hansch suggested that parabolic relationships (Hansch and Clayton, 1973) account for the dependence of transport rates on lipophilicity. For compounds of an optimum log $P_{ow(max.)}$, which is specific for each biological system, the transport rate will be maximal, yielding a maximum concentration of the chemicals at the target site within a given time. Any change in log $P_{ow}$ then results in a decreased transport rate and hence decreased biological activity.

$$\log 1/C = a \log P_{ow} + b (\log P_{ow})^2 + c$$

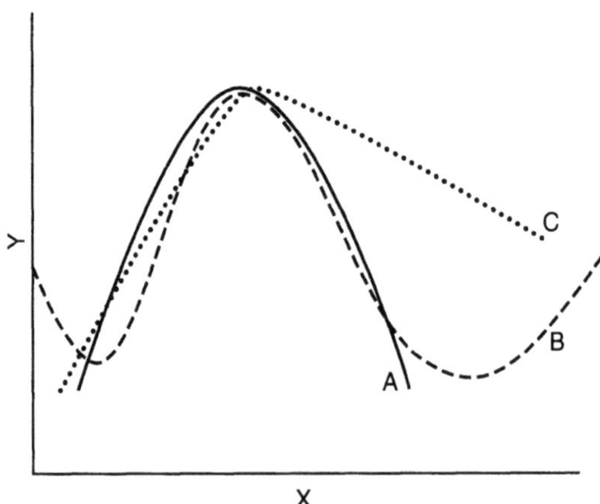


**Figure 3.5**  Examples on non-linear regression functions (A, parabolic; B, polynomial; C, bilinear).

For 16 different sets of hypnotics tested in a variety of ways on mice, rats, rabbits and guinea pigs, an average log $P_{ow(max.)}$ of 2 (range 1.5–2.7) was found (Hansch 1968a). The log $P_{ow(max.)}$ value indicates the optimum lipophilic character for a set of compounds that coincides with least restrictions to their random walk through biological tissue to the sites of action. The further away from equilibrium the activity is recorded, the lower log $P_{ow(max.)}$ mostly will be.

Sustaining the transport model, further non-linear representations of the observed structure–activity relationships were derived. According to the McFarland model (Seydel and Schaper, 1979; Kubinyi 1993), the probability of a drug reaching the receptor after passing several membranes depends on its lipophilicity in a symmetrical manner with linearly increasing and decreasing sides of the curve:

$$\log 1/C = a \log P_{ow} - 2a \log (P_{ow} + 1) + c$$

This function yields an improved description especially of the ascending left-hand side, where the parabola results in systematic deviations due to its regular curvature. However, the practical use of this function is minor, because it always results in log $P_{ow(max.)}$ values of zero, even when different numbers of barriers are considered.

The protein-binding model of Franke yields a complex function between the logarithm of the biological activity and some suitable hydrophobic parameter consisting of a linear part that passes into a parabola if steric hindrance occurs (Franke and Schmidt, 1973).

$$\log 1/C = a (\log [P_{ow} > Px])^2 + b \log P_{ow} + c$$

The point at which the linear course turns into curvature is denoted log $Px$, which does not coincide with the maximum of the function. The expression $[P_{ow} > Px]$ takes the value of zero if log $P_{ow} < \log Px$ or the value of the difference (log $P_{ow} - \log Px$) if log $P_{ow} > \log Px$.

The bilinear model of Kubinyi (1977, 1993) was derived from a reconsideration of the McFarland probability model and takes into account the different volumes of the aqueous and the organic target phases in biological systems:

$$\log 1/C = a \log P_{ow} - b \log (\beta P_{ow} + 1) + c$$

The term $a$ gives the slope of the left-hand ascending side of the curve and $(a - b)$ that of the right-hand descending side. The non-linear parameter $\beta$, which must be estimated by a stepwise iteration procedure, relates to the volume ratio of the aqueous and lipid phases in the system. Setting $\beta = 1$ and $b = 2a$ produces the original McFarland model. Kubiny's bilinear model can be derived from kinetically controlled model systems as well as from equilibrium models, indicating that it is valid under diffusion control as well as under equilibrium or pseudo-equilibrium conditions. For many data sets, the bilinear function aptly fits the experimental observations. Difficulties in calculations may arise from unbalanced data sets, which often occur in environmental sciences. Because of steady-state testing conditions, the log $P_{\text{ow(max.)}}$ is generally high (log $P_{\text{ow}}$ 5–7) and only few precise measurements are available to support the exact course of the right-hand side of the curve (Figure 3.6).

As for the linear models, the non-linear relationships are also restricted to predictions within the parameter range covered, and extrapolations outside this domain – as evident, for example, for fourth-order polynominal functions (which are of questionable meaning for QSARs) – are hazardous.

### 3.2.2 MULTIVARIATE STATISTICS

The handling of large data arrays on a multitude of endpoints and parameters and the extraction of the information they contain (as in a hazard profile) requires the application of multivariate statistics. In environmental studies these are chiefly used in two fields:

direct comparison of an array of test parameters;
derivation of powerful QSARs based on the information from several tests.

For series of compounds, the data obtained by several testing procedures constitute a multidimensional parameter space in which the inherent hidden

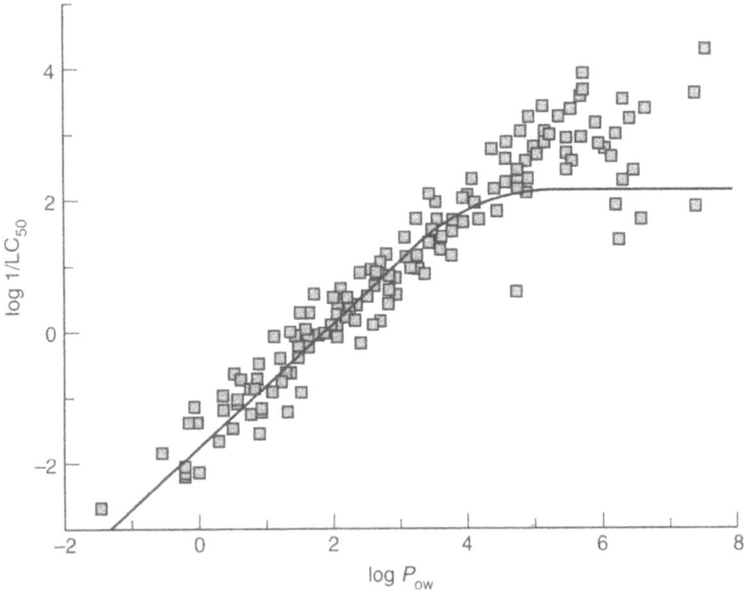

**Figure 3.6**  Example of a non-linear log $P_{\text{ow}}$-dependent model (Veith, Call and Brooke, 1983) in comparison to experimental data on fish toxicity (log $1/LC_{50}$ in mmol/l). The bilinear function provides a good description of the activity data for the compounds with log $P_{\text{ow}} <$ log $P_{\text{OW(max.)}}$, whereas the model is less well defined for the highly lipophilic substances because their data are extremely variable.

intercorrelations are not easily detected. Regression analysis of such data sets yields only some aspects of the information. To recognize the extent and the nature of the basic properties encoded in such data sets, principal component analysis (PCA) or factor analysis (FA) can be used to analyse data matrices with measurements for chemicals obtained in a series of test systems simultaneously (Wold *et al.*, 1984; Thielemans and Massart, 1985; Schaper and Kaliszan, 1987; Jolliffe, 1986; Johnson and Wichern, 1988; Martens and Naes, 1989; Eriksson and Hermens, 1995). Practical applications include the analysis of molecular descriptors and various activity data to reveal mutual relationships and possible redundancies between parameters (section 1.6 and Chapter 8). By explaining the variance/covariance structure of the data matrix through linear combinations, the dimensionality of the data set can be reduced. Hypothetical variables, the principal components (PC) corresponding to the information content of groups of the original variables, are extracted. The PCs are orthogonal vectors, calculated by linear combination of the original (scaled) variables, and they are supposed to represent the substantial, but unobservable, properties. The first PC is determined to be the linear function representing the largest portion possible of the total variance of the data; that is, the function most correlated with the original variables (Figure 3.7).

From the remaining variation in the data cloud a second PC is extracted that is not correlated with the first PC and explains more of the remaining variance than any further PC could. This procedure is then continued until $n$ orthogonal PCs are calculated from the $n$ variables. Geometrically, the PCs correspond to a new coordinate system obtained by rotating the $n$-dimensional data cloud, with the original variables serving as the starting coordinate axes. The relative placement of the data points in the multidimensional space is not altered during the rotation. The new axes represent the orthogonal directions with maximum variability and provide a simpler and more comprehensive description of the covariance structure of the data set. This corresponds to a classification of the original variables into clusters with respect to their intercorrelations. All variables in a particular group are highly correlated among themselves, but have relatively small correlations with variables in the other groups. Each group corresponds to an underlying principal factor (i.e. PC) that is responsible for the observed correlations.

For comprehensive analysis, only those PCs that represent a significant amount of the total information (i.e. variance) will be selected. Further PCs, assumed to contain little or no substantial information, will then be ignored. This rejection of variables corresponds to the objective of PCA to reduce the dimensionality of the data matrix. Two criteria should be used to select significant PCs:

The eigenvalue should be > 1: if it is not, the PC represents less information than corresponds to one original variable. Such minor variance may correspond to experimental data scatter, which can (should) be ignored in further analysis.

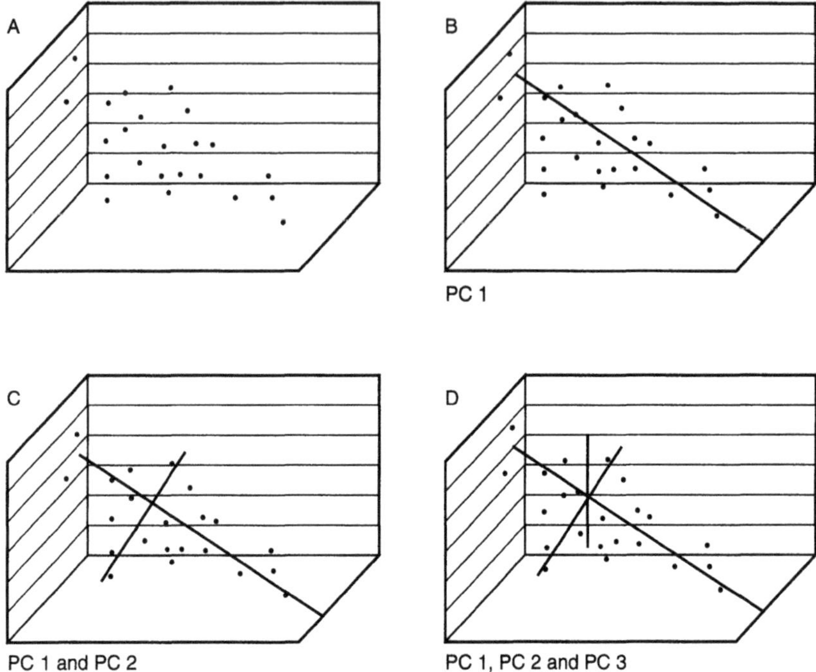


**Figure 3.7** Derivation of principal components (PC) in a three-dimensional parameter space. A: data cloud – the vectors in B, C and D correspond to the first, second and third PC, respectively; B: the first PC represents the maximum variance in the data cloud; C: the second PC is orthogonal to PC1 and represents the second largest variance in the data cloud; D: the third PC is orthogonal to PC1 and PC2 and represents the third largest variance in the data cloud.

The total % variance explained should not exceed the accuracy of the underlying data. If, for example, toxicity data are subject to 20% variability, it is inappropriate to extract more PCs than correspond to 80% (max. 85%) of the total variance (proportion of information). The remainder is, for example, experimental data scatter.

The number of extracted PCs should be limited to those with the largest eigenvalues that represent the substantial variance of the original data. In many data sets on (eco)toxicological data, these two rules of thumb will result in the selection of two or three PCs representing about 80–90% of the total variance. A further indication of whether the selected PCs are appropriate to represent the underlying data set is obtained from the PC loadings. The loadings are specified in terms of eigenvalue/eigenvector pairs. For the $n$ original variables with respect to all possible PCs, they represent the correlations between the PCs and the original variables. The larger the absolute value ($\leq 1$ for scaled variables) of a PC loading, the more information from the original variable is contained in the PC. Only if considerable loadings ($> 1/n$) are observed, can the PC be regarded as significant. From the loadings variables matrix those original variables can be identified that contribute to the same PC(s) and thus can be assumed to contain similar information. (For examples of PCA application see section 1.6 (relationships between descriptors of chemical structures) and Chapter 8 (comparison of toxic effects in several biotests).)

The placing of the original variables' vectors in the PC plot visualizes the intercorrelations between the test systems (Figure 3.8A).

The cosine of the angle between variable vectors corresponds to the respective correlation coefficient (i.e. the smaller the angle between the vectors, the higher the intercorrelation). When more than one significant PC underlies the data set, there is always some inherent ambiguity associated with the model. Even when the loadings are, in general, different, the PCs may have identical statistical properties. To improve the interpretation of the extracted PCs, the variables' vectors can be rotated in the component space. The variance of the squared PC loadings is hence maximized, without changing the placement of the vectors relative to each other (Figure 3.8B). Ideally, the rotation reveals patterns of loadings such that each variable loads highly on a single PC and has small to moderate loadings on the remaining PCs.

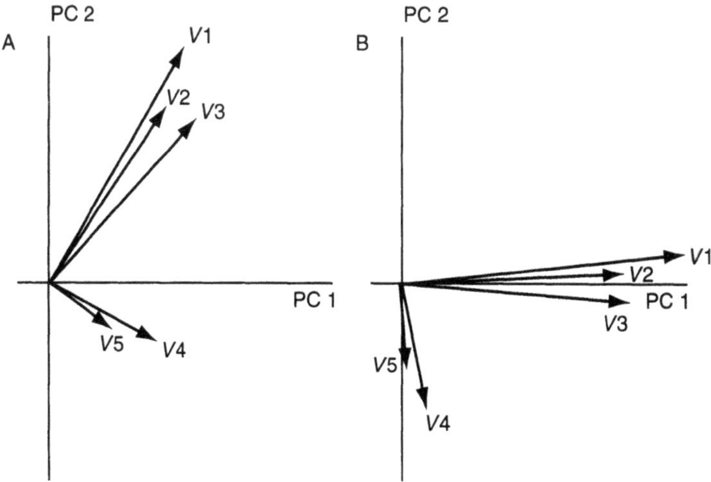


**Figure 3.8** Principal component loadings of five variables (*V1–V5*) before (A) and after (B) VARIMAX rotation. The rotation facilitates the interpretation of the PCA results by clustering the variable vectors with regard to the principal components. The rotation does not change the relative placement of the vectors, hence the angles between the vectors remain constant and indicate the degree of correlation between the original variables (*V1–V5*).

Depending on the number of PCs considered, the pattern of the rotated vectors may vary significantly. If there is only one dominant PC, it will generally be obscured by any orthogonal rotation; hence then rotation is not advisable.

From the PC loadings and the specific variances, PC scores representing the relative ranking of the individual compounds with respect to the principal components are calculated. These quantities can be used for diagnostic purposes as well as for input to a subsequent analysis to understand, for example, the collinear ranking of compounds by various tests. Like the original test results, from which they represent a concentration, the PC scores can be subject to QSAR analysis. This procedure, which is aimed at recognizing the compound-specific features underlying the activity observed in different tests, may yield similar results as directly deriving QSARs for the individual test parameters and then comparing the respective functions (for an example see Chapter 8). Although the interpretation of QSARs based on PC scores may be less obvious, and they are inconvenient for predictive applications, their advantage is the use of the condensed information obtained from a wide data basis with, for example, experimental scatter eliminated. The underlying QSARs may be easier to recognize and are not biased by outlying test results, hence the QSARs can be more sound. The common principle (interactions) encoded in the individual test results can be identified and characterized, such as the relative influence of the physico-chemical properties of the compounds in several toxicity tests, and collinearities can be recognized. Thus they can form the foundation for the derivation of simpler models for the individual endpoints.

Partial least squares (PLS) analysis allows the simultaneous investigation of the relationships between a multitude of activity data ($Y$ matrix) and a set of chemical descriptors ($X$ matrix) through latent variables (Wold *et al.*, 1984; Geladi and Kowalski, 1986; Hellberg, 1986; Geladi and Tosato, 1990). The latent variables correspond to the component scores in PCA and the respective coefficients to the PCA loading vectors. The PLS model can also be applied when the number of (collinear) descriptors exceeds the number of compounds in the data set. The main difference between PCA and PLS concerns the criteria for extracting the principal components and the latent variables, respectively: PCA is based on the maximum variance criterion, whereas PLS uses covariance with another set of variables ($X$ matrix).

In environmental studies, multivariate statistics are equally applicable to exposure- and effects-related data. With respect to the objective of data interpretation they can be used for:

evaluation of test systems reflecting similar effects;
comparison of testing conditions to design more informative testing schemes;
comparison of species sensitivity with a view to considering various modes of
    action;

evaluation of hazard profiles;
experimental design.

The particular merit of multivariate statistics is that it increases the maximum amount of information from a set of compound-specific data from a variety of tests. It makes evident that a large number of analogous experiments cannot provide more information than would be accessible from a fraction of the planned experimental effort. The recommendation for careful design of any experimental investigation, so that the observations and results provide substantial information with the minimum of experimental effort, does not apply only from the statistician's point of view; it is also concerned with the aspect of limited resources (time- and cost-effectiveness) and, in the field of (eco)toxicology, with animal welfare.

### 3.2.3 CLASSIFICATION METHODS

The analysis of semiquantitative activity data is not possible through classical regression techniques. However, chemical properties that are critical for active and inactive substances, respectively, can be identified by classification methods (Dunn and Wold, 1990; McFarland and Gans, 1990). Pattern-recognition techniques can be used to identify the relationships between groups of compounds as well as classify individual chemicals. The statistical methods used include PCA, FA and PLS (Derde and Massart, 1982; Wold *et al.*, 1983). The so-called supervised classification techniques use prior knowledge of the classes of some compounds (the training set) as a basis for assigning further compounds to these classes. To classify compounds in different classes, regions of pattern space occupied by compounds of known activity must be defined. The efficacy of any classification model for predictions strongly depends on the appropriate selection of descriptors. Those should not only facilitate the ability of the discriminant function to differentiate between classes but should also contain information about class membership *per se*, allowing the deduction of a physico-chemical basis for the class memberships.

Cluster analysis is an exploratory data analysis technique aimed at grouping items (e.g. chemicals and their properties) into clusters of similar items according to their position in the multidimensional parameter space. The various clustering methods differ mainly in how they calculate the distance from a point to the cluster: it may be to the nearest point, the most distant point or the centroid of the cluster, with the result that the respective clusters will have different shapes.

Discriminant analysis (DA) yields a function that separates the members of different classes based on either physico-chemical descriptors or indicator variables, such as the presence or absence of defined substructures (Figure 3.9).

The compounds are assigned location coordinates in the $n$-dimensional space according to their values in the descriptors included in the model and

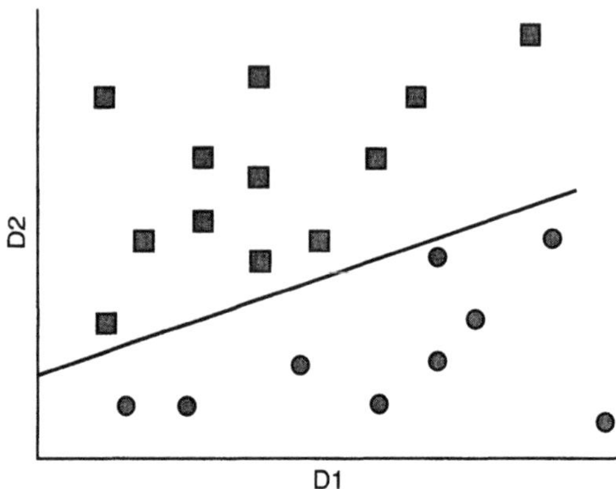

**Figure 3.9** Example of a linear discriminant analysis classifying two categories of compounds (squares and circles, respectively) on the basis of two descriptors (D1 and D2).

the number of classes as defined by the activity data. When only two groups are to be discriminated, multiple regression can be used with the activity data included as an indicator variable (e.g. readily degradable compounds: 1; non-readily degradable compounds: 0). Discriminant function values $> 0$ will then indicate readily degradable and values $< 0$ non-readily degradable compounds. The larger the absolute value in this case, the larger the distance from the discriminant function and hence, by implication, a higher degree of certainty in the classification.

Successful discriminant analysis is based on the assumption that the data in each class are normally distributed and all classes have the same covariance matrix (McFarland and Gans, 1990). Discriminant analysis is extremely sensitive to collinearities among descriptors and the ratio of the number of chemicals in the data set to the number of descriptors in the model should exceed 10:1. For the training-set selection, the data quality, the statistical significance, the type of descriptors and the limitations of the model range, the same restrictions apply for discriminant analysis as explained for classical regression analysis (section 3.2.1):

The activity can be unambiguously assigned to one class.

The compounds are evenly distributed over the different activity classes.

The data on physico-chemical descriptors cover several orders of magnitude.

The descriptor data are evenly distributed over the entire range of the model.

The range covered by each parameter exceeds significantly the variability in the parameter for the individual compounds.

The number of data points per independent $X$ variable is $\geq$ 10-15.

The descriptors used simultaneously in the model are not intercorrelated.

Each relevant substructure accounted for by an indicator variable occurs in a sufficient number of compounds.

Each substructure accounted for by an indicator variable occurs in combination with different other substructures.

For the model to have a comprehensive predictive power, a wide variety of substructures accounted for by an indicator variable should be considered.

The model has been validated with a set of compounds not used for the derivation of the model.

In addition to the discriminant function, to use the model for predictions further information is needed about:

the compound class(es) covered;
the mode(s) of action accounted for;
the range of parameter(s) covered;
the recognized outliers (i.e. (classes of) compounds for which the model is not valid).

If operated in accordance with its possibilities and limitations, discriminant analysis can provide classification models for dichotomous endpoints in environmental sciences, such as biodegradability (section 4.8) and mutagenicity (section 5.7).

A currently popular approach to classification and pattern-recognition problems involves neural networks. Neural networks are mainly used as (non-)linear approximations to multivariable functions or as classifiers (Ripley, 1993). Principally, the technique is intended to mimic the computational properties of the brain, which is highly parallel in its operation. Artificial neural networks (Figure 3.10) consist of units with some of the properties of real neurons.

Each of the units has multiple inputs, which may be partly excitatory and partly inhibitory. The units usually obtain a weighted sum of all the inputs, which, as a single output, is passed down the axon analogue elements. The output scalar compares to the average spike rate of a neuron (Crick, 1989).

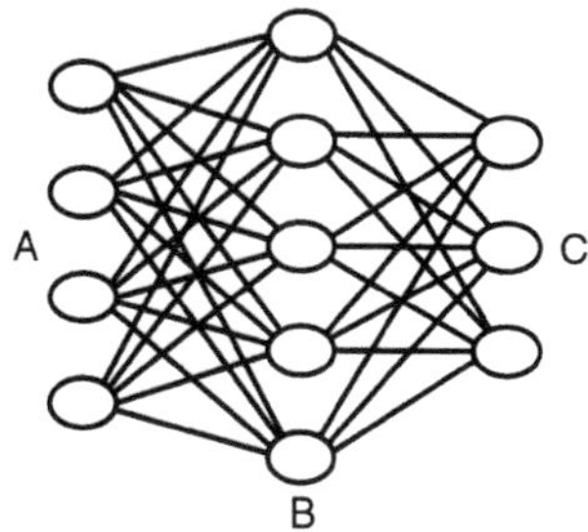


**Figure 3.10**   Principal architecture of a three-layer artificial neural network (Aoyama and Ichikawa, 1991). A: input layer with the number of input "neurons" corresponding to the number of parameters plus 1; B: hidden layer with an arbitrary number of "neurons"; C: output layer with the number of output "neurons" corresponding to the number of categories in the respective classification problem.

Although the respective neural net software is mostly only a simulation of parallel processing implemented on serial computers, the claim of neurophysiological analogy is emphasized by the use of biological terminology for conventional computing operations. Neural networks are in fact mathematical models of mutually interconnected functions arranged in one or several (usually three) layers (Aoyama and Ichikawa, 1991; Rojas, 1993), which are supposed to produce a classifying output for a given multivariate input array. Composite functions are the processing elements ('neurons'). The input arguments are summarized by integration functions and then weighted for their contributions to the overall discriminant function. The respective coefficients of the transfer functions, corresponding to the coefficients in regression analysis, are iteratively adjusted ('training', 'learning'), generally by least-squares procedures. The speed of convergence ('learning rate') is dominated by the network topology, although no principles are yet established about how to design the network architecture for a given problem, except by trial and error. The final output provides the classification of the input into any of the considered categories, with the number of classes being given by the number of 'neurons' in the output layer. Based on training sets of data with known classification, the weight factors of the transfer functions are calculated in a forward manner to reproduce the desired results ('supervised learning'). The most popular equations ('learning rules', learning laws') to minimize the error function (i.e. the divergences between the calculated output and the actual classifications) for multilayer neural networks are encoded in the backpropagation algorithm. The weights of the transfer functions are iteratively adjusted until the deviations between the desired and the currently calculated output are sufficiently small. The only necessary operations are vector matrix, matrix vector and vector vector multiplications – linear recombinations as, for example, in principal component analysis. The major affliction of the backpropagation method is its liability to get stuck in local minima of the error function rather than obtaining its global minimum, especially when the function is evaluated for output values other than zero and one (Rojas, 1993). Practically, backpropagation is a statistical method using least-squares procedures for the fitting of functions, corresponding to multiple (non-)linear (weighted) regression with normalized descriptors. If linear associators are used, neural network calculations are identical with multiple linear regression, whilst sigmoidal activation functions correspond to logistic regression. Comparing the performance of neural networks with modern statistical techniques for the classification of an actual data set revealed no superiority of various neural networks (Ripley, 1993) and several methodological shortcomings have to be realized:

Neural networks tend to have very large numbers of parameters relative to the number of training cases (Ripley, 1993).

The uncertainty in the parameters (weights) and in the predictions are hardly quantified (Ripley, 1993).

The neural network capability with regard to the description of the training-set data ('recognition') as well as its prediction power strongly depends on the transformation method, the initial starting weights, the speed of convergence and the number of iteration cycles (Schüürmann and Müller, 1994) as well as the design of the training set and validation set of chemicals (Rorije, Wezel and Peijnenburg, 1995); overfitting of the initial data set consequently results in poor predictions for further cases.

Different neural network algorithms give highly variable results (Ripley, 1993).

Because of their black-box character, neural network models are hardly interpretable in mechanistic terms.

The advantage of neural networks in directly accounting for non-linear relationships is opposed by their inadequate yield of stable and interpretable QSAR models for the investigation and prediction of environmentally relevant parameters. The essential step of identifying the relevant descriptors for a QSAR model (i.e. determining the rate-limiting processes resulting in the observed activity) is detached from scientific considerations and transferred to the computation system. Throwing the book of parameters at a set of activity data is liable to produce irrelevant chance relations, especially when dealing with ill-defined endpoints such as biodegradability. Hardly transparent relationships between some structural parameters and ambiguous activity data cannot be the objective of QSAR modelling. The use of neural networks for QSAR problems in environmental sciences may hence lead to fake solutions, if this method is used as a substitute for real statistics.

## 3.3 OUTLIERS

Outliers – compounds that do not fit a given model – are a frequent phenomenon in QSAR studies. By means of a numerical QSAR, chemicals may be identified that are in fact peculiar. The failure of a QSAR to account for all data points among a series of compounds can be a blessing in disguise by turning the attention to facts contradicting the presumed hypothesis about an underlying mode of action and thus leading to a better understanding of the interaction investigated.

When outliers are observed, and they always occur when the limits of a QSAR model are exceeded, it is poor practice to omit them (as often indicated in the literature by: 'Statistics are significantly improved upon exclusion of compound $X$'). An explanation for this deviation shall instead be attempted:

The selected structural parameters do not adequately describe the physico-chemical conditions/properties in the system.

There is a change in mode of interaction.

Multiple modes of interaction contribute to the observed effects to different extents.

The activity results from a combination of specific and non-specific effects, with different dependence on structural features.
Solubility limits are exceeded.
The conformations of the compounds change.
Tautomerism of the compounds occurs and is not accounted for by the descriptors.
The test compounds are not stable during the experiment.
There is competition with natural substrates or xenobiotics for binding sites.
Modifications in the structure/conformation of the interaction site occur.
The time dependency of the activity/mode of interaction changes among test compounds in relation to the time of experimental observation.
Differences in the mode of degradation/excretion/pharmacokinetics occur.

Unexplained outliers (i.e. compounds that should, but do not, fit the QSAR model according to the presumed mode of activity), make the relationship likely to be a chance correlation for the selected set of chemicals with, therefore, negligible predictive power. However, explainable outliers are valuable indicators of the limits of a given QSAR model, and serve to guide the selection of the appropriate QSAR for reliable predictions.

## 3.4 VALIDITY ASSESSMENTS OF QSARS

Models in QSAR analysis are local, and hence are valid only for limited changes in chemical structures and biological activity. Outside this range, the models' validity is not substantiated. A validity assessment of a QSAR evaluates the accuracy and precision of the estimates with due regard to the experimental variability in the respective endpoint. It is trivial, but still needs to be remembered, that QSAR predictions can never be more exact than the underlying experimental data. The criteria for assessing the validity of a QSAR model concern:

The statistical validity of the model
The range of the model:
    compound classes
    range of descriptors.
The reliability of predictions.

Erroneous predictions by QSARs may result from:

the selection of an inappropriate QSAR model;
the QSAR model is not statistically sound;
extrapolation outside the parameter range covered by the QSAR model;
the QSAR model is based on an inappropriate set of compounds;
the test conditions for the activity parameter were not identical.

The standards regarding the statistical validity of a QSAR model (section 3.2) relate to the goodness of fit of the equation to the training-set data. For the validity assessment of QSARs, it is worth while examining a graphical representation of the models in relation to all data points, including the outliers. In this way, the statistical quality and the range of the model can be evaluated easily. However, the statistical significance is a necessary, but not suficient, precondition for the predictive applicability of a QSAR model.

The range of a QSAR model determines the cases when it can be used for predicting data for untested compounds. The respective application criteria are guided by the range of chemicals underlying the model. Because each QSAR accounts only for homologous interactions, reliable predictions are only feasible for substances acting by the same mode(s) of action. The assignment of chemicals to the respective class(es) is generally based on traditional structural similarities and the definition of analogous derivatives of some parent compound(s). These inclusion rules can never be exhaustive, and they eventually have to be supplemented by a list of those chemicals (classes) that are excluded from the coverage of the model domain (outliers that were recognized during the application of the model to a wide variety of compounds with known activity).

In like manner, the range in the descriptor variables covered by the training-set chemicals (Figure 3.11) determines the reliability of the predictions.

Only within these limits of the descriptor values can the relationship be assumed to hold. Beyond this domain, the model may reveal a different type of relationship of the activity on the structural descriptors, which evidently must result in erroneous predictions. A striking example of the pitfalls of inapt extrapolations can be demonstrated with the application of a fourth-order polynominal log BCF/log $P_{ow}$ QSAR (Connell and Hawker, 1988) outside its parameter range (log $P_{ow}$ 2.6–9.8). Because of the curvature of the model, higher BCF estimates result for compounds with log $P_{ow} = 0$ than for

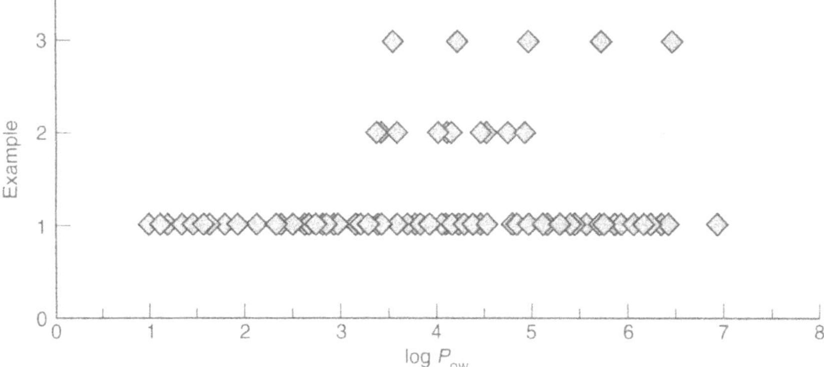


**Figure 3.11**  Examples of the parameter range covered and the distribution of the test compounds in the property domain for QSARs describing the log $P_{ow}$ dependence of bioconcentration factors (Sources of data: 1, Veith and Kosian, 1983; 2, Oliver, 1984; 3: Könemann and Leeuwen, 1980).

those with log $P_{ow}$ = 5. The calculated log BCF is 4.8 and 3.8, respectively, which is evidently nonsense for hydrophilic compounds.

Cross-validation is one method to check the soundness of a statistical model (Cramer, Bunce and Patterson, 1988; Eriksson, Verhaar and Hermens, 1994). The data set is divided into groups, usually five to seven, and the model is recalculated without the data from each of the groups. Consecutively, predictions are obtained for the omitted compounds and compared to the actual data. The divergences are quantified by the prediction error sum of squares (PRESS: sum of squares of predicted minus observed values), which can be transformed to a dimensionless term ($Q^2$) by relating it to the initial sum of squares of the dependent variable ($\Sigma(\Delta Y)^2$):

$$Q^2 = \frac{1 - \text{PRESS}}{\Sigma(\Delta Y)^2}$$

$Q^2$ represents the cross-validated explained variance, the counterpart of which is the explained variance $r^2$. Cross-validation is an important step in the validity assessment of statistical models, but overall it is not sufficient to test the predictive power of a QSAR model, because it remains within the generally homologous series of the test compounds.

The best and the only thorough test of the validity and application range of a QSAR model is the external validation. For a substantial number of diverse compounds, for which the activity is reported, predictions are calculated and compared to the respective measured data. Only in the case of adequate agreement between experimental and predicted values can validity be assumed. From the degree of deviation between these two data for different chemicals, the range of reliable applicability as well as the limitations for each individual QSAR can be concluded. Only after external validation can a QSAR model be recommended for predicting, within its limits, environmentally relevant parameters of chemicals.

# PART TWO

## *Application of QSARs in Environmental Sciences*

One of the applications of QSARs in environmental sciences is as an instrument to supplement the data sets on potentially hazardous substances when there are insufficient experimental data. For priority-setting purposes, there are significant data gaps: < 35% of the required effects data have been measured for the approximately 2000 HPV (high-production volume) chemicals (Zandt and Leeuwen, 1992). The situation is expected to be even worse for those existing chemicals that are produced and marketed at < 1000 tons per year. Because it will be impossible to complete the basic testing of all chemicals that may be released into the environment within the next few decades, QSARs provide one option to help out in the meantime and to direct limited resources for experimental assessments most efficiently towards chemicals that may pose a major hazard to humans and the environment. The potential of new compounds for environmental impact can be assessed from QSAR estimates already in the pre-production phase. Here, QSARs can guide the design of chemicals with lesser side-effects by optimizing their environmentally relevant properties.

The data sets required for the assessment of environmental hazard and risk resulting from the use of chemicals include information on persistence, accumulation, mobility and ecotoxicity. The basic concept of environmental risk assessment is a comparison of the levels of exposure and toxicity (Part 3). An evaluation is made of the probability that the exposure level of potential contaminants may exceed effective, toxic concentrations in the environmental compartment of concern. The severity of the impact is also taken into consideration. The respective exposure and effects concentrations are determined from experimental or, if not available, QSAR data (Nendza, Volmer and Klein, 1990). For that, compound-specific data are required in addition to information on the release rates and the environmental scenario of concern:

Exposure-related parameters:
    partitioning between hydrophilic/lipophilic phases
    water solubility

vapour pressure
Henry's law constant
soil sorption
abiotic degradation
biodegradation
bioconcentration.
Effects-related parameters:
toxicity to fish
toxicity to *Daphnia*
toxicity to algae
toxicity to mammals
mutagenicity.

# 4     *Exposure-related parameters*

Exposure has been defined as the 'concentration of toxic materials in space and time at the interface with target populations' (Travis *et al.*, 1983). The respective parameters required for environmental hazard assessments relate to the spatial and temporal abundance of the contaminants in the various compartments of the ecosystem and hence their bioavailability. Reference environmental descriptions must include the biota exposed to the released chemicals and the hydrological, topographical, geological and meteorological characteristics of the environment that affect the transport and transformation of the contaminants. The key step in exposure assessment is the use of transport and transformation models to quantify the movement of contaminants from the source through the environment to the target populations.

Processes such as accumulation, soil sorption, abiotic (hydrolysis and photolysis) and biotic (microbial) transformation, propagation and distribution will contribute to the effective exposure level of the chemicals (Figure 4.1).

Several models of varying complexity have been developed to calculate and predict the distribution of chemicals in the environment (OECD, 1989a, 1993c). Most of them are derived from the Mackay model (Mackay 1979; Mackay and Paterson, 1981, 1982, 1990; Mackay, Paterson and Shiu, 1992) to estimate the environmental compartment (air, soil, water and biota) in which the chemical is most likely to be found. Based on the concept of fugacity, models were derived for four levels of increasing complexity. Level I

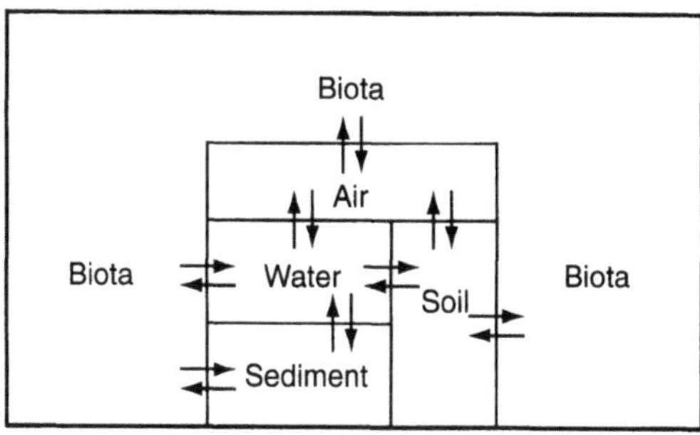


**Figure 4.1** The environmental compartments air, water, soil, sediment and biota are interconnected and the distribution of the contaminants between them determines the effective exposure level of environmental contaminants at the target sites.

assumes equilibrium between all compartments; level II includes transformation and steady-state input of the contaminant; level III is extended to slow transport processes between the compartments; and level IV allows variation of fugacity, input rate and concentration with time.

The compound-specific data required for exposure assessments comprise the 1-octanol/water partition coefficient (log $P_{ow}$), water solubility ($S_w$), vapour pressure ($p_v$), Henry's law constant (H, H$'$), soil sorption coefficient ($K_{oc}$), hydrolysis half-life time, photolysis half-life time and information on biodegradability (OECD, 1993c). These parameters generally relate to steady-state conditions – conditions that are rarely met in the real environment. The experimental data underlying the QSAR models are preferably determined by standardized protocols, but, even then, the absolute values are of variable reliability and precision, which clearly affects the accuracy of the predictions based on the acquired QSARs. The endpoints discussed in the following sections were selected because of their consideration in regulatory evaluation schemes in, for example, the EU (EEC, 1990).

For an important physico-chemical property of compounds, namely the melting point, currently no validated QSAR models of sufficient application range are available. The presumably general predictive estimation methods for this parameter, such as within the CHEMEST programme (EPA 1986a), cannot be recommended for use due to deviations of $\pm$ 80 °C and more between measured and calculated melting temperatures (Lynch *et al.*, 1991; OECD, 1993a). For input in subsequent hazard-evaluation procedures, therefore, experimental melting points should be used. For specific chemical classes, descriptive QSARs have been derived, such as for anilines (Dearden, 1991), but these cannot be extrapolated to other compound classes.

## 4.1 PARTITIONING BETWEEN HYDROPHILIC AND LIPOPHILIC PHASES

No substance released into the environment will remain at the locus of emission, but will spread into all other compartments (biotic and abiotic) according to its respective partition coefficients until ultimately a steady state has been achieved. The interphase partitioning between an aqueous and a non-aqueous phase – for example, water/lipid (biota), water/soil, water/sediment – is generally expressed by the 1-octanol/water partition coefficient (log $P_{ow}$), which assumes 1-octanol to be an acceptable surrogate for the various organic phases in biota and the environment. The log $P_{ow}$ is the key parameter in studies of the distribution of organic chemicals between organic and aqueous phases, which determines the transport to and the interactions with the active sites (Figure 4.2).

It has been related to water solubility, soil/sediment adsorption and bioconcentration. Compounds with low log $P_{ow}$ values ($\leq$ 2) are considered comparatively hydrophilic because they show high water solubility, low soil/sed-

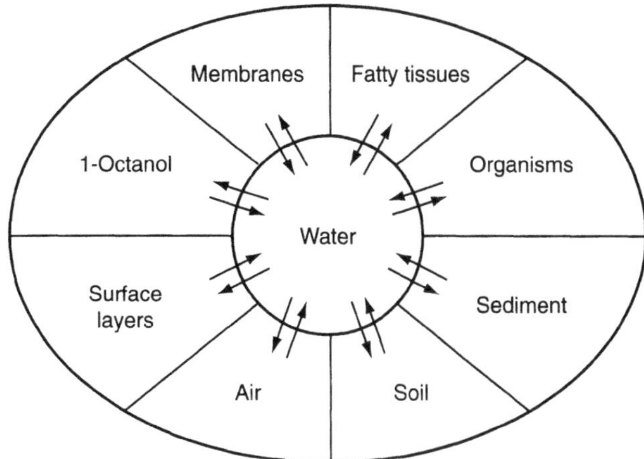



**Figure 4.2** Partitioning between water and the various non-aqueous phases: the mutual distribution processes contribute to the effective exposure levels of environmental contaminants, hence to their bioavailability.

iment adsorption and low bioconcentration. Conversely, log $P_{ow}$ values > 3 characterize more lipophilic substances that are of lower water solubility, higher soil/sediment adsorption and higher bioconcentration. Estimation of distribution-related parameters directly from log $P_{ow}$ is described in later chapters. For an extended treatise of the log $P_{ow}$ parameter, which occurs in environmental QSAR modelling as an endpoint by itself and as an operational descriptor of chemical structures, see section 1.1.

## 4.2 WATER SOLUBILITY

Water solubility is a major property for assessing the fate and exposure of chemicals in the environment. It determines a substance's distribution by the hydrological cycle and mobility, and hence affects a contaminant's concentration in the environmental media. Most of the highly water-soluble substances have low log $P_{ow}$ values, low adsorption to soil and sediment, low bioconcentration and they tend to be readily biodegradable. Water-soluble substances may gain easy access to humans and other living organisms. Knowledge of a chemical's solubility in water is hence a prerequisite for testing biological degradation, bioaccumulation and aquatic ecotoxicity.

Water solubility ($S_w$) is defined as the maximum amount of a compound that will dissolve in pure water at a specified temperature. Addition of further amounts of the chemical, which may be either solid or liquid, will result in a two-phase system at the defined temperature: the saturated aqueous solution and the solid or liquid phase of the undissolved substance (Lyman, Reehl and Rosenblatt, 1990). Under these conditions, water solubility can also be regarded as a parameter describing the partitioning of a compound between an aqueous phase and the pure compound phase. The process of dissolution is determined by interactions in the pure solute phase (e.g. lattice energy), the size of the solute relative to water, solute/solvent interactions and it is a function of temperature (Schwarzenbach, Gschwend and Imboden, 1993). The solubility of surfactants is given by the critical micelle concentration (CMC), only below which surfactants are present in the aqueous solution as individually dissolved molecules. In contrast to liquids and solids, the water solubility of gaseous compounds is of lesser environmental relevance. Water solubility of gases is usually measured at standard conditions (1 atm, 25 °C), which are rarely met in nature. Hence, for gases at lower partial pressure, the Henry's law constant is instead used to describe the concentration of the compound in water in relation to its concentration in air.

All organic chemicals are soluble in water at least to some extent, even though the corresponding concentrations may be very small; for example, $5.64 \times 10^{-9}$ mol/m$^3$ (2.4 ng/l) for 1,2,3,4,6,7,8-heptachlorodibenzo-*p*-dioxin (Shiu *et al.*, 1988). On the other end of the solubility scale are compounds that are infinitely miscible with water such as ethanol. Most environmentally relevant compounds are water soluble between 1 mg/l and 100 g/l, thus covering more than five orders of magnitude. Therefore no single experimental method can be used reliably to obtain data on water solubility and, depending on the experimental procedure, differing numerical values may result for the same compound. In principle, an excess of the chemical is added to pure water and allowed to equilibrate at constant temperature. To reach equilibrium may take several days for very hydrophobic (i.e. low solubility) compounds. The solution is then filtered and/or centrifuged to remove undissolved material before the concentration of the solution is measured (Lyman, Reehl and Rosenblatt, 1990). OECD (1981c) recommended the column elution method for non-volatile compounds of low water solubility (< 10 mg/l). Examples of ranges in water solubility (Dannenfelser *et al.*, 1991) are given in Figure 4.3.

The evident variability in the measured $S_w$ values reported in the literature may be attributed to several factors:

Purity of the test compound: pure substances have to be used because even small amounts of soluble impurities can cause large errors in the measured aqueous concentrations.

Attainment of equilibrium: the required time may take several days.

Suitability of analytical method: a specific set up has to be developed for each chemical species individually.

Stability of the test compound in water: the substance must not degrade during the experiment; losses in the concentration of the test compound may also occur by evaporation or adsorption to glassware.

Undissolved particles: insufficient separation of the phases may lead to apparently increased water solubility.

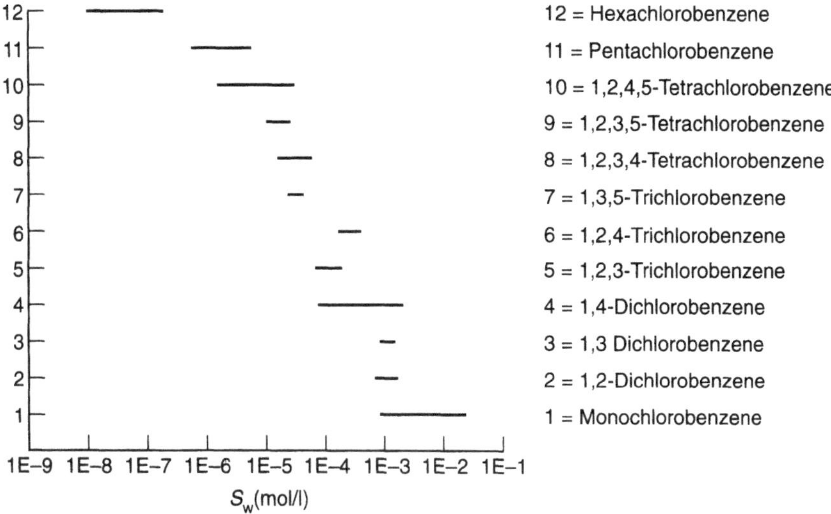

**Figure 4.3** Examples of the ranges in reported water solubility data for chlorobenzenes (Dannenfelser *et al.*, 1991). Reproduced from Nendza and Hermens (1995) with kind permission from Kluwer Academic Publishers, Dordrecht.

Surface-active materials: surface-active compounds may act as solubilizers and greatly affect the amount of test compound in the water phase.

Temperature: water solubility may increase or decrease with changes in the temperature, depending on the nature of the chemicals and the temperature range involved.

Salinity of the test medium: the presence of dissolved salts or minerals generally decreases the solubility of solutes, but also the reverse has been reported; e.g. the increased solubility of benzene in the presence of NaCl (Herington and Kynaston, 1952).

pH and buffer capacity of the water phase: pH conditions affect the solubility of organic acids and bases.

Cosolute effects: the presence of further organic solutes, as is the case in 'real' waters, may significantly alter the solubility of the individual compounds.

Suspended organic matter: soil and sediment components (e.g. humic acids) may result in increased apparent solubilities by serving as a sink compartment due to sorption processes.

In spite of the uncertainties inherent in the experimental data, measured values of water solubility are recommended for use in subsequent hazard assessments if they were obtained by a standard protocol. QSAR estimates may need to be used to fill the data gaps and to supplement the sound experimental data that are available for only a small fraction of existing chemicals.

QSAR models for water solubility are based on the observation that the water solubility of liquid compounds is highly correlated with their 1-octanol/water partition coefficient (Figure 4.4).

The relationship is quite close in the lower log $P_{ow}$ range (0–5), but reveals marked variability for highly lipophilic compounds (log $P_{ow}$ > 5). In the latter range, the underlying experimental data on log $P_{ow}$ as well as those on water solubility are subject to considerable imprecision, which may, at least partly, account for the observed data scatter over two orders of magnitude. As a result, QSAR predictions of the water solubility of highly lipophilic compounds can be regarded only as rough estimates.

For the water solubility of compounds with log $P_{ow}$ < 5–6, numerous linear log $P_{ow}$-dependent equations have been derived for diverse data sets (Table 4.1).

Theoretically, the slope of the respective correlations should be –1 if the activity coefficients in water and in water-saturated 1-octanol were equal. These QSARs are in fact obviously similar in their slopes, which range between –0.9 and –1.5, but vary considerably in their intercepts, ranging between –0.2 and 2.2, depending on the class of chemicals. Practically, this corresponds to a parallel shift of the functions towards higher or lower absolute $S_w$ values for the different classes, which to some extent may reflect class-specific differences in the experimental procedures used to obtain the underlying data (Figure 4.5).

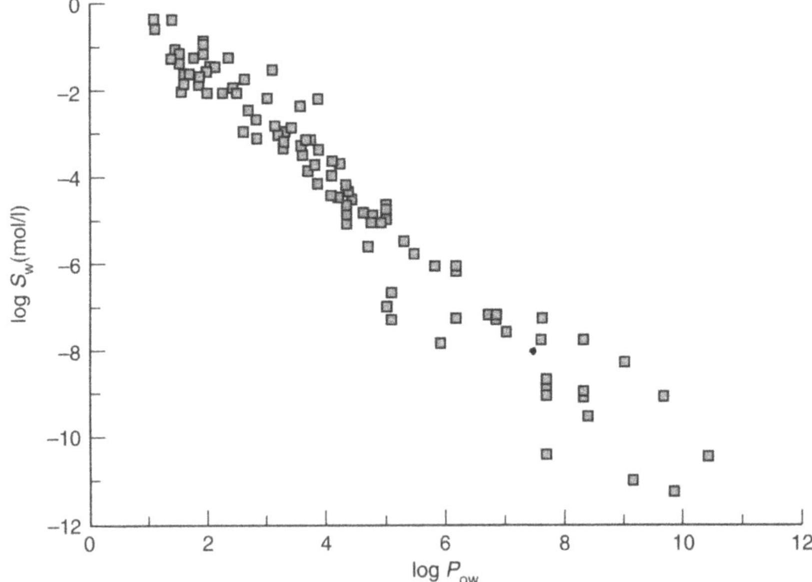

**Figure 4.4**  Relationship between water solubility in mol/l and log $P_{OW}$ (data from Müller and Klein, 1992). Reproduced from Nendza and Hermens (1995) with kind permission from Kluwer Academic Publishers, Dordrecht.

**Table 4.1** Examples of QSAR models for estimating water solubility: linear log $S_w$/ log $P_{ow}$ correlations ($S_w$ in mol/1).

| Model | Chemical class | | $r$ | $n$ | Descriptor range |
|---|---|---|---|---|---|
| 1 | Aromatic P-containing pesticides | $\log S_w = -1.49 \log P_{ow} + 1.46$ | 0.98 | 34 | 1–7 |
| 2 | Phosphoric acid esters | $\log S_w = -0.93 \log P_{ow} + 2.2$ | 0.95 | 11 | 0–5 |
| 3 | Alcohols | $\log S_w = -1.11 \log P_{ow} + 0.93$ | 0.97 | 41 | 0.6–2.8 |
| 4 | Ketones | $\log S_w = -1.23 \log P_{ow} + 0.72$ | 0.98 | 13 | 0.3–2.8 |
| 5 | Esters | $\log S_w = -1.01 \log P_{ow} + 0.52$ | 0.99 | 18 | 0.2–4.7 |
| 6 | Ethers | $\log S_w = -1.18 \log P_{ow} + 0.94$ | 0.94 | 12 | 0.8–2 |
| 7 | Alkyl halides | $\log S_w = -1.22 \log P_{ow} + 0.83$ | 0.93 | 20 | 1.4–3 |
| 8 | Alkynes | $\log S_w = -1.29 \log P_{ow} + 1.04$ | 0.95 | 7 | 2–4 |
| 9 | Alkenes | $\log S_w = -1.29 \log P_{ow} + 0.25$ | 0.99 | 12 | 1.7–3.7 |
| 10 | Aromatics | $\log S_w = -1.00 \log P_{ow} + 0.34$ | 0.98 | 16 | 0.9–3.7 |
| 11 | Alkanes | $\log S_w = -1.24 \log P_{ow} - 0.25$ | 0.95 | 17 | 2.3–4 |
| 12 | Diverse liquids | $\log S_w = -1.34 \log P_{ow} + 0.98$ | 0.93 | 156 | 0–5 |
| 13 | Liquids, solids | $\log S_w = -1.38 \log P_{ow} + 1.17$ | 0.97 | 300 | 0–8 |
| 14 | Diverse liquids | $\log S_w = -1.02 \log P_{ow} + 0.52$ | 0.92 | 111 | –1–5 |
| 15 | Diverse liquids | $\log S_w = -1.16 \log P_{ow} + 0.79$ | 0.97 | 156 | 0–5 |
| 16 | Pesticides, halogenated benzenes, PAHs, P-containing pesticides | $\log S_w = -0.92 \log P_{ow} + 1.18$ | 0.86 | 90 | –4–6.5 |

$r$ = correlation coefficient; $n$ = number of compounds analysed.
Sources of models: 1, Chiou *et al.* (1977); 2, IUCT (1992); 3–12, Hansch, Quinlan and Lawrence (1968); 13, Isnard and Lambert (1989); 14, Valvani, Yalkowsky and Roseman (1981); 15, Müller and Klein (1992); 16, Kenaga and Goring (1980).

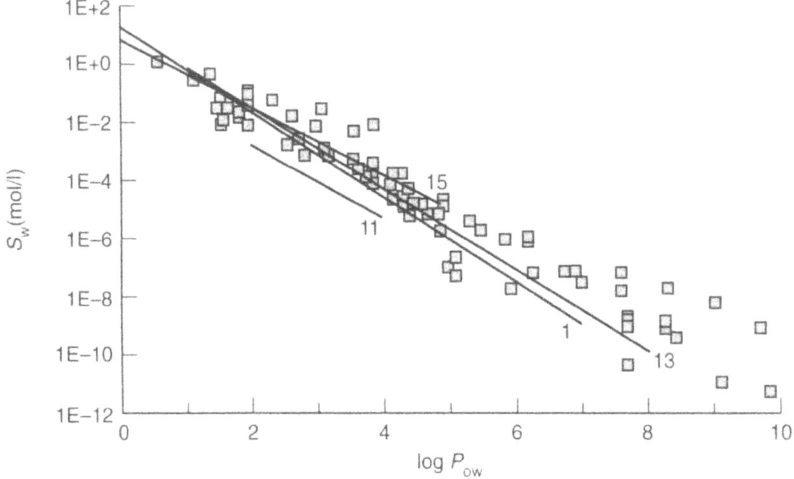


**Figure 4.5** Comparison of selected QSAR models relating log $S_W$ in mol/l to log $P_{OW}$ with experimental data for water solubility (Müller and Klein, 1992); for identification of the QSAR functions, see Table 4.1.

The differences are levelled out when using large diverse data sets to derive QSARs by regression techniques, which produce intermediate slopes and intercepts. As a consequence, $S_w$ estimates by equations 1–16 (Table 4.1) differ by about two log units (i.e. by factor 100) for compounds within the common parameter range covered (log $P_{ow}$ 0–5). Some uncertainties also arise from the neglect of the temperature dependence of $S_w$ by mixing input data obtained at different temperatures and the fact that these models are partly based on experimental as well as on calculated partition coefficients. Equation 15 copes with some inconsistency by using exclusively calculated partition coefficients to re-analyse the solubility data first used to derive equation 12. The result is a similar function, but with significantly improved statistics.

Validation with an extensive experimental database confirmed the principal applicability of the linear log $P_{ow}$-dependent water solubility QSARs for liquids, but significant misfits, up to two orders of magnitude, were detected for solids. In contrast to liquids, the water solubility predictions for solids are limited to few chemical classes, even when a correction is made for lattice energy via melting point (IUCT, 1992). Without this additional term, the models would refer to (hypothetical) subcooled liquids, but not to the true solid state. The supplement of QSARs based on log $P_{ow}$ by melting point ($T_m$) is to account for the fact that the crystal interaction energy of solids has no effect on their partitioning, but considerably affects their solubility (Yalkowsky, Valvani and Mackay, 1983). The melting point obviously compensates these effects only partly, such that these QSARs, which vary significantly for different classes of chemicals (Table 4.2), are generally restricted to small compound classes, such as PAHs and halogenated benzenes (Valvani and Yalkowsky, 1980). The model by Isnard and Lambert (1989) is based on a larger data set of various compounds ($n = 300$), but it contains a considerable fraction (> 50%) of liquids. These two-parameter QSARs can be applied for liquids by setting the melting point ($T_m$ in °C) to 25 °C. The extension of models based on log $P_{ow}$ and $T_m$ with correction factors for some compound class-

**Table 4.2** Examples of QSAR models for estimating water solubility: linear log $S_w$/ log $P_{ow}$ correlations ($S_w$ in mol/l; $T_m$ in °C), including melting-point correction.

| Model | Chemical class | | $r$ | $n$ | Descriptor range |
|---|---|---|---|---|---|
| 17 | PAHs | $\log S_w = -0.88 \log P_{ow} - 0.01\, T_m - 0.01$ | 0.99 | 32 | 3–7 |
| 18 | Halogenated benzenes | $\log S_w = -0.99 \log P_{ow} - 0.01\, T_m + 0.72$ | 0.995 | 35 | 2–6.5 |
| 19 | Halogenated aromatics | $\log S_w = -1.37 \log P_{ow} - 0.0094\, T_m + 1.64$ | – | 15 | – |
| 20 | Liquids, solids | $\log S_w = -0.94 \log P_{ow} - 0.01\, T_m + 0.32$ | 0.98 | 162 | 2–8 |
| 21 | Liquids, solids | $\log S_w = -1.26 \log P_{ow} - (0.0054\, T_m - 25) + 1.0 -$ | | 300 | 0–8 |

$r$ = correlation coefficient; $n$ = number of compounds analysed.
Sources of models: 17, 18, Valvani and Yalkowsky (1980); 19, Brüggemann and Altschuh (1991), 20, Yalkowsky, Valvani and Mackay, (1983); 21, Isnard and Lambert (1989).

es and the molecular weight (Meylan, Howard and Boethling, 1996) yields statistically improved equations but the mechanistic relevance of these terms, especially the molecular weight, remains unclear.

Models based on descriptors other than log $P_{ow}$ provide alternatives for $S_w$ estimations, and also serve to cross-check the predictions obtained, especially when there is reasonable doubt about the correctness of the respective log $P_{ow}$ values (Table 4.3).

QSARs based on molecular surface area or on solvatochromic parameters (Taft *et al.*, 1985) are not generally applicable, but have to be restricted to chemicals homologous to the base set. The use of topological descriptors for environmental parameters is questionable *per se*. Likewise a QSAR based on connectivity indices supplemented by substructure contributions accounting for polarity (Nirmalakhandan and Speece, 1988a, 1989; Speece, 1990) revealed only moderate predictive power. Even for compounds like dibenzodioxins, for which a special indicator variable is included, the estimates are excelled by those from log $P_{ow}$-dependent QSARs (Müller and Klein, 1992). According to thermodynamic considerations, $S_w$ data are also accessible from the aqueous activity coefficients $\gamma_w$ of the solute (Tewari, Miller and Wasik, 1982; Arbuckle, 1983; Chen, Holten-Anderson and Tyle, 1993; Myrdal and Yalkowsky, 1994). The calculations generally coincide with experimental data within $\pm$ 1 log units, but reveal increased errors for highly lipophilic compounds (log $P_{ow} > 5$). The application of this approach is further limited by missing parameter values for various environmentally relevant compounds in schemes for calculating activity coefficients, such as UNIFAC (Fredenslund, Gmehling and Rasmussen, 1977; Hansen *et al.*, 1991). The estimation of $S_w$ from the free energy of solvation and the molecular contact surface area (Schüürmann, 1995) has not yet been validated for diverse compounds.

**Table 4.3** Examples of QSAR models for estimating water solubility: log $S_w$ correlations with various parameters ($S_w$ in mol/1).

| Model | Chemical class | | $r$ | $n$ | Descriptor range |
|---|---|---|---|---|---|
| 22 | Not for stronger hydrogen-bonding donor solutes | $\log S_w = -3.36\ V + 0.46\ \pi^* + 5.23\ \beta + 0.55$ | 0.99 | 93 | – |
| 23 | Diverse | $\log S_w = 1.63\ ^0\chi - 1.37\ ^0\chi^v + 1.00\ \Phi + 1.56$ | 0.99 | 470 | – |
| 24 | Diverse | $\log S_w = x^w/(1 - x^w)$ | – | – | – |

$r$ = correlation coefficient; $n$ = number of compounds analysed; $V$ = molar volume; $\pi^*$ = solute/solvent dipolarity/polarizability; $\beta$ = hydrogen bond acceptor basicity; $^0\chi$, $^0\chi^v$ = zero-order connectivity indices; $\Phi$ = polarizability = $\Sigma$ (–0.94 (#Cl) – 0.36 (#H) – 0.77 (#double bonds) – 2.62 (#F) + 1.47 (#I) – 1.24 ($I_{alkane/alkene}$) + 1.01 ($I_{ketone/aldehyde}$) + 0.64 (#NH$_2$) – 1.70 (#NO$_2$) – ($I_{dibenzodioxin}$)); $x^w$ = mole fraction = $x^{oc}$ ($\gamma^{oc}$ ($\gamma^{oc}/\gamma^w$) ($x^w$ = mole fraction in the water phase; $x^{oc}$ = mole fraction in the organic phase; $\gamma^w$ = activity coefficient in the water phase; $\gamma^{oc}$ = activity coefficient in the organic phase).
Sources of models: 22, Taft *et al.* (1985); 23, Speece (1990); 24, Chen, Holten-Anderson and Tyle (1993).

To avoid major inadequacies, the appropriate QSAR model for predictive purposes has to be carefully selected, with special consideration to the chemical class accounted for and the parameter range covered. If available, a QSAR specifically derived for the chemical class of the compound under investigation should be used; models with very few compounds should be avoided. If a sound specific model is not at hand, equations 14 and 15, alternatively 20 or 21 (Tables 4.1, 4.2), are recommended for estimating $S_w$ values of liquid compounds with log $P_{ow} < 6$; in a validation exercise with 266 diverse liquids, the QSAR 15 yielded mean square residuals of only 0.219 log (mol/l) units (Müller and Klein, 1992). For compounds of higher lipophilicity, it is less reliable, because this parameter domain is not covered by the model and the corresponding log $P_{ow}$ values – measured or calculated – tend to be subject to significant errors. Accordingly, log $P_{ow}$-dependent water solubility models may be applied to compounds with log $P_{ow} > 5$ only to obtain rough estimates. The prediction of $S_w$ data for solids is generally worse than that for liquids. Because of the large database involved, the use of models 20 and 21 is recommended for solubility estimation of solids, when no reliable compound class-specific QSAR is available (OECD, 1993).

## 4.3 VAPOUR PRESSURE

The residence time of chemicals in soil and water is, among other factors, determined by the volatility of the substances – the tendency to evaporate into the air compartment. Partitioning and transport of substances between environmental media are thus affected by vapour pressure ($p_v$). Highly volatile chemicals have the potential for rapid, long-distance dispersion in the atmosphere, and can be taken up by terrestrial animals by inhalation or skin absorption, by terrestrial plants through stomatal pores or the cuticle, as well as into water bodies.

Vapour pressure ($p_v$) represents the partial pressure of a compound above the pure solid or liquid phase at thermal equilibrium; it corresponds to a steady state with a continuous exchange, but no net transfer, of molecules between the two phases. From thermodynamic considerations, the vapour pressure of a chemical is determined by its enthalpy of vaporization ($\Delta H_v$) and the temperature ($T$) as described by the Clausius–Clapeyron equation:

$$\frac{dp_v}{dT} = \frac{\Delta H_v}{RT^2}$$

Because $\Delta H_v$ is not directly accessible, the total enthalpy change may be aggregated from the individual processes involved (Yalkowsky and Mishra, 1991), which include the heat capacities and enthalpies of melting and boiling as well as the melting point ($T_m$) and the boiling point ($T_b$), with the melting being relevant only for compounds that are solid at the given temperature. With regard to solids, an extrapolation below the melting point accounts only for the subcooled liquid and not for a true solid state.

For illustrative purposes, vapour pressure may be portrayed as solubility in air. This parameter is strongly dependent on the ambient temperature; if measured at different temperatures, the logarithm of $p_v$ can be linearly related to the reciprocal temperatures (K). For most liquids, vapour pressure ranges between $10^{-3}$ and $4 \times 10^5$ Pa at room temperature. It is experimentally accessible using a (mercury) manometer to measure the pressure established in the gas phase above the pure compound at defined temperatures. For volatile chemicals ($p_v > 100$ Pa), measured data are generally accurate, whereas for low-volatility compounds ($p_v < 100$ Pa), the experimental results may scatter by one order of magnitude (Schwarzenbach, Geschwend and Imboden, 1993).

No validated QSARs are available to predict $p_v$ directly from chemical structure, but there are several methods for calculating $p_v$ based on derivations of the Clausius–Clapeyron equation (Table 4.4).

For conventional reasons, $p_v$ is obtained in most models in atm (1 Pa $= 0.98 \times 10^{-5}$ atm). Models of vapour pressure have mainly been applied in process engineering and resulted in assessments with highest precision for high-temperature ranges between the boiling point and the critical temperature. Less attention has been paid to conditions relevant in environmental

**Table 4.4** Examples of QSAR models for estimating vapour pressure $p_v$.

Model  Chemical class

| | | | |
|---|---|---|---|
| 1 | Liquids | $\ln p_v$ (atm) | $= K_F (8.75 + R \ln T_b) (T_b{-}C)^2/(0.97\ RT)\ [1/(T_b{-}C){-}1/(T{-}C)]$ |
| 2 | Liquids, (solids) | $\ln p_v$ (atm) | $= K_F (8.75 + R \ln T_b)/(0.97\ R)[1{-}(3{-}2T^*)^{m*}/T^*{-}2m(3{-}2T^*)^{m-1}\ \ln T^*]$ |
| 3 | Liquids, solids | $\ln p_v$ (atm)  $\ln \Delta p_{(s)}$ | $= K_F \ln(RT_b)/0.97)[1{-}(3{-}2T^*)^m/T^*{-}2m^* (3{-}2T^*)^{m*-1}\ \ln T^*]+\ln\Delta p_{(s)}$  $= 0.6 \ln (RT_m)\ [1{-}(3{-}2T^{**})^{m*}/T^{**} - 2m^* (3{-}2T^*)^{m*-1}\ \ln T^{**}]$ |
| 4 | Liquids, solids | $\log p_v$ (Pa)  $\Delta S_v$  $\Phi (K, T_b, T)$ | $= 0.434\ (\Delta S_v/R)\ T_b\ \Phi (K, T_b, T) + f (T_m) + 5.006$  $= 1.03\ [10.60 + 3.66 \log T_b + (1/MW)\ (0.0935\ T_b + 1.035 \times 10^{-3}\ T_b{}^2 - 1.345 \times 10^{-6}\ T_b{}^3)]$  $= (1 + K)\ (1/T_b - 1/T) + (K/T_b)\ \ln (T_b/T)$ |
| 5 | Liquids, solid hydrocarbons, halogenated hydrocarbons | $\ln p_v$ (atm) | $= -(4.4 + \ln T_b)\ (1.803\ (T_b/T{-}1) - 0.803 \ln (T_b/T) - 6.8\ (T_m/T - 1)$ |
| 6 | Liquids, solids | $\ln p_v$ (atm) | $= -([T_m - T]/T)\ (8.5 - 5.0 \log \sigma + 2.3 \log \o) - ([T_b - T]/T)\ (10 + 0.08 \log \o) + ([T_b{-}T]/T - \ln[T_b/T])\ (-6 - 0.9 \log \o)$ |

$K_F$ = compound class-specific constant (Table 4.5, default value 1.06); $R$ = gas constant (cal/mol K); $T$ = ambient temperature (K); $T_b$ = boiling point (K); $T_m$ = melting point (K); $T^* = T/T_b$; $T^{**} = T/T_m$; $C = 0.19\ T_b - 18$; m = specific factor for liquids (m = 0.19) or solids (if $T^* > 0.6$, m = 0.36; if $0.6 > T^* > 0.5$, m = 0.8; if $T^* < 0.5$, m = 1.19); m* $= -0.2575\ T^{**} + 0.4133$; $K = 0.803$; MW = molecular weight; $f(T_m) = 2.9532\ (1 - T_m/T)$ for $T_m > T$; $f(T_m) = 0$ for $T_m < T$; $\ln \Delta p_{(s)}$ (equation 3) = term for solids only; $(T_m/T{-}1)$ (equation 5) = term for solids only; $\sigma$ = rotational symmetry number; $\o$ = conformational flexibility.

Sources of models: 1, 2, Grain (1990); 3, Lyman (1985); 4, Altschuh, Brüggemann and Karcher (1993); 5, Mackay *et al.* (1982); 6, Mishra and Yalkowsky (1991).

hazard and risk assessment, thus the methods are less accurate at ambient temperatures between $-10\ ^\circ$C and $40\ ^\circ$C. As a result of this extrapolation from the boiling point to environmental temperatures (generally $25\ ^\circ$C), the estimates for volatile compounds ($p_v > 100$ Pa) are significantly better than for those of low vapour pressure ($p_v < 100$ Pa) (Lyman, 1985; Grain, 1990; Müller and Klein, 1993).

A prerequisite for applying $p_v$ estimation techniques is the knowledge of the compound's boiling point ($T_b$) and for solids also of the melting point ($T_m$). The methods described by Grain (1990), Lyman (1985) and Altschuh, Brüggemann and Karcher (1993) are of general applicability and they are not restricted to particular chemical classes. In Table 4.4, model 1 is derived from the Antoine equation, which describes the temperature dependence of vapour pressure; model 2 is based on the Watson correlation, which describes the temperature dependence of the heat of vaporization; model 3 constitutes an extension of 2. These three models additionally use a class-specific constant ($K_F$) as input, which is assumed to describe the polarity of the compounds (Table 4.5).

The $K_F$ values (Grain, 1990), which range between 0.97 and 1.38, were derived for monofunctional compounds, but they also apply to polyfunctional compounds if the respective highest $K_F$ value is used. For compound classes not considered, a default $K_F$ value of 1.06 is recommended, but the lack or ambiguity of specific $K_F$ values for some classes may result in erroneous estimates. Models 1–4 are equally well suited for liquid compounds, whereas for solids methods 3 and 4 are preferable. The fifth function (Mackay *et al.*, 1982), in contrast to models 1–4, is only applicable for liquid and solid hydrocarbons and halogenated hydrocarbons; these compound classes are also covered by the other models. Model 5 has been revised on a thermodynamic basis by Mishra and Yalkowsky (1991), who introduced terms on the rotational symmetry and the conformational flexibility of the molecules to extend the application range to diverse liquids and solids. Calculations based on the free energy of solvation and the contact surface area (Schüürmann, 1995) have been limited to substituted benzenes.

Calculated $p_v$ values may be subject to considerable uncertainty; the best procedure is therefore to apply all suitable models and then to decide on the most reasonable estimates. A comparison of the results obtained by the different methods will also reveal erroneous predictions. As a whole, these models result in acceptable estimates for $p_v$ values $> 1$ Pa; the predictions are worse for compounds with $p_v < 1$ Pa (Figure 4.6).

Major uncertainties have to be considered especially when an estimate from the Meissner method (Rechsteiner, 1990) is used instead of an experimental boiling point. Furthermore, these errors are propagated when, for example, calculated $p_v$ values are used together with calculated $S_w$ data to estimate the Henry's law constant.

**Table 4.5** Class-specific $K_F$ values for $p_v$ estimations (Grain, 1990).

| Compound class | $K_F$ factor |
|---|---|
| *Hydrocarbons* | |
| $n$-Alkanes | 0.97–1.00 |
| Branched alkanes | 0.99 |
| Mono-, di-olefins | 1.00–1.01 |
| Cyclic alkanes | 1.00 |
| Alkyl derivatives of cyclic alkanes | 0.99 |
| *Halides* | |
| Monochlorides | 1.01–1.05 |
| Monobromides | 1.01–1.04 |
| Monoiodides | 1.01–1.03 |
| Polyhalides with hydrogen | 1.01–1.05 |
| Mixed halides without hydrogen | 1.01 |
| Perfluorocarbons | 1.00 |
| *Carbonylic compounds* | |
| Esters | 1.01–1.14 |
| Ketones | 1.01–1.08 |
| Aldehydes | 1.01–1.09 |
| *Nitrogen compounds* | |
| Primary amines | 1.05–1.16 |
| Secondary amines | 1.03–1.09 |
| Tertiary amines | 1.01 |
| Nitriles | 1.01–1.07 |
| Nitro compounds | 1.01–1.07 |
| *Sulphur compounds* | |
| Mercaptanes | 1.01–1.05 |
| Sulphides | 1.01–1.03 |
| *Alcohols* | |
| One OH-group | 1.22–1.31 |
| Two OH-groups | 1.33 |
| Three OH-groups | 1.38 |
| *Other oxygen compounds* | |
| Aliphatic ethers | 1.01–1.03 |
| Oxides | 1.01–1.08 |

## 4.4. HENRY'S LAW CONSTANT

The transport of a chemical between air and water compartments is affected by its volatilization as well as its water solubility. For the water phases not only surface waters but also, for example, the water in moist soils, rain and mist have to be considered. The extent and rate of such transfer (evaporation rates) between media are important factors with regard to the mobility of a chemical contaminant and hence for the exposure concentrations.

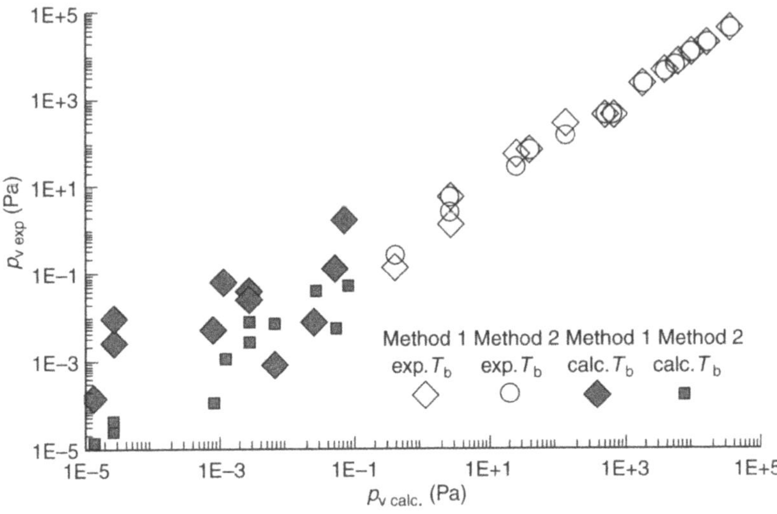


**Figure 4.6**   Comparison of calculated vapour pressure ($p_{v\,calc.}$) with experimental ($p_{v\,exp.}$) data (Grain, 1990). Method 1: estimates based on the Antoine equation using either experimental (exp. $T_b$) or calculated (calc. $T_b$) boiling points. Method 2: estimates based on the Watson correlation using either experimental (exp. $T_b$) or calculated (calc. $T_b$) boiling points. Considerable discepancies between calculated and experimental $p_v$ occur for compounds of low vapour pressure (Pa) using calculated $T_b$ data. Reproduced from Nendza and Hermens (1995) with kind permission from Kluwer Academic Publishers, Dordrecht.

Henry's law describes the partitioning of a chemical between a gaseous phase (e.g. air) and a liquid phase (e.g. water), stating that the solubility of a gaseous compound in that liquid is proportional to its partial pressure above the solution. The proportionality factor obtained for equilibrium conditions is represented by the Henry's law constant. It is expressed either as H (Pa m$^3$/mol), the ratio of the partial pressure in the vapour phase ($p_v$ in Pa) and the concentration in water ($S_w$ in mol/m$^3$), or as H$'$ (dimensionless), the ratio of the concentrations (mol/m$^3$)in air and water:

$$H = \frac{\text{vapour pressure } (p_v)}{\text{water solubility } (S_w)}$$

$$H' = \frac{\text{equilibrium concentration of chemical in air}}{\text{equilibrium concentration of chemical in water}}$$

Because the concentration of chemicals in the air and their partial pressure are related by the ideal gas law, H and H′ can be converted according to:

$$H = H' \, RT$$

where $R$ denotes the gas constant and $T$ the ambient temperature (K). If for the partial pressure a value of 101 325 Pa (= 1 atm) can be assumed, the Henry's law constant corresponds to the reciprocal of the water solubility, which evidently is an unlikely situation under most environmental conditions (Lyman, Reehl and Rosenblatt, 1990). Direct measurement of the air/water partitioning is practical only for compounds that are not very volatile (i.e. with low $p_v$), otherwise the Henry's law constants are estimated from the ratio of the vapour pressure and the water solubility of the chemicals.

The $p_v/S_w$ ratio of compounds is the method of choice to calculate the Henry's law constant. Assuming constant activity coefficients in the liquid phase over the entire range of solubility, this relationship is applicable if the total pressure is near 1 atm and if there is no association of molecules in the vapour phase (Lyman, 1985). The input parameters $p_v$ and $S_w$ have to refer to the same state of the chemicals; that is for solid compounds they both have to be obtained either for the solid or for the subcooled state. This indirect approach can obviously not be applied for chemicals completely miscible with water (with infinite water solubility) nor is it suitable for compounds of high water solubility, because significant method errors occur. If limited to substances with relatively low water solubility ($S_w < 1$ mol/l), the $p_v/S_w$ ratio provides fairly accurate estimates of the Henry's law constant (OECD, 1993a). However, the resulting data for Henry's law constant are subjected to the variability in the $p_v$ and $S_w$ values used as input and the respective uncertainties caused by propagated errors have to be taken into account.

To substantiate the estimates obtained by the $p_v/S_w$ ratio, further predictions by a different method may be acquired and the results compared. Specifically for hydrocarbons, halogenated hydrocarbons, acids, esters and alcohols, Nirmalakhandan and Speece (1988b) derived a QSAR based on polarizability, first-order connectivity indices and indicator variables for some substructures. Hine and Mookerjee (1975) described two methods for estimating the Henry's law constant directly from the molecular structure, using either group contributions for about 70 fragments or bond contributions for 34 different bond types. The authors rate the group-contribution method to give more reliable results than the bond-contribution method. Both methods are basically applicable to monofunctional hydrocarbons; multifunctional compounds may be subject to variant deviations. The bond-contribution approach has been extended by Meylan and Howard (1991), who also provide a computer program for doing the respective calculations. Problems still arise for highly chlorinated aliphatic pesticides (e.g. aldrin, dieldrin and lindane), and when multiple polarizable groups occur. For simple compounds, the fragment/bond methods should provide sufficiently reliable results; however, because the methods use class-specific correction terms in addition to the bond/group contribution, they are likely to fail when applied to compounds with unconsidered structure fragments. Based on thermodynamic considerations, the Henry's law constant may also be estimated from boiling points and aqueous activity coefficients (Myrdal and Yalkowsky, 1994) or the free energy of aqueous solvation (Schüürmann, 1995), but validation of this method had been limited to homologous (chloro)benzenes.

## 4.5 BOILING POINT

The boiling point serves as an indicator of a chemical's physical state at ambient conditions. Mostly it is not used as an autonomous parameter in environmental hazard assessments, but for estimating further properties such as the vapour pressure. The boiling point is defined as the temperature at which the vapour pressure of a liquid is equal to the pressure of the atmosphere on the liquid (Rechsteiner, 1990). The measurement of boiling points is in principal simple, requiring only a calibrated thermometer, but the results may be severely affected by:

purity of the test compound: impurities may alter the boiling temperature;
decomposition of the test compound;
ambient pressure.

The calculation of boiling points with sufficient accuracy is feasible only for some compound classes. The quite old, though still the most-recommended method (Meissner, 1949; Rechsteiner, 1990) for monofunctional hydrocarbons, alcohols, aldehydes, ketones, carboxylic acids, ethers, esters, amines and nitriles is based on molar refractivity, parachor and a class-dependent variable (Table 4.6).

This model is highly empirical in nature and the two interrelated polarizability descriptors (section 1.2.2) can be only vaguely related to dispersive interactions in the liquid phase. The estimates for the respective monofunctional compounds are generally within ± 5–10% of the experimental data, but larger deviations occur for polyfunctional substances. For a set of heterogeneous compounds, a mean method error of 21% (°C) has been reported (Lynch *et al.*, 1991). Significant outliers comprise (IUCT, 1992):

sulphur-containing compounds except thiols
nitro compounds
isocyanates
amides
hydrazines
anhydrides
halogenated aromatic hydrocarbons
carbamates.

**Table 4.6** Example of a model for estimating boiling point $T_b$ (K) (Rechsteiner, 1990).

| Chemical class | |
| --- | --- |
| Acids, alcohols, amines, esters, ethers, mercaptanes, hydrocarbons, ketones, aldehydes, nitriles | $T_b = (637\,(MR)^{1.47} + B)\,/\,parachor$ |

| | *B* |
| --- | --- |
| Acids (monocarboxylic) | 28 000 |
| Alcohols (monohydroxy) including phenols, cresols, etc. | 16 500 |
| Primary amines | 6 500 |
| Secondary amines | 2 000 |
| Tertiary amines | –3 000 |
| Esters (monocarboxylic acid/monohydroxy alcohol) | 15 000 |
| Esters (dibasic acid/monohydroxy alcohol) | 30 000 |
| Ethers, mercaptanes | 4 000 |
| Acetylenic hydrocarbons | –500 |
| Aromatic hydrocarbons | –2 500 |
| Paraffinic, naphthalenic hydrocarbons | –2 500 |
| Olefinic hydrocarbons | –4 500 |
| Ketones, aldehydes | 15 000 |
| Monochlorinated normal paraffins | 4 000 |
| Nitriles | 20 000 |

$MR$ = molar refractivity; $B$ = class-specific constant; parachor = the product of the molar volume and a surface tension term.

Because of the considerable uncertainties in the estimated boiling temperatures, measured values, which for many chemicals are in the literature, are recommended as input parameters for, for example, vapour pressure computations, so as not to increase the propagated errors unnecessarily.

## 4.6 SOIL SORPTION

The sorption to soil components is a determinant factor for the mobility of contaminants, accounting for their distribution among soil, sediment and water phases. The extent to which chemicals partition between the solid and solution phases in soil, or between water and sediment in aquatic ecosystems, determines the likelihood of the contaminants leaching through the soil or being immobile. The soil sorption hence influences the elution of compounds into groundwater bodies as well as their availability for transformation by soil microbes, their volatilization from soil surfaces and their bioavailability for exposed organisms.

The extent of sorption to soils is governed by a variety of physico-chemical properties of both the soil and the contaminant. The soils' heterogeneous chemistry and physics resulting from the varying proportions of the major components – mineral and organic matter, water, air and (micro)organisms – account for the differences in the binding capacity of different soils. The relevant soil parameters include:

organic carbon content
size of particles
porosity
clay content
biological activity and biomass
humidity
pH value
cation exchange capacity
temperature.

Sorption occurs if the free energy of interaction between the soil components and the chemical is negative. The underlying processes of physisorption and chemisorption may be caused by (Oepen, Kördel and Klein, 1991):

van der Waals interactions
hydrophobic bonding
hydrogen bonding
charge-transfer interactions
ligand exchange and ion bonding
direct and induced ion-dipole and dipole–dipole interactions
covalent binding.

The sorption of non-polar substances, generally to the organic matter of the soil, can be regarded as a distribution process between the polar phase of the soil water and the organic phase of the soil components. The equilibrium constant of this partitioning between solid and solution phases constitutes the adsorption coefficient ($K$) for soil and sediment:

$$K = \frac{\text{concentration of chemical sorbed to soil}}{\text{mean concentration of chemical in water}}$$

The partition coefficient in a soil/water system depends on concentration, thus the equilibrium concentrations in both phases have to be determined for varying amounts of contaminant. The relationship between sorbed and dissolved fractions constitutes the sorption isotherm. With regard to soil type, the occupation of binding sites and the chemical investigated, several mathematical models have been derived to account for the dominating processes. The Langmuir isotherm and the Freundlich isotherm presume low concentrations adsorbed to homogeneous soil surfaces at steady state. This simplification neglects solvation effects in organic portions of the soil as well as diffusion into pores. The empirical Freundlich isotherm, even though it has no theoretical foundation, generally provides a sufficiently accurate description of contaminant sorption to soils (Oepen, 1990):

$$\frac{x}{m} = K_f\, C_e^{\,1/n}$$

with $x/m$ representing the chemical concentration in soil ($\mu$g/g), $C_e$ the chemical concentration in the aqueous phase at equilibrium ($\mu$g/ml), $K_f$ the respective equilibrium constant (Freundlich adsorption constant) and the compound-dependent exponent $1/n$ the intensity of sorption (ideally $1/n = 1.0$, but mostly it ranges between 0.7 and 1.3; Lyman, 1990). The resulting adsorption coefficients are considered independent of the chemical concentration applied.

Adsorption coefficients may be determined experimentally by batch equilibrium studies (e.g. OECD, 1983). Dispersions with a defined soil/solution ratio, containing each of several initial concentrations of the chemical, are agitated until equilibrium is achieved. The phases are separated by centrifugation and the compound's concentration is determined in the aqueous fraction. The reduction in concentration of the dissolved chemical in water is used as a measure of sorption. A desorption test is conducted consecutively and the adsorption isotherms are determined. Variability in the measured soil sorption coefficients may arise from:

Inhomogeneity of the soil: the different soil components reveal different sorption capacities.

Attainment of equilibrium: the required time may take several days.

Analytical method: a specific set up has to be developed for each chemical species individually.

Stability of the test compound in water: the substance must not degrade during the experiment; losses in concentration of the test compound may also occur by evaporation or adsorption to glassware.

Stability of the test compound in soil: the substance must not be biodegraded during the experiment.

Surface-active materials: solubilizers may greatly affect the concentration of the test compound in the aqueous phase.

pH and buffer capacity of the water phase: pH conditions influence the sorption of organic acids and bases.

Cosolute effects: the presence of further organic or inorganic solutes, as is the case in 'real' environments, may significantly alter the sorption of the individual compounds.

Incomplete phase separation: sorption to suspended soil/sediment fractions may enhance the apparent chemical concentration in the aqueous phase.

The sorption capacities of soils vary considerably and the sorption coefficients measured for the same compound with different soils may range over several orders of magnitude. Extrapolation of sorption coefficients obtained with one soil type to another soil type is therefore almost impossible. Even a normalization to the organic carbon fraction (% OC), the principal interaction site for hydrophobic compounds, reduces the variance of sorption coefficients measured in different soils (1–10% OC) only to one order of magnitude, and for more polar chemicals the normalized $K_{oc}$ values may still vary by a factor of 100–1000 (Lambert, Porter and Schiefestein, 1965; Lambert, 1968; Hamaker and Thompson, 1972).

$$K_{oc} = K_f \, (100/\%OC)$$

Soil sorption data normalized for either the organic carbon content or organic matter content can be converted according to:

$$\log K_{oc} = \log K_{om} + 0.2365$$

As a consequence of the multifunctional nature of the sorption parameter $K_{oc}$, recent developments are directed to testing schemes using HPLC with various stationary phases. The basic principle is to determine separately the different interactions involved in sorption on standardized material, and then to recombine the data from the individual measurements for a comprehensive sorption parameterization. Until the experimental sorption data will be determined by a stringently standardized method, and the $K_{oc}$ values refer to a consistently defined endpoint, sorption modelling must remain a formalization of empirical relationships. Accordingly any QSAR approach to soil sorption is principally restricted by the problem of determining precisely which activity is accounted for:

sorption *to which soil components*?
sorption *by which interaction processes*?

If soil sorption is regarded as a partitioning process, characterized by the distribution coefficient $K_d$, it sets the basis for potential relationships to the corresponding descriptors of the chemicals, such as log $P_{ow}$, water solubility, molar refractivity and simple connectivity indices (Table 4.7).

These properties are generally highly intercorrelated for the compound sets analysed, and the respective QSARs thus represent the linear increase in soil sorption with increasing hydrophobicity of the chemicals. The models derived for non-ionic lipophilic aromatic hydrocarbons reveal the best statistical quality, corresponding to the presumed uniform mode of interaction with the soil components. However, the existence of a correlation between the distribution coefficients for two systems does not necessarily imply that the mechanisms controlling phase distribution are the same. The log $P_{ow}$-dependent models yield slopes close to unity (0.8–1) for non-polar aromatics, but lesser lipophilicity contributions (i.e. slopes of approximately 0.5) for presumably more polar pesticides. This finding coincides with results obtained with other endpoints. The marked differences in the intercepts of these QSARs may reflect either class-specific sorption mechanisms or peculiarities in the procedures for measuring the respective underlying data sets. Using HPLC capacity factors as soil sorption predictors (e.g. Kördel, Stutte and Kotthoff, 1993; Müller and Kördel, 1996) has the advantage of simulating more realistically

**Table 4.7** Examples of QSAR models for estimating soil sorption: $\log K_{oc}$ correlations with various parameters.

| Model | Chemical class | | $r$ | $n$ | Descriptor range |
|---|---|---|---|---|---|
| 1 | Pesticides | $\log K_{oc} = 0.52 \log P_{ow} + 1.12$ | 0.95 | 105 | –0.6–7.4 |
| 2 | Pesticides | $\log K_{oc} = 0.54 \log P_{ow} + 1.38$ | 0.86 | 45 | 2–6 |
| 3 | Aromatics, PAHs | $\log K_{oc} = 0.83 \log P_{ow} + 0.29$ | 0.95 | 20 | 1–6 |
| 4 | Aromatic herbicides | $\log K_{oc} = 0.94 \log P_{ow} - 0.01$ | 0.97 | 19 | – |
| 5 | Aromatic hydrocarbons | $\log K_{oc} = 0.99 \log P_{ow} - 0.35$ | 1.00 | 5 | 2–5.2 |
| 6 | Phenols, benzonitriles | $\log K_{oc} = 0.57 \log P_{ow} + 1.08$ | 0.87 | 24 | 0.5–5.5 |
| 7 | Anilines | $\log K_{oc} = 0.62 \log P_{ow} + 0.85$ | 0.91 | 20 | 1–5.1 |
| 8 | Dinitroanilines | $\log K_{oc} = 0.38 \log P_{ow} + 1.92$ | 0.91 | 20 | 0.5–5.5 |
| 9 | Esters | $\log K_{oc} = 0.49 \log P_{ow} + 1.05$ | 0.87 | 25 | 1–8 |
| 10 | Pesticides | $\log K_{oc} = -0.55 \log S_w^1 + 3.64$ | 0.84 | 106 | $\log S_w$:–4–6.3 |
| 11 | Aromatic hydrocarbons | $\log K_{oc} = -0.59 \log S_w^2 - 0.20$ | 0.97 | 10 | $-\log S_w$:3.4–10 |
| 12 | Chlorinated hydrocarbons | $\log K_{oc} = -0.56 \log S_w^3 + 4.28$ | 0.99 | 10 | – |
| 13 | PCBs, chlorobenzenes | $\log K_{oc} = 0.08\, MR - 0.27$ | 0.99 | 15 | – |
| 14 | Esters | $\log K_{oc} = 0.06\, MR - 0.27$ | 0.95 | 10 | – |
| 15 | Amines | $\log K_{oc} = 0.09\, MR - 0.68$ | 0.96 | 8 | – |
| 16 | Phenols, polycyclics | $\log K_{oc} = 0.67\,^1\chi^v + 0.37$ | 0.97 | 32 | 0.9–8.0 |
| 17 | PAHs | $\log K_{oc} = 1.03\,^2\chi^v + 0.76$ | 0.99 | 8 | 1.1–4.7 |
| 18 | PAHs, benzenes, halogenated phenols | $\log K_{oc} = 0.55\,^1\chi + 0.67$ | 0.97 | 37 | 2.3–10 |
| 19 | Hydrophobic chemicals (only C, H, halogen) | $\log K_{oc} = 0.52\,^1\chi + 0.70$ | 0.98 | 81 | 1–11 |
| 20 | Anilines, benzenes heterocyclics | $\log K_{oc} = 0.53\,^1\chi + 2.09\,^{\Delta 1}\chi_x^v + 0.64$ | 0.97 | 56 | $^1\chi$:1.4–10 |
| 21 | Non-polar, polar compounds | $\log K_{oc} = 0.53\,^1\chi + \Sigma P_f N + 0.62$ | 0.98 | 189 | $^1\chi$:1–13 |

$r$ = correlation coefficient; $n$ = number of compounds analysed; $S_w$ = water solubility ($S_w^1$, mg/l; $S_w^2$, mole fraction; $S_w^3$, $\mu$mol/l); MR = molar refractivity; $^n\chi$ = connectivity indices, $^{\Delta 1}\chi^v$, $\Sigma P_f N$ = polarity correction terms.
Sources of models: 1, Briggs (1981); 2, 10, Kenaga and Goring (1980); 3, Hodson and Williams (1988); 4, Brown and Flagg (1981); 5, 11, Karickhoff (1981); 6, 9, 19, Sabljic *et al.* (1995); 12 Chiou, Peters and Freed (1979); 13, 16, Koch and Nagel (1988); 14, 15, Oepen (1990); 17, Sabljic and Protic (1982); 18, Sabljic (1987a); 20, Bahnick and Doucette (1988); 21, Meylan, Howard and Boethling (1992).

the distribution between a stationary solid phase and a liquid mobile phase, but also the disadvantage of requiring experimental descriptor data. The use of water solubility as a $K_{oc}$ predictor paraphrases the $\log P_{ow}/\log K_{oc}$ functions. The $\log K_{oc}/MR$ correlations obtained might indicate some contributions of the compounds' polarizability to sorption behaviour, as could be anticipated, for example, for esters (model 14 in Table 4.7) or amines (model 15), but the derived models do not allow rationalization of these effects for the different chemical classes. For dissociable compounds, the soil pH significantly influences their adsorption, hence a correction for the dissociated fraction will provide more realistic estimates (Gestel, Ma and Smit, 1991). Concerning the connectivity models, it is first of all striking that different indices perform alike on similar sets of compounds. This indicates no unique dependency between soil sorption and the structural features encoded in these descriptors. To account for polarity effects on adsorption, Meylan, Howard and Boethling (1992) supplemented a linear QSAR based on the first-order connectivity index by substructure indicators for the occurrence of polar fragments in the structures (model 21). Their negative correction terms throughout reflect the assumption of reduced soil sorption of such chemicals, whereas substructures causing enhanced soil sorption are not accounted for. These polarity correction factors decrease the estimated log $K_{oc}$ values by 0.12 (nitrogen on noncyclic aliphatic carbon) to 2.0 (carbamate) log units (nitro: −0.63, ketone: −1.25, ester: −1.31, aliphatic alcohol: −1.52, carboxylic acid: -1.75). Like any substructure model, this QSAR is limited to compounds with fragments considered in its derivation, otherwise the estimation results correspond to those obtained with the simple linear QSARs.

Because of the considerable variability in experimental sorption data, the available correlations cannot be validated generally and they have to be assumed satisfactory only for the defined series of compounds, mostly persistent hydrophobic organics; they are not recommended for application to compounds of different chemical classes. Accordingly, analogous problems, as when selecting the 'correct' $K_{oc}$ value from a variety of experimental data, arise also when choosing the appropriate QSAR for predictions. Comparison of the predictive power of soil sorption QSARs (Müller and Kördel, 1996) revealed that models based on experimental HPLC partition coefficients estimate $K_{oc}$ values with higher accuracy than those based on calculated descriptors. The second-best parameters were calculated log $P_{ow}$ values, which also gave acceptable results, whereas connectivity indices and molecular fragment indicators proved to be much less useful. Whenever available, a chemical class-specific model should be applied, taking advantage of the presumably similar mode of sorption within the class. If no definite QSAR for the respective class exists, a log $P_{ow}$-dependent model may be used for non-polar compounds, provided that sorption depends mainly on van der Waals or hydrophobic interactions. To substantiate the respective estimates, it is best to make predictions by using other suitable models and then to decide on the most reasonable estimates. By comparing the results obtained by the different methods, erroneous predictions may be recognized. The simplistic approach of log $K_{oc}$ being approximately equal to log $P_{ow}$ (OECD, 1992c) is recommended only for a rough first approximation. Major problems remain for polar compounds, for which other modes of sorption (e.g. ionic or ligand exchange interactions) can be assumed and where ionization effects may be relevant. Appropriate QSARs for reliable predictions of $K_{oc}$ for polar compounds are not yet available.

## 4.7 ABIOTIC DEGRADATION

The residence time of a defined chemical in the environment is determined by degradation processes, among which abiotic reactions in water and air phases may contribute significantly to the breakdown of the structures. In this context the terms 'degradation' and 'removal' should not be confused, because degradation does not remove any material and the resultant degradation products remain present. It always has to be remembered, therefore, that degradation may in some cases yield even more toxic metabolites (toxification) and the fate and effects of these need to be considered in a meaningful hazard assessment for the parent compound. Abiotic degradation processes do not usually achieve a complete breakdown of the structures (mineralization), and most degradation experiments monitor only the disappearance of the parent compound without regard to the reaction products.

To evaluate the persistence of compounds adequately, the various degradation pathways (Schwarzenbach, Gschwend and Imboden, 1993) that may be undergone have to be considered in combination. The extent, the rates and the byproducts of the individual processes should preferably be integrated. The predominant abiotic degradation reactions comprise reactions with reactive species in the atmosphere (so-called photodegradation), hydrolysis and photolysis (Figure 4.7).

These reactions may take a considerable time with half-lives ($t_{1/2}$) of the parent compounds ranging from seconds to years. Pseudo first-order

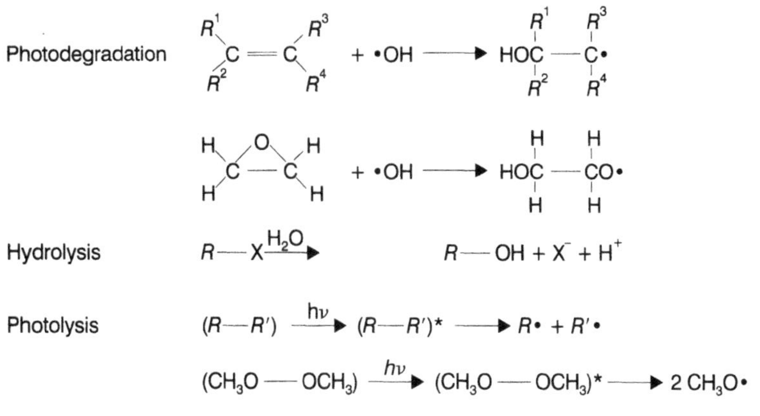

**Figure 4.7**   Examples of abiotic degradation reactions.

kinetics (section 2.1) are usually applied to describe the time course of the transformations (i.e. the reaction rates depend solely on the concentration of the contaminants, whereas the concentrations of the environmental reaction partners (e.g. water and radicals in the atmosphere) are assumed sufficiently large to be taken as constant). Besides the concentration of the compounds, the ambient temperature is of major influence on the transformation rates, which generally increase exponentially with increasing temperatures.

All the various modes of degradation of the chemicals have to be considered potentially relevant for the persistence of compounds in the environment. The degradation rates, either determined in the laboratory under ideal conditions or estimated from structural information, can only indicate which of several possible transformation pathways is most likely to occur. Because the degradability assessed this way may not correspond to the degradation occurring in the environment, the extrapolation from laboratory data to the field is extremely difficult.

Under given local conditions, different degradative reactions, abiotic and also biotic, at varying rates may predominate, yielding different metabolites of variant fate. The assessment of the transformation of chemicals can be valid, therefore, only for a stringently defined environmental scenario. An extrapolation to more general conditions is feasible only with regard to comparing different compounds on a relative scale – for example, to identify the substances with the least likelihood of persistence. These inherent uncertainties have to be accounted for when evaluating the degradability of chemicals, no matter the provenance of the data.

### 4.7.1 ABIOTIC DEGRADATION IN THE AIR COMPARTMENT (ATMOSPHERE)

The atmospheric concentration of contaminants is determined by evaporation from terrestrial and aqueous compartments, removal by wet and dry deposition, and transformations of the chemicals. The reactions accounted for by photodegradation represent only one possible primary step in the transformation of the chemicals and the products of this reaction may undergo further degradation by any (a)biotic pathway. Photolytic degradation is generally related to the fate of contaminants in the atmosphere, although similar processes may also occur in water bodies or on soil surfaces along with hydrolysis and microbial degradation.

The (indirect) photodegradation – reaction with reactive species formed by photochemical processes – has been recognized as the major transformation pathway for chemicals in the troposphere. The electrophilic addition of tropospheric radicals constitutes the principal degradation pathway. The relevant reactive species are hydroxyl radicals (OH•) and ozone ($O_3$) during day time and $NO_3$• radicals at night. The hydroxyl radicals result from reactions of oxygen atoms with water vapour, photolysis of $HNO_2$ and reactions of $HO_2$ (a product of the photolysis of aldehydes and ketones) with NO; ozone results from the photolysis of $NO_2$; the $NO_3\bullet$ radicals result from reactions of $NO_2$ with ozone. Because of their photochemical instability, the $NO_3\bullet$ radicals exist only during night time in sufficient concentrations to react with alkenes and phenols, whereas the reaction rates with other organic compounds are insignificant. Ozone transforms alkenes and alkynes through addition to the multiple bond, but hardly reacts with alkanes and aromatic compounds. In contrast, the hydroxyl radicals react with almost any kind of aromatic compounds by abstraction of hydrogen atoms from C–H and O–H bonds, by addition to double bonds and by addition to aromatic rings. This degradation pathway is the dominant one controlling the atmospheric persistence of chemicals, because the other transformations occur at much lesser rates: up to 90% of organic compounds are transformed more rapidly by reactions with hydroxyl radicals than by any other process. Only a few compound classes are inert under tropospheric conditions, especially the perhalogenated alkanes.

The measurement of rate constants in the gas phase is restricted because of the difficult, time- and cost-consuming experiments, which are generally limited to compounds of at least some water solubility ($S_w > 1$ $\mu$mol/l) and volatility ($p_v > 1$ Pa); hence QSARs may be used to obtain estimates. Estimation of the reaction rate constants for the OH• or $NO_3\bullet$ radicals from the corresponding vertical ionization energies (Güsten, Klasinc and Maric, 1984; Sabljic and Güsten, 1990) still requires experimental input data. Some recent models use calculated quantum-chemical descriptors (Klamt, 1993; Medven, Güsten and Sabljic, 1996). The widely recommended method of estimating rate constants for the reaction of organic compounds with hydroxyl radicals has been derived by Atkinson (1987, 1988). The underlying principle is to account for molecular fragments that are likely to be attacked by reactive species. The total reaction rate constants ($k_{tot.}$) for the hydroxyl radicals are calculated from rate constants of four important types of reactions: H atom abstraction from C–H and O–H bonds ($k_{H-abst.}$), addition of hydroxyl radicals to C–C multiple bonds ($k_{add.(C=C)}$), addition to aromatic rings ($k_{add.(aromat.)}$) and reactions with N, S or P ($k_{N,S,P}$):

$$k_{tot.} = k_{H-abst.} + k_{add.(C=C)} + k_{add.(aromat.)} + k_{N,S,P}$$

The rates of the four types of reactions are estimated using additivity fragment constants. By considering exclusively substructures that are liable to transformations, ignoring persistent moieties, this approach is principally assumed to correspond to the worst case. If none of the substructures considered in the model is present in the molecules, zero degradation is assumed. The factors (group contributions) relate to the respective reaction rates of the degradable fragments with reactive species, and degradation as a result of further functional groups is neglected.

For the calculation of the rate constants for hydroxyl radicals and ozone, a computer program is available (AOP, 1990). In a validation study, this program was used for calculating rate constants of 369 compounds and to compare those with experimental data (Müller and Klein, 1991). Considering that almost all of the available experimental data were included in the model derivation, the experimental and calculated rate constants differed by more than a factor of two for only 34 compounds (9.2%). The agreement between experimental and calculated rate constants was generally satisfactory (deviations < factor 2) for:

C H compounds (not aromatic)
C H O compounds
C H N compounds (not $NO_x$)
C halogen compounds (< 3 halogens per C atom)
C H S compounds
C H O S compounds.

whereas major discrepancies occurred for:

–NO compounds
–$NO_2$ compounds
–$NO_3$ compounds
C H P compounds
C H O P S compounds
aromatic compounds.

The AOP computer program is an extension of the Atkinson method, because it is extended by fragment values that were not given by Atkinson. Although some assumptions are not validated, the good overall agreement of experimental and calculated data justifies the application of this program for the estimation of rate constants (OECD, 1993a).

Direct photolysis, occurring in air and water phases, involves reactions mediated by the absorption of solar radiation. Chemicals consisting of conjugated $\pi$ electron systems (e.g. aromatic rings and conjugated double bonds, so-called chromophores) may absorb light in the wavelength range 290–600 nm and be promoted to an excited state, which may result in a fragmentation of the compound. A prerequisite for the derivation of predictive models for the rate of direct photolysis is to quantify the rate of light absorption and the quantum yields ($\Phi$). Although the net rate at which an aqueous solution absorbs light can be estimated by computerized calculations (Schwarzenbach, Gschwend and Imboden, 1993), the data necessary for the calculation of net quantum yields are available only from experiment. The simplistic, but not recommended, assumption $\Phi = 1$ (i.e. 100% efficiency) results in the maximum photolysis rate and may significantly underrate the persistence of the contaminants. Currently no validated QSARs are available to predict direct photolysis rates.

## 4.7.2 ABIOTIC DEGRADATION IN THE WATER PHASE

The persistence of contaminants in aqueous compartments is dependent among other factors on the chemical reactions between the contaminants and water. Abiotic degradation is of minor relevance for chemicals in natural aquatic sediments, because their transformation is dominated by biodegradation processes (section 4.8). Sediment and particulate matter in water bodies, however, may influence greatly the efficacy of abiotic transformations by altering the truly dissolved (non-sorbed) fraction of the compounds – the only fraction available for reactions (Weber and Wolfe, 1987). Among the possible abiotic transformation pathways, hydrolysis has received the most attention, although only some compound classes are potentially hydrolysable, such as alkyl halides, amides, amines, carbamates, esters, epoxides, and nitriles (Harris, 1990; Peijnenburg, 1991). The carbon atom bound to a heteroatom is attacked by a nucleophile, generally $H_2O$ or $OH^-$, resulting in the bond cleavage between the carbon atom and the leaving group (e.g. the hydrolysis of an ester bond yields the corresponding acid and the alcohol). The hydrolysis rate of carboxylic acid esters is dependent on the ambient pH values due to changing contributions of the acid catalysed, neutral and base catalysed mechanisms, with the $OH^-$ (base)-catalysed reaction being predominant at pH $\geq$ 6–7 (Schwarzenbach, Gschwend and Imboden, 1993).

Estimation methods for the hydrolysis rates of several types of carboxylic acid esters, carbamates, aromatic nitriles and phosphoric acid esters have been reported. Hydrolysis rates are subject to substituent effects, and consequently linear free-energy relationships (LFERs), as represented by Hammett or Taft correlations, have hence been applied to their estimations. Reviews (e.g. Harris, 1990; Peijnenburg, 1991) reveal that QSARs are available for only a few compound classes (Table 4.8) and are mostly based on limited sets of experimental data.

The alkaline hydrolysis rates depend on the electronic and steric features of the leaving group (alcohol moiety), and generally decrease with size and branching of the alcohol and increase with electron-withdrawing groups on the alcohol. The respective QSARs are strictly limited to homologous compounds due to the class-specific sensitivity of the chemical reaction centre, as indicated by the greatly varying slopes and intercepts of the functions. The same effects are described by model 8 in Table 4.8, where the electronic substituent constant is replaced by a quantum-chemical descriptor. Before this equation can be used to predict absolute $K_{hyd.(OH)}$ values, the calculated net charges on phosphorus have to be scaled with the original data used to derive the model. Models 9 and 10 use rate constants corrected for the sorption to sediments based on organic carbon content to account for the fraction of the compounds not available for the reactions. The available hydrolysis QSARs

**Table 4.8** Examples of QSAR models for estimating the alkaline hydrolysis rate: $\log K_{hyd.}$ correlations with various parameters ($K_{hyd.}$ in $mol^{-1}s^{-1}$).

| Model | Chemical class | | $r$ | $n$ |
|---|---|---|---|---|
| 1 | Benzoic esters | $\log K_{hyd.} = 1.17\,\sigma + 2.26$ | 0.996 | 18 |
| 2 | Phosphoric acid esters | $\log K_{hyd.} = 1.4\,\Sigma\,\sigma - 0.47$ | 0.995 | 4 |
| 3 | Phthalate esters | $\log K_{hyd.} = 4.59\,\sigma* + 1.52\,E_S - 1.02$ | 0.986 | 5 |
| 4 | $N$-Methyl-$N$-phenyl-carbamates | $\log K_{hyd.} = -0.26\,pK_{a(alcohol)} - 1.3$ | 1.0 | 3(!) |
| 5 | $N$-Phenylcarbamates | $\log K_{hyd.} = -1.3\,pK_{a(alcohol)} + 13.6$ | 0.99 | 20 |
| 6 | $N$-Methylcarbamates | $\log K_{hyd.} = -0.91\,pK_{a(alcohol)} + 9.3$ | 0.99 | 6 |
| 7 | $N,N$-Dimethyl-carbamates | $\log K_{hyd.} = -0.17\,pK_{a(alcohol)} - 2.6$ | 0.89 | 7 |
| 8 | Phosphoric acid esters | $\log K_{hyd.} = -9.65\,\rho + 2.85\,E_{S(alcohol)} + 4.89$ | 0.95 | 19(?) |
| 9 | Aromatic nitriles ($m$, $p$-subst.) | $\log K_{corr.} = 0.54\,\log P_{ow} + 0.57\,\sigma - 5.28$ | 0.925 | 17 |
| 10 | Aromatic nitriles ($o$-subst.) | $\log K_{corr.} = -0.46\,\log P_{ow} + 1.26\,\sigma - 4.56$ | 0.981 | 7 |

$r$ = correlation coefficient; $n$ = number of compounds analysed; $\rho$ = net charge on P; $K_{corr.}$ = hydrolysis rate corrected for the fraction of compound sorbed to sediment.
Sources of models: 1, 2, Harris (1990); 3, Wolfe Steen and Burns (1980); 4–7, Wolfe Zepp and Paris (1978); 8, Johnson *et al.* (1985); 9, 10, Peijnenburg *et al.* (1993).

are limited in their application range but are of acceptable validity when operated in accordance with their restrictions. Other abiotic degradation reactions (e.g. reductive dehalogenation) have also been modelled using LFER-type correlations (Peijnenburg *et al.*, 1991) or quantum-chemically derived parameters (Rorije *et al.*, 1995).

Several QSARs, developed for different purposes, are not applicable for environmental hazard assessment as the underlying transformation rates were obtained not in water but in a mixture of solvents. Further development is needed, therefore, especially with respect to validation and extension of the existing QSARs, the derivation of new QSARs for extra compound classes and the consideration of the various mechanisms of hydrolysis.

## 4.8 BIODEGRADATION

The persistence of xenobiotics in the environment is determined substantially by the ability of ambient microorganisms to utilize the chemicals as nutrient sources. This may reduce the concentration of the parent compound either through the formation of – potentially hazardous – metabolites (primary biodegradation) or through the complete mineralization of the substance to produce carbon dioxide, water and mineral salts (ultimate biodegradation). Biodegradation can be an effective mechanism for tranforming organic compounds in water, soil and sediment. The extent of biodegradation depends on:

structure and concentration of the compound
the compound's toxicity towards microorganisms
water solubility of the compound
presence of other xenobiotics
exposure time
soil type
water content
oxygen concentration
nutrients
temperature
pH
ionic strength
diversity of the microbial community
adaptability of the microbial community to xenobiotics.

Biodegradation of organic chemicals is a complex multistep process involving uptake, intracellular transport and enzymatic reactions inside the microorganisms. Depending on the ambient conditions, different modes and rates of biodegradation may predominate. These are influenced by factors related to the chemical (substrate), the microorganisms and the environment (Scow, 1990; Schwarzenbach, Gschwend and Imboden, 1993; Pedersen *et al.*, 1994). The high adaptive and mutational capacity of microorganisms may, in suitable environmental conditions, result in the transformation of almost any substance. The different microbial communities in different environments may render a chemical biodegradable at one site but not at another due to the different degradative capacities in different soils. Bacteria have a variety of enzyme systems, but specific enzymes for transforming xenobiotics are not present except in a few species of microorganism. The required enzymes are either always present at certain concentrations and activities (constitutive enzymes) or have to be induced, expressed or transferred by plasmids during an adaptive lag phase (inducible enzymes). Primary metabolic reactions are mediated mostly by non-specific enzyme systems, which catalyse oxidation, reduction and hydrolysis, with the different transformations occurring either consecutively or simultaneously (competitive) at different sites of the substrate.

The initial point of attack for oxidations is generally that moiety in a chemical with the most readily available electrons, such as:

double bonds
$\pi$ electron systems of aromatic rings
non-bonded electrons of sulphur or nitrogen
$\sigma$ electrons of C–H bonds.

Moieties containing electron-withdrawing heteroatoms are primarily subject to reductions, for example:

carbonyl groups
nitro groups
sulphoxide and sulphone groups
halides.

Hydrolysis preferably occurs on moieties with unsaturated C, P or S atoms with a multiple bond to another more electronegative atom, such as:

esters
amides
alkyl halides
carbamates
ureas
(thio)phosphates.

For large compounds with a molecular weight > 500, which cannot react with the intracellular bacterial enzymes because their transfer through membranes is hindered, biodegradation is generally negligible, hence abiotic degradation (e.g. hydrolysis) may be the only degradation pathway.

In different environmental conditions, a given chemical may be biodegraded by different pathways, resulting in different degrees of persistence. The actual biodegradability depends on the rates of the concurrent transformation reactions. Only if a reaction occurs at a sufficiently high rate can it contribute significantly to the breakdown of a structure. The higher the rate, the higher is the probability for this transformation to be relevant in the metabolic pathway of the respective chemical.

A uniform experimental assessment scheme cannot reflect the varying ambient (field) situations and transformation processes. A (defined) test system needs clear evaluation criteria, but the various test procedures do not agree on a definite endpoint. Accordingly, it has to be recognized that biodegradability is not a uniform principal property of chemical contaminants and that biodegradability is not a well-defined parameter.

Various testing protocols for the experimental determination of biodegradability have been developed. The procedures (Table 4.9) differ considerably in:

the treatment and concentration of the test compounds
the kind of inoculum used (e.g. pure cultures, surface water, sewage water, soil)
size and density of the inoculum
period of adaptation
incubation time.

Biodegradation is a strongly time-dependent phenomenon, and the period allowed for acclimation of the microbes to a substrate significantly affects the experimental results. The rate of adaptation is influenced by the dosage, either continuous or intermittent, of the xenobiotic. Because the concentrations of compounds in laboratory tests are generally much higher than those in the environment, toxicity towards the bacterial inoculum may result in

**Table 4.9** Examples of variables in some biodegradation tests. Reproduced with permission from OECD).

| Test | Measured variable | Inoculum (CFU/ml) | Test compound concentration (mg/ml) | Duration (d) | Path-level |
|---|---|---|---|---|---|
| OECD[a] | DOC | $10^7$–$10^8$ | 5–40[b] | 28 | 70% |
| STURM[a] | $CO_2$ | $10^7$–$10^8$ | 10 + 20 | 28 | 60% |
| AFNOR[a] | DOC | $10^7$ | 40[b] | 28 | 70% |
| Closed Bottle[a] | BOD | $10^4$–$10^6$ | 2–10 | 28 | 60% |
| MITI[a] | BOD | $10^7$–$10^8$ | 100 | 28 | 60% |
| Zahn-Wellens[a] | DOC | $10^5$–$10^6$ | 50–400[b] | 28 | 70%/20% |
| Babeu and Vaishnav (1987) | BOD | $10^5$–$10^6$ | 0.4–3.2 | 5–20 | 16% |
| Pitter (1976) | COD | 100 mg DW/l | 200 | 5 + 20[c] | 90%, 15 mg $g^{-1}$ $h^{-1}$ |
| Bridie *et al.* (1979) | BOD | 100 mg DW/l | 30 | 5 | – |
| Urano and Kato (1986) | BOD | 30 mg DW/l | 100 | 14 | 40%/25% |
| Kondo *et al.* (1988) | SAM | < 3 mg DW/l | 0.1–1000 | 3 | 50%/15% |

BOD = biological oxygen demand; DOC = dissolved organic carbon; COD = chemical oxygen demand; $CO_2$ = carbon dioxide production; CFU = resultant number of colony-forming units; DW = dry weight; SAM = specific analytical method.
[a] Test procedure in accordance with OECD test guidelines (OECD, 1984a, 1989b); [b] test concentration in mg DOC mg $^{-1}$ $l^{-1}$; [c] test duration with adaptation time.
Sources: Degner (1991); IUCT (1992); OECD (1993b).

apparently reduced biodegradation. Further discrepancies in test results are due to the various parameters measured (BOD, COD, $CO_2$ production, DOC, etc.) and the different path-levels evaluated: path-levels from 15% to 90% are used to classify degradable and non-degradable substances. Within a tiered assessment scheme, OECD (1984a, 1989b) established three categories of degradation testing, resulting in a classification of compounds as readily degradable, inherently degradable or persistent. To interpret the results of laboratory tests correctly, it is essential to understand the mechanisms and rate-limiting processes of the respective biodegradation. The experimental conditions may be as important as the structure of the test compound for the outcome of a biodegradation test. A comparative assessment of experimental biodegradation data obtained by different test procedures (Degner, 1991; OECD, 1993b) revealed that the different test conditions result in different classifications of biodegradability: in other words, a compound evaluated readily biodegradable in one test may be non-degradable in another. Some variability in the test results can be related to the concentrations of the test compounds, the inoculum size and the adaptation and incubation periods. Nevertheless, the major proportion of the variability could not be explained from such factors. For an adequate judgement of the relevance of QSAR estimates of biodegradability, it is necessary to rationalize the inherent diversity of the endpoints and parameters concerned. Additionally, the currently used test methods result in the dichotomous categorization of the biodegradability of chemicals as either 'yes' or 'no', which impedes the classification of substances of intermediate degradability. It must always be realized, therefore, that the major problem of biodegradation modelling is not the availability of appropriate descriptors or statistical techniques, but the ambiguity of the underlying activity data.

Because of the severely empirical nature of biodegradability assessments, therefore, anybody who aspires to evaluate and model biodegradation data such as $BOD_5$ is strongly recommended first to conduct such measurements (and try to replicate them) before proposing a superficially precise QSAR, using elaborate parameters and statistics. There is substantial variability of at least ± 20% of the input activity data (OECD,1984a); for example, the official German procedure (DEV, 1987) uses for calibration purposes a solution of glucose and glutamic acid with a theoretical oxygen demand of 306.7 mg/l, for which the acceptable $BOD_5$ ranges from 180 to 230 mg/l.

In spite of the considerable uncertainties about the relevance of the different activity data sets, numerous QSARs on biodegradation have been reported. However, the significance of the respective estimates cannot be expected to exceed that of the underlying data and they should be regarded as indicators of probabilities towards higher or lower biodegradability. Accordingly, any QSAR approach to biodegradability is principally restricted by the problem of determining precisely which activity is accounted for:

transformation *by which reaction*?
transformation *to what extent*?

Numerous QSAR models for estimating biodegradation have been derived, with reviews given by, among others, Kuenemann and Vasseur (1988), Kuenemann, Vasseur and Devillers (1989), Parsons and Govers (1990), Scow (1990), OECD (1993b), Peijnenburg (1994) and Peijnenburg and Karcher (1995). The available models are all based on the assumption that structural features of a compound may indicate its biodegradability:

Acyclic compounds:
   chain length
   degree of branching
   saturation state of the carbon chain
   oxidation state of the terminal groups.
Aromatic compounds:
   kind of substituents
   number of substituents
   position of substituents
   number of rings.

Additionally, physico-chemical properties may affect the degradability of the chemicals, relating to processes such as transport into the microbial cell. Electronic parameters can be used to explain different transformation mechanisms caused by the different polarity of the chemicals. The electron density on the aromatic ring, which is dependent on the functional groups, is determinative for the ring cleavage of an aromatic system. However, because of the multitude of processes involved in biodegradation, no single descriptor model can accurately predict the biodegradability of a broad range of chemicals. If the reaction mechanism (i.e. the explicit reaction equation and stoichiometry) of the degradation process is known, thermodynamic descriptors may be used; these have the advantage of allowing for variable environmental conditions, such as the presence of water, concentrations of reactants and products, temperature and redox conditions (Govers *et al.*, 1995).

QSAR models on biodegradation can be categorized with respect to the underlying data set and the statistics used for analysis:

Models for homologous series of substances using simple regression methods. Quantitative substructure-based models derived from multivariate statistics (e.g. discriminant analysis), which may be applicable to a variety of chemical classes.
Use of classification criteria (e.g. associated with substructures) to discriminate degradable/non-degradable compounds qualitatively.

To test the applicability of published QSARs concerning biodegradation, Degner (1991) compiled > 70 models and compared the respective estimates with experimental data. The validation exercise was founded on a set of measured data for > 600 diverse chemicals, uniformly obtained by the MITI procedure (OECD, 1984a, 1989b). In spite of the shortcomings of this test, the available data pool proved useful. The MITI degradation test takes an intermediate position in terms of degradative capacity and can be regarded as representative of degradation tests for category 1 (ready biodegradability). These data do not comply with all QSARs because of the different endpoints modelled. However, this validation study provides a realistic approximation of the predictive power of the various models to discriminate readily degradable and non-readily degradable compounds. Even if a sufficiently large uniform data set on other biodegradation parameters became available, the principal results of this study would probably not change drastically.

From the multitude of QSAR models for estimating biodegradability, only a few provide an adequate level of agreement between calculated and experimental data. Classifications were considered adequate if > 75% of the MITI data could be discriminated into readily degradable and non-readily degradable substances for any arbitrary path-level (for details of this validation study for the individual QSAR models see OECD, 1993b). The reasons for misclassifications by many models can be ascribed partly to the inconsistent data material, the endpoint inhomogeneity and the selection of limited sets of homologous test compounds to derive the respective QSARs. Many of the models could not be validated with the MITI data as they apply to chemical classes whose compounds are either all readily degradable or all non-readily degradable in the MITI degradation test, or the models were developed for chemical classes that were not included in the MITI data set. From the literature, 64 regression models for specific compound classes were retrieved, of which 35 could be tested with the MITI data, but only seven QSARs were successfully validated (Table 4.10).

These models were derived with four to eight homologous substances and, because of their specificity, are suitable for application to corresponding substances only. The number of validated QSAR models for specific compound classes is hence too low to make predictions solely on this basis; in the MITI data set they were applicable for estimating the biodegradability of only 3% of the chemicals.

For the classification of larger data sets, more general group-contribution models have been derived on the basis of substructure indicators, which may

**Table 4.10** Examples of validated QSAR models for estimating biodegradability: log COD/log $k$ correlations with various parameters.

| Model | Chemical class Restrictions | | $r$ | $n$ | Validation test (%) | Path-level at validation |
|---|---|---|---|---|---|---|
| 1 | Alicyclic alcohols, ketones $P_{ow}$: 1.27–6.72 | $\log (COD_{rate}) = -0.51 \log P_{ow} + 2.53$ | 0.98 | 4 | 100 $(n = 5)$ | 15[a] |
| 2 | Disubstituted phenols only:-OH-OCH$_3$-CH$_3$, -Br,-Cl | $\log (k_{rate}) = -1.36\ Y_{vdw} - 9.3$ | 0.98 | 8 | 100 $(n = 14)$ | $2.2 \times 10^{-12}$[b] |
| 3 | Disubstituted phenols only:-COOH,-OH, -CH$_3$,-NO$_2$,-Cl | $\log (COD_{rate}) = -0.32\ \sigma + 1.43$ | 0.95 | 7 | 77 $(n = 14)$ | 21[c] |
| 4 | *ortho*-substituted phenols only:-COOH, -OH,-CH$_3$,-NO$_2$,-Cl | $\log (COD_{rate}) = -0.43\ \sigma_o + 1.70$ | 0.98 | 5 | 83 $(n = 6)$ | 30[c] |
| 5 | *meta*-substituted phenols only:-COOH, -CH$_3$,-OH,-Cl | $\log (COD_{rate}) = -0.616\ \sigma_m + 1.72$ | 0.94 | 4 | 80 $(n = 5)$ | 32[c] |
| 6 | *para*-substituted anilines only:-CH$_3$, -NH$_2$,-NO$_2$,-Cl | $\log (COD_{rate}) = -0.78\ \sigma_p + 1.04$ | 0.94 | 5 | 80 $(n = 4)$ | 14.5[c] |
| 7 | Aliphatic cyclic compounds with > 1 substituent | $\log (COD_{rate}) = -0.293\ ^0\chi^v + 3.216$ | 0.970 | 6 | 100 $(n = 4)$ | 8.5[a] |

$r$ = correlation coefficient; n = number of compounds analysed; $COD_{rate}$ = chemical oxygen demand (mg g$^{-1}$ h$^{-1}$); $k_{rate}$ = transformation rate of substrate ($10^{-12}$ l organisms$^{-1}$ h$^{-1}$); $Y_{vdw}$ = van der Waals radius; $\chi$ = connectivity index.
[a] mg COD g$^{-1}$ h$^{-1}$; [b] organisms$^{-1}$ h$^{-1}$; [c] mg g$^{-1}$ h$^{-1}$.
Sources of models: 1, 7, Vaishnav, Boethling and Babeu (1987); 2, Paris *et al.* (1983); 3–6, Pitter (1985). For details of validation see Degner (1991), IUCT (1992), OECD (1993b)

be applicable to a variety of chemical classes. Factors are used to quantify the contributions of certain substructures to the degradability of the compounds. Increasing biodegradability has been demonstrated in the presence of functional groups such as carboxyl-, hydroxyl- and methyl groups and decreasing degradability in the presence of nitro-, amino-, cyano- and halogen groups (Kawasaki, 1980; Kobayashi, 1981; Paris, Wolfe and Steen, 1982; Pitter, 1985). The respective criteria may be combined by rules of logic to support expert judgement of the biodegradability of environmental contaminants. Non-linear substructure models are supposed to account also for the contribution of group interactions to the chemicals' degradability. They are usually derived by neural network techniques, but their application is currently limited by their insufficient reproducibility and the lack so far of thorough validation for predictive purposes. As a principal limitation, all substructure models are not applicable to compounds with structural elements not in the original data set. Niemi *et al.* (1987) developed a model for estimating biodegradation based on the presence or absence of functional groups for 261 chemicals using cluster analysis, stating the probability for correct classification to be 92%. This and other models unrelated to specific chemical classes (Geating, 1981; Mudder, 1981; Babeu and Vaishnav, 1987; Boethling and Sabljic, 1989; Desai and Govind, 1990; HDI, 1990; Howard *et al.*, 1992) were subjected to a validation exercise with the MITI data. They mostly yielded < 70% conform classifications of the data, except the substructure model by Niemi *et al.* (1987), which provided 76% coincident predictions (OECD, 1993b). A feature of all these models, which gives cause for concern, is a marked difference in recognition of degradable and non-degradable compounds. Although they correctly classify 50–83% (mean 73%) of the degradable substances as being degradable, the predictions are false for most of the non-degradable chemicals, the success rate being as low as 10% for some models (range 10–70%, mean 47%). This imbalance indicates the major problems with the application of biodegradation models: predictions that compounds are readily degradable have a reliability of < 50% of being true, hence they are useless and should be discarded. Only if a compound is predicted non-degradable by such models, is there a good probability that it really is non-degradable and the predicted result may be used with some confidence. The rejection of QSAR-predicted ready degradability as being untrustworthy is based on the rationale that incorrect predictions for readily degradable compounds as persistent may be filed as overprotective assessments, but the opposite classification of persistent chemicals as readily degradable may result in substantial hazard for people and the environment.

Based on the information gained during the validation exercise, especially the necessity of stringent application criteria for QSARs on biodegradation, Degner *et al.* (IUCT, 1992; OECD, 1993b) concluded that there cannot be one universal model for predicting the diverse degradation processes, but rather a set of QSARs supplemented with rules to guide the selection of the

**Table 4.11** Examples of substructure models for estimating the biodegradability $B$ of specific compound classes ($B > 0$: readily degradable, $B < 0$: non-readily degradable).

$$B = \Sigma \, (n_i \cdot \text{factor}) + \text{intercept}$$

| Model | Chemical class | Substructure | Factor |
|---|---|---|---|
| 8 | Acyclic compounds *except* phosphoric acids, hydrazines, disulfides, tertiary amines, compounds with non-terminal heteroatoms (e.g. ethers and esters), compounds with triple bonds | $-C=O$ <br> $-CH_2-NH_2$ <br> $-CH_3$ <br> $-OH$ <br> $-$Halogen <br> Intercept | + 0.07 <br> + 0.14 <br> $-0.11$ <br> $-0.13$ <br> $-0.20$ <br> +0.39 |
| 9 | Acyclic compounds *except* hydrazines, disulphides, compounds with non-terminal heteroatoms (e.g. ethers and esters), compounds with triple bonds | $-CH_2-NH_2$ <br> $-CH_3$ <br> $-C=O$ <br> $-OH$ <br> $-$Halogen <br> $-P(R_1)$ <br> $-N(CH_2-R_1)_3$ <br> Intercept | +2.81 <br> +0.0025 <br> +0.00092 <br> $-0.0022$ <br> $-0.0046$ <br> $-0.37$ <br> $-0.68$ <br> +0.009 |
| 10 | Acyclic compounds | > 1 halogen atom <br> Tertiary butyl <br> Tertiary amine with > 2 $-\!\!-CH_2$ (non-amide) <br> Atoms other than C, H, N, O, P, S, halogen <br> 2 terminal isopropyl groups <br> Phosphoric compound <br> Azo group <br> Disulphide <br> Hydrazine <br> $C \equiv C$ triple bond | $-1$ <br> $-1$ <br> $-1$ <br> <br> $-1$ <br> <br> $-1$ <br> $-1$ <br> $-1$ <br> $-1$ <br> $-1$ <br> $-1$ |
| 11 | Monocyclic aromatic compounds only: mono- and disubstituted carbocyclic compounds | Aryl-C(O)R$_2$ <br> Aryl-C(O)OR$_2$ <br> Aryl-OH <br> Aryl-CH$_2$(R$_2$) <br> Aryl-NHC(O)R$_1$ <br> Aryl-NH$_2$ <br> Aryl-NO$_2$ <br> Aryl-halogen <br> Aryl-SO$_3$H <br> Intercept | +0.22 <br> +0.22 <br> +0.15 <br> +0.14 <br> +0.03 <br> $-0.19$ <br> $-0.38$ <br> $-0.32$ <br> $-0.16$ <br> +0.08 |
| 12 | Monocyclic aromatic compounds only: carbocyclic compounds | Aryl-C(O)R$_2$ <br> Aryl-C(O)OR$_2$ <br> Aryl-OH <br> Arul-CH$_2$(R$_2$) <br> Aryl-NHC(O)R$_1$ <br> Aryl-NH$_2$ <br> Aryl-NO$_2$ <br> Aryl-halogen <br> Aryl-SO$_3$H <br> Aryl-(R$_3$) <br> Intercept | +0.09 <br> +0.09 <br> +0.003 <br> +0.003 <br> $-0.052$ <br> $-0.34$ <br> $-0.55$ <br> $-0.48$ <br> $-0.32$ <br> $-0.50$ <br> +0.38 |

$n_i$ = occurrences of any specified substructure i; $R_1$ = unspecified fragment; $R_2$ = $-$H or unbranched alkylchain without multiple bonds; $R_3$ = fragment with non-terminal heteroatoms or branched alkylchain. Sources of models: Degner (1991); IUCT (1992); OECD (1993b).

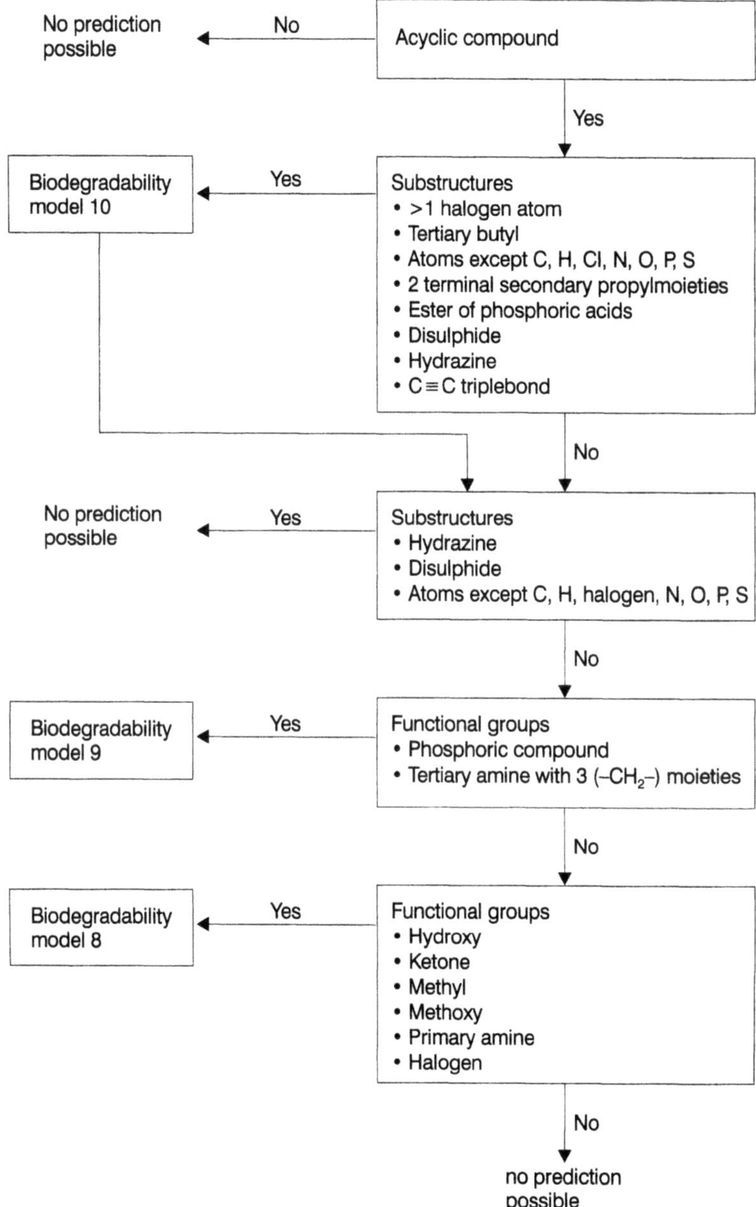



**Figure 4.8**  Flow chart of hierarchic QSAR models for predicting the biodegradability of acyclic compounds (Degner, 1991; IUCT, 1992; OECD, 1993b). Reproduced with permission from OECD.

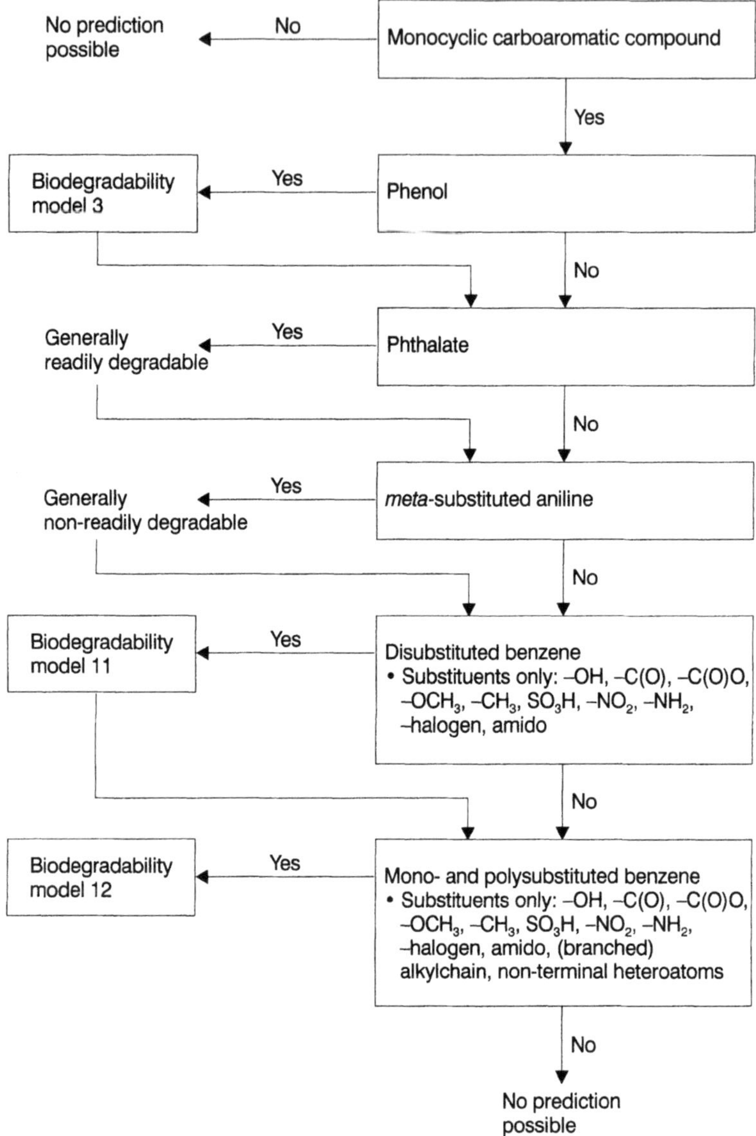


**Figure 4.9**    Flow chart of hierarchic QSAR models for predicting the biodegradability of monocyclic carboaromatic compounds (Degner, 1991; IUCT, 1992; OECD, 1993b). Reproduced with permission from OECD.

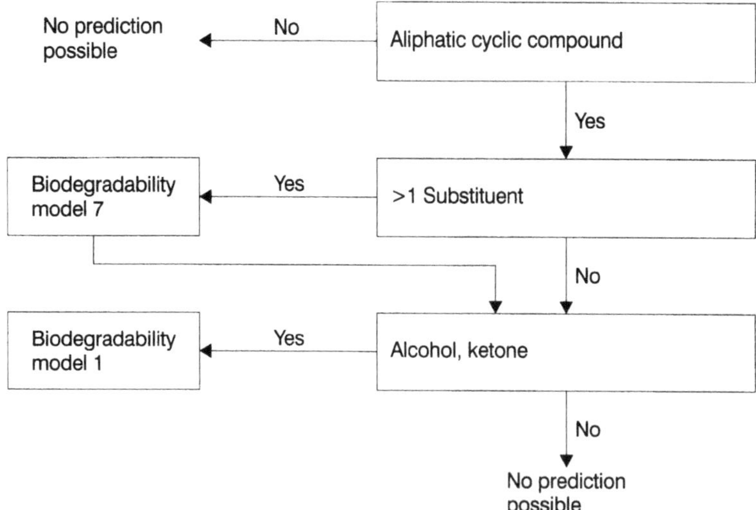



**Figure 4.10** Flow chart of hierarchic QSAR models for prediciting the biodegradability of alicyclic compounds (Degner, 1991; IUCT, 1992; OECD, 1993b). Reproduced with permission from OECD.

appropriate model. Starting out from the potentials and limitations identified for the substructure QSARs examined, confined models were derived for specific chemical classes. For acyclic compounds and monocyclic carboaromatic compounds several particular QSARs based on substructure indicators revealed sufficient predictive power if operated in concordance with the respective restrictions (Table 4.11).

In combination with other validated QSARs (Table 4.10), hierarchic models are recommended for estimating the biodegradability of three major classes of environmental contaminants: acyclic compounds, monocyclic carboaromatic compounds and aliphatic cyclic compounds (Figures 4.8–4.10).

At present it is not possible to establish schemes for further compound classes because of the lack of consistent experimental data.

The principle of the approach using hierarchic models is that a set of discriminant criteria can be used to identify the most appropriate QSAR for a given compound. First, the compounds have to be assigned to a chemical class based on structural characteristics. This categorization of substances according to their parent structure and substructures is intended to group compounds with a similar degradation pattern. Within the classes defined in this manner, biodegradability may then be related to the same structural descriptors. Incorrect classification of the compounds (i.e. if an unsuitable QSAR model is applied) may result in substantial errors in predicting biodegradability, hence the restrictions for the individual models have to be respected. The hierarchic models should be followed completely to ensure that not only the first applicable QSAR but the most appropriate QSAR is identified. If more than one QSAR is regarded suitable, all of them should be used for predictions and then a decision made about the most reasonable estimates. By comparing the results obtained by different methods, erroneous predictions will also be recognized.

The drawback of the hierarchic system of validated QSARs is that for a rather large proportion of chemicals no reliable predictions can be made. A re-validation procedure of these QSARs with a further set of > 300 experimental data revealed that for 53% of the chemicals recommended QSARs were available and that > 95% of the predictions (degradable or non-degradable) were correct.

## 4.9 BIOCONCENTRATION

The accumulation of chemicals in biotic and/or in abiotic compartments of the ecosphere such as organisms, water bodies, sediment and soil is of major concern for environmental hazard assessment. The uptake of dissolved contaminants into the biophase occurs mostly by direct adsorption, but also along the trophic web. The internal concentration in the body may increase by accumulation to a level that causes toxic effects, even if the external concentration remains below the critical limit. Even exposure for a short time may produce high internal concentrations that persist in the organism much longer than in the surrounding water. Because of their elevated and lasting level in biotic compartments, substances that are accumulated may evoke potentially chronic effects, not only in the organisms directly exposed but also in species at higher levels in the foodchain, including humans. Bioaccumulation is therefore an important link between the pollution of surface waters and human exposure to xenobiotic substances.

Accumulation is the general term for any phenomenon associated with increasing the concentration of chemicals in a compartment relative to the surrounding phases. With regard to organisms, the accumulation processes are defined according to the mode of uptake of contaminants:

*bioaccumulation*: uptake from the environment via any possible pathway;
*biomagnification*: uptake via the foodweb resulting in increased concentrations in higher trophic levels;
*bioconcentration*: uptake from the surrounding phase via adsorption, lipid diffusion, etc.

The potential of chemicals to bioaccumulate is generally characterized by the bioconcentration factor BCF, which serves as a measure of the compounds' concentration in the organism concurrent with ambient concentrations under steady-state conditions (e.g. for aquatic environments):

$$BCF = \frac{\text{concentration of chemical in organism at equilibrium}}{\text{mean concentration of chemical in water}}$$

The definition of bioconcentration as a partitioning process between the outer environmental phase (e.g. the water body) and the inner biophase (e.g. a fish) implies that the chemical transfers into the organism until a steady state is reached; that is, the rate of uptake into, and the rate of loss from, the body are equal. The BCF, regarded as an equilibrium constant, may be obtained either from the ratio of the corresponding concentrations in the water and in the biophase, or from the ratio of the respective uptake and elimination rate constants in the water/biophase system, assuming a first-order, one-compartment model with lipid diffusion (section 2.1) to be the relevant transport process (Hansch, 1969; Hamelink, Waybrant and Ball, 1971; Hunn and Allen, 1974; Isensee and Jones, 1975; Kanazawa, Isensee and Kearney, 1975; Bahner *et al.*, 1977; Clayton, Pavlou and Breitner, 1977; Hamelink and Spacie, 1977; Macek, Petrocelli and Sleight, 1979; Zitko, 1980; Mackay 1982; Walker, 1987; Nagel and Loskill, 1991; Pedersen et al. 1994, Berg *et al.*, 1995). Any assessment of bioconcentration does not concern only the parent toxicants but also has to take into account any accumulation of their degradation products resulting from metabolism, biodegradation, hydrolysis, photolysis, etc.

Most methods for experimental BCF determinations represent an assessment of the potential for accumulation. They do not account for specific environmental conditions, such as differences in the species exposed or the environmental bioavailability of the chemicals. The tests generally use fish as a model organism to serve as a predictor for bioconcentration in other aquatic species as well, and for bioaccumulation/biomagnification along aquatic foodwebs. Test guidelines for BCF in fish are available (e.g. OECD, 1981a), where the fish are exposed to the chemicals and from the concentrations in fish and water the BCF is obtained. The tests vary with respect to:

exposure regime (static, semi-static, flow-through)
duration of the test (OECD 305A–D: 1–8 weeks; OECD 305E: 8 h–90 d)
sampling frequency
requirements concerning uptake/elimination phase and steady state
requirement for measuring the lipid content of the fish
method of calculating the BCF value.

Several factors may contribute to the substantial variability observed in measured BCF values, which may range over several orders of magnitude for the same compound; for example, for pentachlorobenzene, BCF values between 900 and 250 000 have been reported (e.g. Bruggeman *et al.*, 1984; Gobas, Shiu and Mackay, 1987; Hawker, 1990). The evident variability in experimental BCF data may arise from:

Species sensitivity: the bioconcentration of xenobiotics in organisms varies with the size, lipid content, age, sex and lifespan of the species.
Purity of the test compound: pure substances have to be used because even small amounts of soluble impurities can cause large errors in the measured bioconcentration.

Attainment of equilibrium: the required time may take several days to weeks.

Analytical method: a specific method has to be developed for each chemical species.

Stability of the test compound in water: the substance must not degrade during the experiment; losses in the concentration of the test compound may also occur by evaporation or adsorption to glassware.

Surface-active materials: the presence of solubilizing agents alters the bioavailability of the test compounds significantly; the apparent increase of the total (but not bioavailable) concentration in the water phase, possibly above the water-solubility limit for chemicals of low water solubility and high lipophilicity, may cause spuriously low BCF values (i.e. experimental artefacts).

pH and buffer capacity of the water phase: pH conditions strongly influence the bioconcentration of organic acids and bases.

Water chemistry: hardness, ionic strength, etc. are determinant especially for the bioconcentration of surfactants.

Cosolute effects: the presence of further organic solutes, as is the case in 'real' waters, may significantly alter the bioconcentration of the individual compounds.

Suspended organic matter: soil and sediment components (e.g. humic acids) may result in decreased bioavailability by serving as a sink compartment due to sorption processes.

Further practical problems in measuring BCF stem from insufficiencies in the chemical analyses. Difficulties arise in the determination of the concentration in the aqueous solution especially for highly lipophilic chemicals with $\log P_{ow} > 6$ and low water solubility $< 10$ μg/l. A major fraction of these compounds will be adsorbed by suspended particles, colloidal organic matter in the water phase (Bruggeman *et al.*, 1984; Schrap and Opperhuizen, 1990) or the glassware. The water analyses that are currently applied can hardly distinguish between the bioavailable and the non-bioavailable chemical in the water, thus resulting in a miscalculation of BCF. The situation becomes even more complex with regard to real environments. Natural freshwater and marine ecosystems contain dissolved and colloidal humic materials that can bind hydrophobic organic contaminants and reduce their bioavailability to varying extents (Donkin, 1994). Ambient microorganisms may contribute to altered concentration levels by degrading the chemicals and producing potentially effective metabolites. Increased concentration levels for substantial periods of time – even decades after the direct input has ceased – may result from the re-mobilization of chemicals from sinks and reservoirs (e.g. sediments), as has been observed for persistent pesticides such as DDT and toxaphene or for PAHs in estuaries and coastal waters.

The processes involved in bioconcentration may require a considerable period of time and only upon continuous exposure, as a result of persistence or continuous release, will chemicals reach the steady state (section 2.1). The attainment of equilibrium, especially for lipophilic compounds, may take up to several weeks or even months and, for animals such as crustaceans with a short lifespan relative to the biological half-lives of the chemicals, the equilibrium may never be reached. Especially in these cases, only BCF determinations by kinetic methods based on uptake and elimination rates will provide reliable experimental data.

Most highly accumulating substances reveal intermediate to high lipophilicity, low water solubility, a low degree of ionization and low degradability – that is, high stability (Esser, 1986; Connell, 1988; Bysshe, 1990; Nendza, 1991b). The chemical properties associated with bioconcentration (Table 4.12) are highly intercorrelated, describing the same principal attributes.

Because permeation through biological membranes is limited for large molecules, compounds exceeding a molecular weight of about 500 (Umweltbundesamt, 1990) or widths of about 10 Å (Opperhuizen *et al.*, 1985; Moser and Anliker, 1991) can be assumed to be hindered from reaching the sites of potential accumulation. As a result, these compounds may reveal lesser bioconcentration comparing to their lipophilicity, but they still do accumulate – as shown by the considerable residues of them, especially in organisms of higher trophic level (e.g. Shaw and Connell, 1982; Biddinger and Gloss, 1984; Berg *et al.*, 1987; Muir, Norstrom and Simon, 1988; Suedel *et al.*, 1994; Hendriks, 1995; Nendza *et al.*, 1997).

The consideration of bioconcentration as an interphase distribution between aqueous and organic phases governed by diffusion processes anticipates a potential parallelism with partitioning processes between other non-miscible phases. Direct relationships between the partition coefficients in different systems (water/organic phases A and B, respectively) were recorded by Collander (1951):

$$\log P_{Aw} = a \log P_{Bw} + c$$

For conventional reasons, the 1-octanol/water system is mostly selected as an easily accessible surrogate. Because the lipid tissue of the fish is the principal site for bioaccumulation and 1-octanol often imitates the properties of lipids satisfactorily, direct correlations are usually observed between log BCF and

**Table 4.12** Structural characteristics of accumulating compounds.

| Property | | Effect on bioconcentration |
|---|---|---|
| Lipophilicity: | with ↑ log $P_{ow}$ between 0 and 6 | ↑ BCF |
| | with log $P_{ow}$ > 6 | ↓ BCF |
| Water solubility: | ↓ solubility in aqueous phases | ↑ BCF |
| Molecular charge: | ↓ degree of ionization | ↑ BCF |
| Molecular size: | with molecular weight > 500 | ↓ BCF |
| | with molecular diameter > 10 Å | ↓ BCF |
| Stability: | with ↓ rate of transformation | ↑ BCF |

$\log P_{ow}$. Numerous QSARs estimating bioconcentration based on lipophilicity have been published (reviews include: Connell, 1988; Bysshe, 1990; Calamari and Vighi, 1990; Nendza, 1991b). In general, the models describe an increase in BCF associated with an increase in $\log P_{ow}$ (Table 4.13).

Analogous relationships between BCF and other descriptors, generally collinear with $\log P_{ow}$, have been derived; such as water solubility (Metcalf *et al.*, 1973, 1975; Chiou *et al.*, 1977; Kenaga and Goring, 1980). The bioconcentration factors for *Daphnia*, molluscs, mussels, algae and microorganisms have also been related to $\log P_{ow}$ or water solubility (Table 4.14).

Models for accumulation in higher plants, such as freshwater macrophytes (Gobas *et al.*, 1991) and terrestrial plant cuticles (Sabljic *et al.*, 1990), and in mammalian tissues (e.g. Tichy, 1987, 1991; Connell, Braddock and Mani, 1993), use water/lipid and air/lipid partitioning descriptors. The linear $\log P_{ow}$/log BCF correlations for aquatic environments, which are assumed to describe the same processes, reveal a wide variation in slopes and intercepts, which has been attributed to the physiological differences of the tested organisms (e.g. varying lipid content and varying metabolic capacities) and to the various classes of chemicals under study. For a comparative assessment of accumulation in different species and on different trophic levels, it is a prerequisite to normalize the data for the lipid content of the organisms (Connell, 1989; Hebert and Keenleyside, 1995; Wezel *et al.*, 1995).

**Table 4.13** Examples of QSAR models for estimating bioconcentration in fish: linear correlations.

| Model | Chemical class | | $r$ | $n$ | Descriptor range |
|---|---|---|---|---|---|
| 1 | Halogenated aromatics | $\log BCF = 0.54 \log P_{ow} + 0.12$ | 0.95 | 8 | 2.6–7.6 |
| 2 | – | $\log BCF = 0.94 \log P_{ow} - 1.50$ | 0.87 | 26 | – |
| 3 | Diverse | $\log BCF = 0.79 \log P_{ow} - 0.40$ | 0.93 | 122 | 1.0–6.9 |
| 4 | Diverse | $\log BCF = 1.00 \log P_{ow} - 1.32$ | 0.97 | 44 (?) | 1.3–6.0 |
| 5 | Chlorobenzenes | $\log BCF = 1.02 \log P_{ow} - 0.63$ | 0.99 | 11 | 3.4–5.5 |
| 6 | Chlorobenzenes | $\log BCF = 0.89 \log P_{ow} + 0.61$ | 0.95 | 18 | 3.4–5.0 |
| 7 | Diverse | $\log BCF = 1.02 \log P_{ow} + 0.84$ $\log S_{oct.} + 0.0004$ $(T_m - 25) - 1.13$ | 0.95 | 36 | 1.45–8.26 |
| 8 | Chlorinated polycyclic hydrocarbons | $\log BCF = 11.26 \text{ SASA} - 25.56$ | 0.93 | 30 | 2.4–2.7 |
| 9 | Chlorinated aromatics | $\log BCF = -0.862 \log S_w +$ $(3.30 + \log L)$ | – | – | 2–6 |

$r$ = correlation coefficient; $n$ = number of compounds analysed; $S_{oct.}$ = solubility in 1-octanol (mol/l); $T_m$ = melting point (°C); SASA = solvent accessible surface area ($10^{-10}$m$^2$); $S_w$ = water solubility (mol/m$^3$); L = fraction of lipid in the organism.
Sources of models: 1, Neely, Branson and Blau (1974); 2, Kenaga and Goring (1980); 3, Veith and Kosian (1983); 4, Mackay (1982); 5, Oliver and Niimi (1983); 6, Chiou (1985); 7, Banerjee and Baughman (1991); 8, Schüürmann and Klein (1988); 9, Hawker and Connell (1991).

**Table 4.14** Examples of QSAR models for estimating bioconcentration in various acquatic organisms and compartments: linear log BCF/log $P_{ow}$ correlations.

| Model | Chemical class | Organism/ compartment | | $r$ | $n$ | Descriptor range |
|---|---|---|---|---|---|---|
| 10 | Diverse | *Daphnia* | $\log \mathrm{BCF} = 0.90 \log P_{ow} - 1.32$ | 0.96 | 22 | 1.8–6.2 |
| 11 | Diverse | Molluscs | $\log \mathrm{BCF} = 0.84 \log P_{ow} - 1.23$ | 0.83 | 33 (?) | 3.4–7.8 |
| 12 | Diverse | Mussel | $\log \mathrm{BCF} = 0.86 \log P_{ow} - 0.81$ | 0.96 | 16 | 1.7–6.2 |
| 13 | Dibenzo-thiophene | Oyster | $\log \mathrm{BCF} = 0.49 \log P_{ow} + 1.03$ | 0.62 | 14 | 4.4–5.9 |
| 14 | Diverse | Algae | $\log \mathrm{BCF} = 0.68 \log P_{ow} + 0.16$ | 0.90 | 41 | 0.6–6.2 |
| 15 | Pesticides | Microorganisms | $\log \mathrm{BCF} = 0.91 \log P_{ow} - 0.36$ | 0.98 | 14 | 3.1–6.9 |
| 16 | Aromatics | Sediment | $\log \mathrm{BCF} = 1.00 \log P_{ow} - 0.21$ | 1.00 | 10 | 2.1–6.3 |

$r$ = correlation coefficient; $n$ = number of compounds analysed.
Source of models: 10, 11, Hawker and Connell (1986); 12, Geyer *et al.* (1982); 13, Ogata *et al.* (1984); 14, Geyer, Politzki and Freitag (1984); 15, Baughman and Paris (1981); 16, Karickhoff, Brown and Scott (1979).

It is well established that the linear log $P_{ow}$/log BCF correlations do not hold for chemicals with log $P_{ow} > 6$, with BCFs that no longer increase in correspondence with their log $P_{ow}$. A maximum range in log BCF values of approximately 6–7 for compounds with log $P_{ow}$ 6–8 is assumed, followed by a plateau or a gradual decrease with further increase in log $P_{ow}$. Before any mechanistic interpretation, however, considerable doubts have to be addressed about the experimental BCF values of superlipophilic chemicals. Experiments have been conducted almost exclusively at concentrations orders of magnitude above the water solubility of the test compounds with the help of solvent carriers. However, the total amount of the test compound in the aqueous phase is not relevant for the accumulation, because only the truly dissolved (i.e. bioavailable) fraction of the chemical can be taken up. As a consequence, BCF values must be far too low if they are calculated with the nominal concentration of the oversaturated aqueous phase. Extrapolations of accumulation data to concentrations below the water solubility revealed substantially higher BCF values than so far reported (Geyer *et al.*, 1992; Franke *et al.*, 1994) and were recently confirmed experimentally (Schmieder *et al.*, 1995). BCF data for superlipophilic compounds that were not determined by a kinetic method at concentrations below their water solubility should be eliminated, therefore, from QSAR analyses. Accordingly, the possible explanations for the non-steadiness of the linear log $P_{ow}$/log BCF correlations relate particularly to deficiencies of the biological test data:

exceeded water solubility of the test chemicals
decreased bioavailability in the water phase
non-attainment of equilibrium
hindered membrane passage
metabolism/degradation
inaccuracies in the estimations/measurements of log $P_{ow}$ and BCF

influence of steric conformations
differences in phase (solvent) properties of natural lipids and 1-octanol.

Investigations of the water to 1-octanol transfer for hydrophobic compounds revealed the rate constants to be essentially independent of log $P_{ow}$ (Hawker and Connell, 1989), indicating that diffusion in the aqueous phase is the controlling factor for these solutes. For an extended log $P_{ow}$ range, a curvilinear relationship has to be expected between the logarithm of the water to 1-octanol transfer rate constants and log $P_{ow}$, as has been also found for water to lipid transfer or uptake to aquatic organisms. These qualitative similarities in mass-transfer kinetics in abiotic and biotic partitioning systems suggest analogous control processes for lipid/water and 1-octanol/water systems, but at different solute $P_{ow}$ values and with different magnitudes of rate constants.

Differences in the thermodynamic properties of lipid/water and 1-octanol/water partitioning processes (e.g. enthalpy changes) have been observed for different types of lipophilic chemicals, indicating that there may not be a unique log $P_{ow}$/log BCF relationship for all contaminants (Opperhuizen *et al.*, 1988). Comparing the phase properties of the fish lipids and 1-octanol towards organic chemicals reveals different structures of the lipid phases. Besides storage lipids in some fish species, fish lipid consists primarily of biological membranes, in which the molecules are predominantly arranged in bilayers. The lipid phase thus has a distinct structure and restricted spatial dimensions. Because the 1-octanol phase is a bulk phase, presumably with little or no structure, organic solutes may display different activity coefficients and partitioning behaviour in 1-octanol than in membranes. The loss of a linear correlation between log $P_{ow}$ and log BCF can then be at least partly ascribed to differences in solvent characteristics between natural lipids and 1-octanol. For molecules less than a certain volume or certain dimensions, 1-octanol reveals a satisfactory surrogate; that is, the activity coefficients in 1-octanol and fish lipid are approximately equal, whereas for larger molecules this similarity breaks down and the activity coefficients in the membrane phase are much larger than in 1-octanol, which then is no longer a satisfactory surrogate and log $P_{ow}$ is no longer a linearly corresponding descriptor. It seems likely that the relatively low solubility of voluminous molecules in membranes as compared to 1-octanol is one of the causes for the loss in linear correlation between log $P_{ow}$ and log BCF. To compensate for the differences in lipid solubility, the inclusion of a parameter for octanol solubility in log BCF/log $P_{ow}$ relationships has been suggested (Banerjee and Baughman, 1991). Furthermore, membrane/water partition coefficients may be used as a more reliable parameter to estimate and correlate the BCFs of organic chemicals in aquatic organisms (Gobas, Shiu and Mackay, 1987).

To describe the reduced bioconcentration of superlipophilic compounds, non-linear QSAR models have been derived (Table 4.15).

**Table 4.15** Examples of QSAR models for estimating bioconcentration in fish: non-linear correlations.

| Model | Chemical class | | $r$ | $n$ | Descriptor range |
|---|---|---|---|---|---|
| 17 | Chlorobenzenes | $\log \mathrm{BCF} = 3.41 \log P_{ow} - 0.26 (\log P_{ow})^2 - 5.51$ | – | 6 | 3.5–6.4 |
| 18 | Chlorinated hydrocarbons | $\log \mathrm{BCF} = 0.0069 (\log P_{ow})^4 - 0.185 (\log P_{ow})^3 + 1.55 (\log P_{ow})^2 - 4.18 \log P_{ow} + 4.79$ | – | 45–46 (?) | 2.6–9.8 |
| 19 | Diverse | $\log \mathrm{BCF} = 0.99 (\log P_{ow} - 1.47\log (4.97 \times 10^{-8} P_{ow} + 1) + 0.0135$ | – | 132 | 1.0–11.2 |
| 20 | Diverse | $\log \mathrm{BCF} = 2.12\, {}^2\chi^v - 0.16\, ({}^2\chi^v)^2 - 2.13$ | 0.97 | 84 | 1.1–11.8 |

$r$ = correlation coefficient; $n$ = number of compounds analysed; ${}^2\chi^v$ = valence-corrected connectivity index. Sources of models: 17, Könemann and Leeuwen (1980); 18, Connell and Hawker (1988); 19, Nendza (1991b); 20, Sabljic (1987c).

A parabolic relationship has been developed by Könemann and Leeuwen (1980) for chlorobenzenes. Connell and Hawker (1988) derived a polynomial $\log P_{ow}$-dependent function to describe BCFs of chlorinated hydrocarbons with a maximum bioconcentration for compounds with $\log P_{ow}$ 6.7. Spacie and Hamelink (1982) proposed sigmoid modelling to account for the fact that the linear correlations also break down for highly hydrophilic compounds. A bilinear function based on $\log P_{ow}$ describes the highest BCF associated with a given lipophilicity – that is, a worst-case estimate of bioconcentration, corresponding to the empirically postulated coincidence of $\log P_{ow}$ and log BCF, with a maximum about $\log P_{ow}$ 7 (Nendza, 1991b). This function was the result of a validation exercise, when published QSARs were compared with experimental BCF data from the literature. The first remarkable observation concerns the substantial scatter in BCF values for compounds of approximately identical lipophilicity by at least two orders of magnitude, although the trend of BCFs increasing with $\log P_{ow}$ is readily evident (Figure 4.11).

Closer inspection of the data revealed clustering with respect to chemical classes, which may reflect either specific accumulation behaviour among chemical classes or characteristics of the experimental assessment by different investigators, which generally studied homologous series of compounds.

Contrasting the experimental BCF data with QSAR functions (Figure 4.12) reveals significant deviations (a) among the curves and (b) between the predicted and measured data.

Considering that the linear models are not applicable in the high lipophilicity range, the functions all result in the same relative ranking of accumulation potential because they depend on the same principal property ($\log P_{ow}$) and they yield estimated average BCFs, which may be exceeded by measured values by more than one order of magnitude. The non-linear functions have a broader application range extending to the domain of superlipophilic compounds. Inspection of the respective residuals ($\log \mathrm{BCF}_{obs.} - \log \mathrm{BCF}_{calc.}$) reveals some overestimates for all regression functions but, more severely,

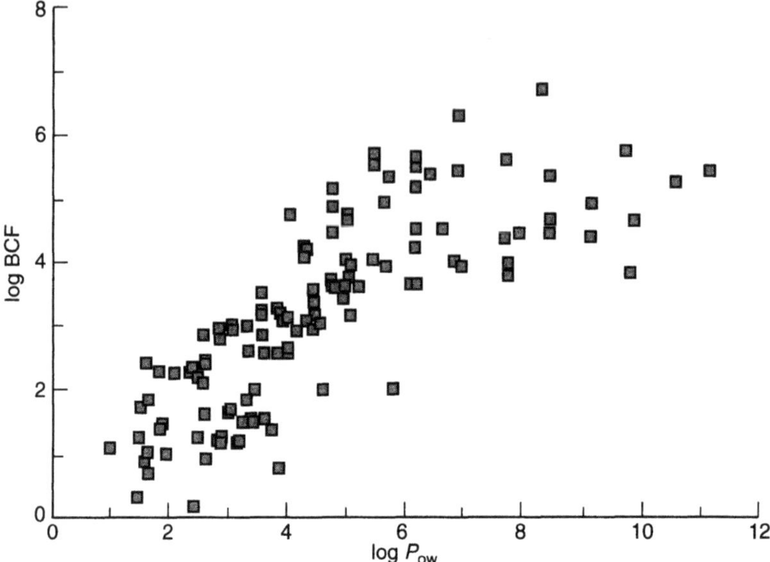

**Figure 4.11** Relationship between experimental bioconcentration data (based on fish lipid content) and log $P_{ow}$. Reproduced from Nendza (1991b) with permission from VCH, Weinheim.

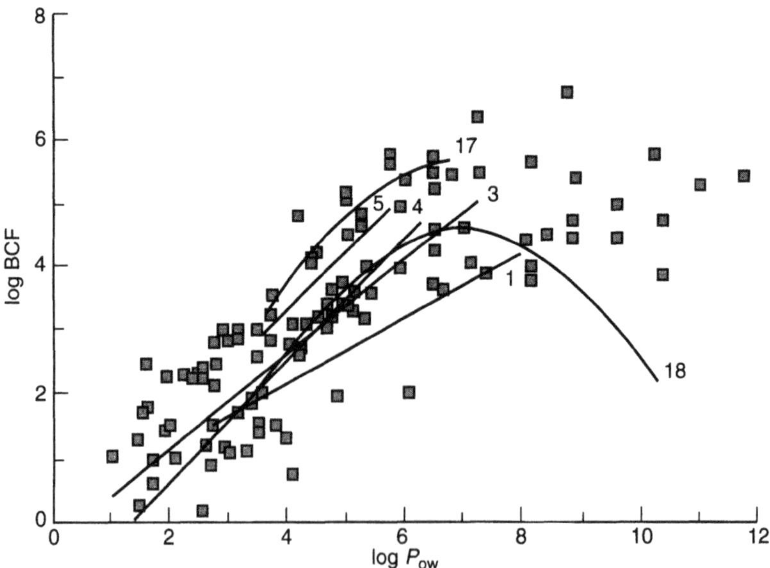



**Figure 4.12** Comparison of selected QSAR models relating log BCF to log $P_{ow}$ with experimental bioconcentration data (Nendza, 1991b); for identification of the QSAR functions see Tables 4.13 and 4.15.

also much underrating of BCFs. In contrast to this approach, it is evident from the data plot of the BCF values (Figure 4.11) that a non-linear function can be constructed based on log $P_{ow}$ and this describes the highest BCF associated with a given lipophilicity (Figure 4.13).

If, then, discrepancies between measured and calculated values occur, the measured BCFs are lower than calculated. For this function, which does not represent a regression on the BCF data but is a description of the hypothetical worst case, no standard statistical parameters can be given:

$$\log \text{BCF} = 0.99 \log P_{ow} - 1.47 \log (4.97 \times 10^{-8} P_{ow} + 1) + 0.0135$$

The bilinear curve resumes a linearly increasing part between log $P_{ow}$ 0 and 6, where the empirically postulated coincidence of log $P_{ow}$ and log BCF is reflected by a near-unity slope (0.99) for the first-order log $P_{ow}$ term and the intercept of about 0. Maximum log BCF values of approximately 7 are obtained for compounds with log $P_{ow}$ between 7 and 8. Compounds that are more lipophilic are expected to be less accumulating, which corresponds to the negative slope derived for the second log $P_{ow}$ term of the bilinear function. If restricted to compounds with log $P_{ow} < 7$, the equation can be simplified thus:

$$\text{only if } \log P_{ow} < 7: \qquad \log \text{BCF} = \log P_{ow}$$

The bilinear derivation, as well as its linear simplification for a narrow application range, yields estimates corresponding to the accumulation poten-

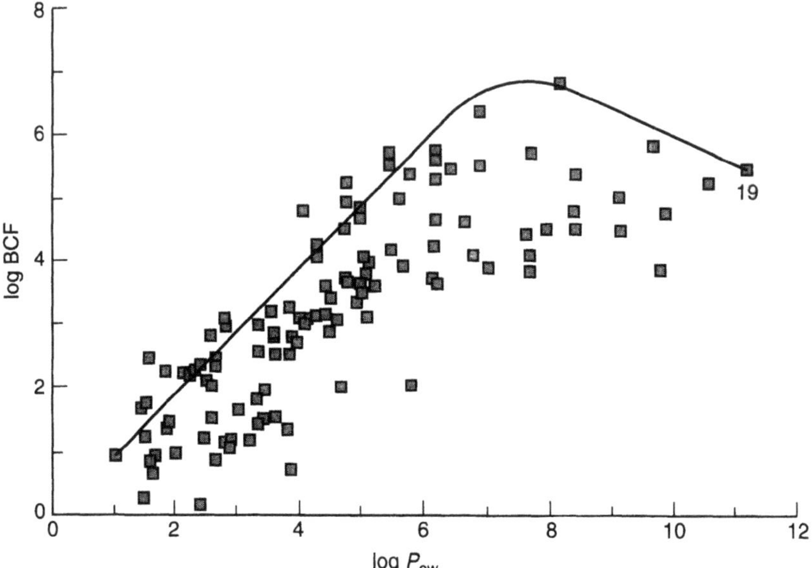


**Figure 4.13**  Comparison of a bilinear 'worst case' QSAR model relating log BCF to log $P_{ow}$ with experimental bioconcentration data (Nendza, 1991b); for details of the QSAR function see Table 4.15. Reproduced with permission from OECD.

tial of contaminants associated with the assumed lipophilicity. This procedure can be justified as a realistic worst-case approach to model the accumulation potential of the prevailing majority of environmental chemicals.

An understanding of the limits of the reliability of the available tools to predict BCF values is of major relevance, because the BCF represents the first parameter where QSARs are applied for legislative purposes. Based on physico-chemical properties (i.e. log $P_{ow}$), no substantial bioconcentration is assumed if log $P_{ow} < 3$ (BCF < 100). Chemicals ranging in log $P_{ow}$ between 3 and 6 are classified as highly accumulating, which eventually results in the demand for testing. Superlipophilic compounds characterized by a log $P_{ow} > 6$ and a molecular weight of > 500 are classified as modestly accumulating. This simplified prediction scheme may substitute (except in Japan) for the experimental determination of a compound's BCF on the basic tier of chemical legislation. From this aspect, it is essential that any user of QSARs on BCF realizes the limitations and restrictions of this approach. The correlations between log BCF and log $P_{ow}$ will be linear as long as the ratio of the respective activity coefficients remains constant. But several factors may cause deviations and apparent loss of the linear log $P_{ow}$/log BCF correlation, often due to variable experimental conditions for the underlying data. The evident variability in parameterization of lipophilicity by log $P_{ow}$ must also be recognized (Schüürmann and Klein, 1988), but this uncertainty is only a minor factor compared to the substantial problems arising from the quantification of the biological endpoint.

Several factors may cause major discrepancies between QSAR-estimated BCFs and experimental values:

Because of the non-linearity of the log $P_{ow}$/log BCF correlations, extrapolations have to be restricted to the parameter range covered by the model with respect to similarity with the structures used for deriving the model and to the structure descriptors, which must not exceed the domain of the underlying data set. The result of undue extrapolation outside the model domain may be erroneous predictions by several orders of magnitude; for example, for highly hydrophilic (log $P_{ow} < 0$) or highly hydrophobic (log $P_{ow} > 7$) compounds the application of a linear log $P_{ow}$/log BCF QSAR will yield false results.

Deviations caused by extreme lipophilicity have been attributed to differences in transfer rates for lipid/water and 1-octanol/water partitioning processes, differences in solvent phase characteristics between natural lipids and 1-octanol, and size-limited diffusion (Opperhuizen *et al.*, 1985, 1988; Gobas, Shiu and Mackay, 1987; Anliker, Moser and Poppinger, 1988; Hawker and Connell, 1989; Banerjee and Baughman, 1991).

Deviations may occur for substances of large molecular diameter due to hindered membrane passage. Lack of permeation was observed for chemicals with effective cross sections > 9.5 Å (Opperhuizen *et al.*, 1985). It was demonstrated from the relationship between the steric configuration of the compounds and the lack of uptake that the type and composition of the organisms membranes can also influence the bioconcentration potential of a chemical. In agreement with the size-limited uptake are results reported for dispersed dyestuffs and pigments (Anliker, Moser and Poppinger, 1988) that revealed no substantial bioconcentration.

Further deviations may result from substructure effects such as for 2,4-dinitro-substituted phenols (Butte, Willig and Zauke, 1987; Deneer *et al.*, 1987; Hauk *et al.*, 1990), which were reported to reduce bioconcentration to less than estimated from log $P_{ow}$. These deviations were only partly explainable from the compounds' reduced apparent lipophilicity due to dissociation and have been related also to differences in lipid-phase activity coefficients (Banerjee and Williams, 1993).

Degradation and metabolic transformation often result in reduced apparent bioconcentration (Spacie, Landrum and Leversee, 1983; Gobas and Schrap, 1990), due not only to the reduction in the concentration of the parent compound but also to the increased polarity of the metabolites formed. The bioconcentration of surfactants as well as, for example, organophosphates (Bruijn and Hermens, 1991a,b) and aromatic amines (Wolf *et al.*, 1992a) is significantly influenced by the (bio)transformation of the parent compounds.

The estimation of BCF values from log $P_{ow}$ is founded on a relatively profound theoretical basis. However, the predictive power of the respective QSARs should not be overestimated and their limitations must be realized. Principally, QSAR predictions of BCF values correspond to the average accumulation observed with the class of compounds under investigation. The assessment of the potential worst case requires a model that reflects the highest accumulation potential associated with the assumed lipophilicity. The respective bilinear QSAR (Table 4.15) formalizes the empirical rules for estimating log BCF (adjusted for the lipid content of the fish) from log $P_{ow}$: the bioconcentration potential corresponds to log $P_{ow}$. Compounds of high lipophilicity with log $P_{ow}$ > 6–7 reveal no further increase in BCF. Factors resulting in less bioconcentration are neglected, as the various contributions are not systematically accountable. This procedure is recommended as a conservative approach to a realistic worst-case assessment of the bioaccumulation potential.

# 5                 *Effects-related parameters*

Asessment of environmental effects relates to the determination of the concentration of toxic material that produces undesirable changes in exposed populations. Unlike human toxicology, the intention of ecotoxicology is not to protect individuals but to secure viable populations. According to legislative and regulative requirements, generally single-species tests are used to measure the direct effects of chemicals or effluents on aquatic organisms. In this way, the relative toxic potencies of chemicals can be determined. Analogously, the relative sensitivity of different species or different life stages within a species can be discerned, whereas the many interactions in the community and the ecosystem that may ultimately control the fate and the effects of contaminants are ignored. However, the intention of single-species testing is not to simulate all interacting processes of ecotoxicological relevance in ecosystems, but merely to provide indicators of potential hazards.

The biological parameters used in environmental effects assessments are generally obtained with several species that may be regarded representative for selected groups of organisms or compartments of the ecosystem:

Aquatic ecosphere:
    algae (primary producer)
    *Daphnia* (primary consumer)
    fish (secondary consumer).
Terrestrial ecosphere:
    plants (primary producer)
    mammals (secondary consumer).
Soil ecosphere:
    microorganisms (primary decomposer)
    earthworm (secondary decomposer).

Indicators of toxicity in aquatic systems are fish lethality ($LC_{50}$ at 96 h (EPA, 1982; ASTM, 1984; OECD, 1989c)), toxicity to *Daphnia* ($EC_{50}$ at 48 h (OECD, 1984c)) and inhibition of algal productivity ($EC_{50}$ at 96 h (OECD, 1984d)). Effects on the terrestrial biota are reflected by reductions in forest and agricultural productivity; phytotoxicity is measured as the inhibition of seedling growth ($EC_{50}$ at 14 d (EPA, 1982)). Toxicity to wildlife and other mammals is extrapolated from oral $LD_{50}$ for small laboratory rodents.

The effects on soil organisms are assessed as toxicity to earthworms (OECD, 1984b) and to microbial communities. The respective endpoints are operationally defined and accessible by measurements.

Toxicity tests are generally classified according to the duration of the test in relation to the generation time of the test organisms used. Ranges in generation time are:

| | |
|---|---|
| bacteria | minutes to hours to days |
| algae | minutes to hours to days |
| crustaceans | weeks |
| fish | months to years. |

In acute toxicity tests, the test organisms are exposed for a relatively short period in relation to their generation time. The acute effects recorded with such tests usually concern the survival of the test species, regardless of how mortality is evoked. In chronic tests, the test organisms are exposed for a significant part of their life cycle or the entire life cycle of one to several generations. The (sub)chronic effects primarily concern sublethal impacts such as reduced reproduction or growth, altered behaviour or development, which ultimately may also lead to increased mortality.

The biological data used to derive and validate QSAR models may have been determined in a standard series of experiments or may have been compiled from a variety of studies. In the latter case, the measurements should have been conducted according to standard protocols and evaluated by consistent criteria. The important test parameters should be identical or comparable throughout any set of data used. If, nonetheless, they vary within the data set, they should be critically evaluated for their effect on the outcome of the biological assays, as such differences may greatly affect the relevance of the derived models.

The available toxicity parameters reflect the impact of xenobiotics on the test organisms in terms of global symptoms (e.g. fish lethality) without due regard to the mode by which the effects are evoked. The mortality may hence result either from non-specific retardation of membrane-bound functions or from interactions with definite targets in the organisms (e.g. enzymes) to produce so-called specific toxicity. The observed effect is nevertheless identical: dead fish. Therefore using such parameters as $LC_{50}$ or NOEC for toxicity ranking among diverse chemicals is quite likely to compare unequal activities. The discrimination of different modes of action may be achieved experimentally, by, for example, monitoring fish acute toxicity syndromes (FATS) (McKim, Bradbury and Niemi, 1987; Bradbury, Henry and Carlson, 1990), by joint toxicity studies (Broderius, Kahl and Hoglund, 1995), by subjecting the chemicals to testing in a battery of *in vivo* and/or *in vitro* assays featuring different targets (Nendza and Wenzel, 1993; Nendza, Wenzel and Wienen, 1995; Wenzel *et al.*, 1997) (Figure 5.1) or, retrospectively, by lateral QSAR comparison of the relevant parameters and their contributions to the overall model.

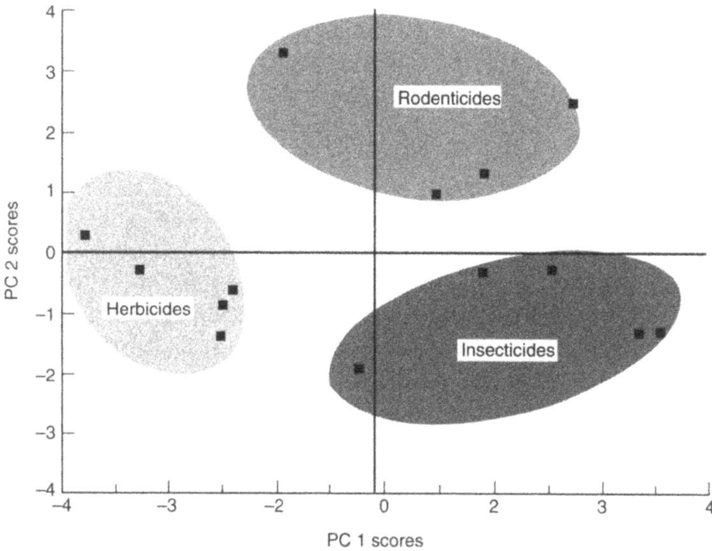


Principal component analysis: % variance explained

| PC | Eigenvalue | Percent (%) | Sum (%) |
|---|---|---|---|
| 1 | 2.16 | 43.2 | 43.2 |
| 2 | 0.96 | 19.1 | 62.3 |
| 3 | 0.92 | 18.3 | 80.6 |
| 4 | 0.56 | 11.1 | 91.7 |
| 5 | 0.42 | 8.3 | 100 |

Principal components: loadings

| | PC 1 | PC 2 | PC 3 |
|---|---|---|---|
| Fish | 0.346 | 0.011 | 0.49 |
| Daphnia | 0.729 | −0.577 | −0.035 |
| Algae | −0.189 | −0.169 | 0.831 |
| Bird | 0.358 | 0.141 | −0.182 |
| Rat | 0.430 | 0.787 | 0.183 |

Principal components: scores

| | PC 1 | PC 2 | PC 3 |
|---|---|---|---|
| H 1 | −1.893 | −0.630 | −0.173 |
| H 2 | −3.720 | 0.316 | −0.967 |
| H 3 | −3.049 | −0.254 | −0.085 |
| H 4 | −2.012 | −0.867 | 0.924 |
| H 5 | −2.040 | −1.360 | 1.342 |
| I 1 | −0.309 | −1.925 | −0.313 |
| I 2 | 1.256 | −0.330 | −0.412 |
| I 3 | 2.090 | −0.284 | −0.954 |
| I 4 | 3.448 | −1.332 | 0.656 |
| I 5 | 3.192 | −1.351 | −2.183 |
| R 1 | 0.664 | 0.956 | 1.287 |
| R 2 | −1.259 | 3.282 | −1.953 |
| R 3 | 1.245 | 1.325 | 1.408 |
| R 4 | 2.386 | 2.453 | 1.252 |

**Figure 5.1** Example of the discrimination of different modes of action based on tests featuring different targets. The toxicity profiles of herbicidal (H 1–5), insecticidal (I 1–5) and rodenticidal (R 1–4) compounds with regard to fish, *Daphnia*, algae, bird and rat revealed highly specific for each of the toxicant classes and can be clearly separated by their scores in a principal component analysis (Nendza and Wenzel, 1993).

Compounds belonging to different classes by their mode(s) of action are toxic in different ways due to different interactions with the organisms and hence they have to be modelled with different QSARs.

With regard to QSAR modelling, the inevitable inhomogeneity of the effects data impedes the derivation of uniform general relationships. The toxicants have to be classified with regard to their potential interactions with biomolecular targets (i.e. their modes of action). These result from, for example, hydrophobic, polar, electrostatic and reactive transient interactions, hydrogen bonding, covalent binding and/or steric fit to the various interaction sites. To reach the site of action, any compound undergoes partitioning. If the lipid diffusion processes are rate-limiting, eventually overruling specific interactions, the observed effect is considered non-specific toxicity and corresponds to the baseline effect, which can be described using a (non-)linear univariate bioactivity/log $P_{ow}$ model with an ascending slope equal to or approaching unity. The intercept of the QSAR equations varies with the sensitivity of the correlated biological response, but both sublethal and lethal endpoints give rise to the same principal relationship. The universality of such models indicates that partitioning is the fundamental process in ecotoxicology. Chemicals with polar moieties such as phenols or anilines, with log $P_{ow} \leq 3$, also produce non-specific effects, but they are more toxic than corresponding to the baseline models and their toxicity is not concentration-additive with non-polar chemicals. The respective compounds are termed polar non-specific toxicants. Further additional interactions between xenobiotics and the biosystem may increase the effects, as can be observed, for example, for uncoupling agents. The considerations outlined results in a model that assumes toxicity to be due to the sum of interactions with the biosystem:

toxicity =    hydrophobic interactions
             + hydrogen bonding
             + polar interactions
             + covalent interactions
             + steric fit to the interaction site
             + reactivity
             + ...

The non-specific effects can be considered as the default mode of action, giving the minimal toxicity of any compound, and may be topped by specific effects that may be defined from clinical symptoms. Reactive toxicity has been ascribed to various interactions caused by a chemical's reactivity. Attempts to link the physico-chemical interactions with the physiological modes of action of toxicants have been successful so far only to a limited extent. The toxicity mechanisms attempt to deduce the compounds' impacts to effects on the biochemical and biomolecular level. However, alike symptoms may result from different molecular interactions. The principal modes of action comprise:

interaction with membranes and alteration of their properties
inhibition of enzymes
interference with specific transport processes
reaction with defined receptors
covalent binding to essential biomolecules.

The species potentially affected by environmental contaminants reveal a great diversity, still the actual impacts generally concern only a limited number of targets at the biomolecular level. With any kind of organism, membranes and their components are of major importance (Seydel *et al.*, 1994). The vital functional units are principally similar in different species, although the set (abundance) of potential targets may vary between species (Table 5.1).

Accordingly, most environmentally relevant chemicals can be classified by their mode of action into relatively few groups. The classes comprise substances evoking the same kind of symptoms, without reference to the underlying molecular mechanisms:

non-specific toxicity (membrane perturbation)
uncoupling of the respiratory chain
inhibition of acetylcholinesterase (AChE)
inhibition of photosynthesis
interaction (alkylation) with thiol moieties (e.g. DNA, proteins)
irritation (reactive cytotoxicity)
neurotoxicity.

With regard to the various target species, chemicals mostly act non-specifically, and further effects become relevant only if specific interaction sites are present in the affected organisms. As a consequence, each class of compounds has a characteristic toxicity profile that depends on the occurring mode(s) of action. The most sensitive species are characterized by the presence of target biomolecules mediating specific effects; for example, algae for inhibitors of photosynthesis and daphnids for inhibitors of acetylcholinesterase (AChE). If such targets are not present in organisms (e.g. photosynthetic systems in fish) non-specific toxicity is usually observed for the respective compounds. The effects are then governed by distribution processes and can be related to the compounds' lipophilicity. As a result, the mode of toxic action cannot be regarded only as a compound-specific property, it also depends on features of the organisms exposed to the chemical and the definite endpoint measured by experiment. The response pattern obtained from a series of tests is therefore specific for the respective mode(s) of action, whereas different modes of action yield different fingerprints (Figure 5.2).

Comparison of compounds from different classes reveals that they may act by the same mode in some organisms, but by dissimilar modes in others. Applying this rationale to QSAR studies implies that the occurrence of outliers cannot be uniform for all models, and that toxicity exceeding the

**Table 5.1** Examples of toxicologically relevant modes of action.

| Mode of action | Site of action | Experimental assay | Model compounds |
| --- | --- | --- | --- |
| Cytotoxicity (incl. reactive compounds, radicals, detergents) | Membranes, filaments, metabolism | Cell culture, multiplication, vitality | Acrolein, 4-nitrobenzaldehyde |
| Non-specific non-polar | Membranes | Cell culture, vitality (neutralred assay) | 1-Octanol, 1,4-dichlorobenzene |
| Non-specific polar | Membranes | Cell culture, vitality (neutralred assay) | Phenol, pyridine |
| Uncoupling | Inner mitochondria ATP-synthesis | $O_2$-consumption mitochondria | 2,4-Dinitrophenol, pentachlorophenol |
| Inhibition of respiration | e⁻-transport respiratory chain | $O_2$-consumption e⁻-transporting particles | Rotenone, amobarbital |
| Inhibition of photosynthesis | e⁻-transport photosynthesis | $O_2$-production, chloroplasts (chlorophyll-fluorescence) | Atrazine, diuron |
| Inhibition of acetylcholinesterase | Neuromuscular end-plate enzymes in receptor complex | AChE activity | Malathion, carbaryl |
| Reaction with thiol-moieties (alkylation) | (Non)-protein –SH DNA | SH-enzymes activity | Folpet, epoxides |
| Mutagenicity | DNA | Bacterial revertants (*Salmonella typhimurium*) | Dimethylsulphate, 4-Nitrochinolinon-N-oxide |

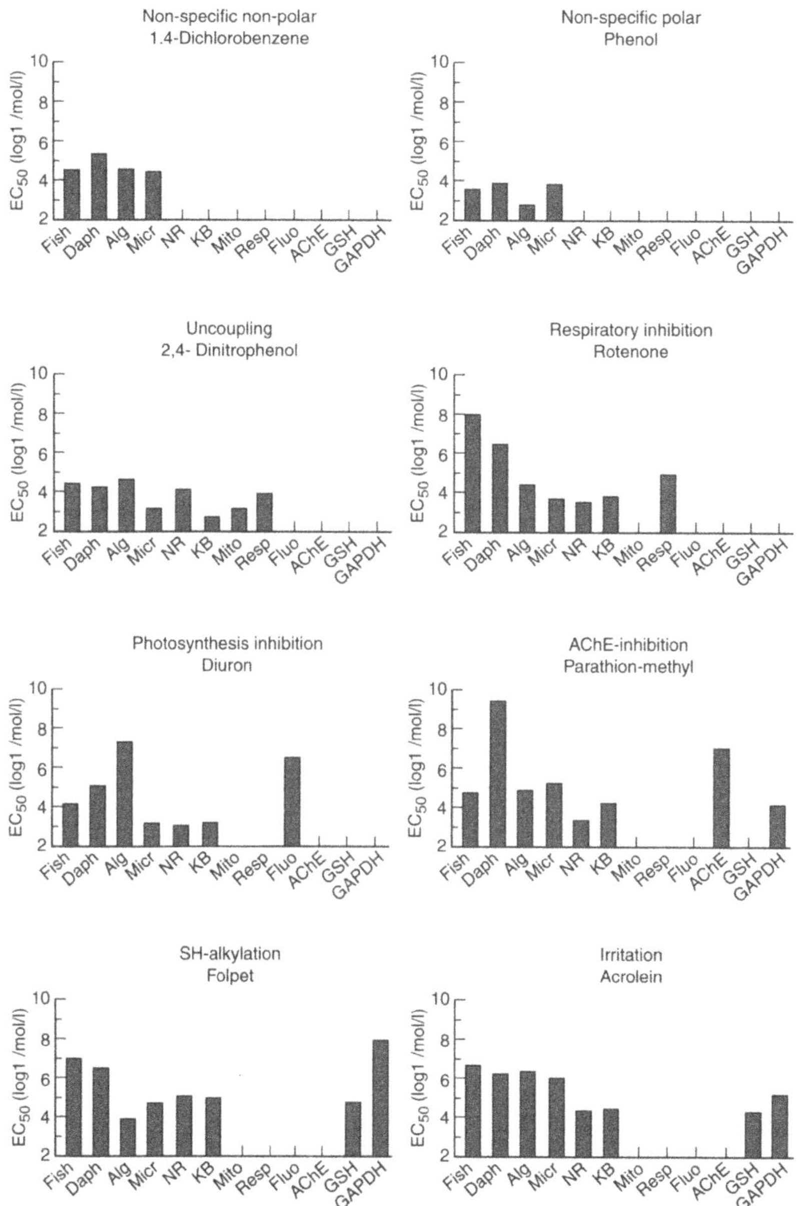



**Figure 5.2** Examples of toxicity profiles of model compounds acting by different modes in different organismic and *in vitro* test systems. Fish = $LC_{50}$ fish; Daph = $EC_{50}$ *Daphnia*; Alg = $EC_{50}$ algae; Micr = $EC_{50}$ Microtox assay; NR = $EC_{50}$ neutralred assay; KB = $EC_{50}$ kenacidblue assay; Mito = LOEC uncoupling of oxidative phophorylation; Resp = LOEC inhibition of respiratory electron transport chain; Fluo = $EC_{50}$ algal flourescence; AChE = $EC_{50}$ acetylcholinesterase; GSH = $EC_{50}$ glutathion; GAPDH = $EC_{50}$ glycerinaldehyde-3-phosphate-dehydrogenase; no bar = no effects $\leq$ 500 mg/l.

lipophilicity-mediated baseline is caused by different structures for different endpoints.

The variability in a chemical's mode of action because of different sensitivities of the targets/organisms is complicated by the fact that compounds assigned a definitive mode of action may nonetheless reveal further interactions with other specific targets. For example, exemplary uncouplers such as pentachlorophenol or respiratory blockers such as rotenone were found to be additionally active also in an acetylcholinesterase inhibition test (Nendza, Wenzel and Wienen, 1995; Wenzel *et al.*, 1997). The discrimination of distinct compound classes thus proves rather difficult (Figure 5.3.).

Especially specific toxicants are rarely of one unique mode of action and may instead interact with several targets to varying extents in different species. A major lesson to be learnt, therefore, is that specific toxicity does not mean only one definite mode of action, but may result from a multitude of concurrent specific interactions yielding the elevated toxicity observed for the respective compounds. Furthermore, the multivariate nature of the (eco)toxicity profiles, which are dependent on the chemicals as well as on the

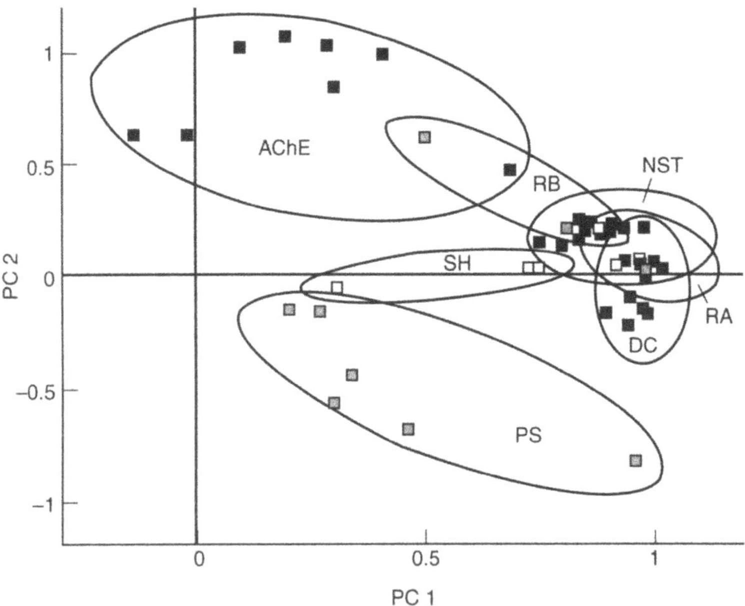


**Figure 5.3** Example of the discrimination of different modes of action based on testing with a battery of specific *in vitro* assays. The overlap of class domains indicates the ambiguity of assignments for chemicals with multiple modes of action. NST = non-specific toxicants; DC = uncouplers; PS = inhibitors of photosynthesis; AChE = AChE inhibitors; RA = reactive compounds; SH = SH blocker; RB = respiratory blocker. Reproduced from Nendza, Wenzel and Wienen (1995) with permission from Gordon and Breach Publishers, Lausanne.

species exposed under the particular conditions, makes it necessary to derive a systematic classification of the liable toxicological manifestations. Such a classification system is an essential prerequisite for process-related QSARs.

Because QSAR models are principally limited to homologous classes of chemicals with regard to the underlying mechanisms, it is a major concern to classify the respective compounds' mode of action correctly before applying predictive models. Only if this task can be fulfilled, generally with little or no toxicological information at hand, can reliable QSAR predictions be made. The selection of an inappropriate QSAR for a particular compound may yield grossly false estimates. Grouping of similar contaminants should ensue from uniform mechanisms of toxic action, but the discrimination criteria (i.e. the prediction of the compounds' mode(s) of toxic action from the chemical structures) is feasible only to yet insufficient extents. Hence the selection of the appropriate prediction model(s) is still based on structural moieties characterizing chemical classes of compounds, which does not necessarily coincide with the grouping by mechanism of toxic action. Until this deficiency can be overcome, there will remain considerable reservations about the reliability of QSAR predictions in ecotoxicology.

For several important parameters relating to the effects of environmental pollutants (especially regarding human health) no validated QSAR models of sufficient predictive capacity are yet available:

subacute mammalian toxicity
sensitization
corrosivity
carcinogenicity
teratogenicity
neurotoxicity
immunotoxicity
endocrine disruption.

## 5.1 TOXICITY TO FISH

Fish represent a high level in aquatic foodwebs and as such are important indicators of the status of aquatic ecosystems. A variety of freshwater fish species are frequently used for testing (estuarine or marine species are rarely considered):

trout (*Salmo gairdneri*)
bluegill (*Lepomis machrochirus*)
guppy (*Poecilia reticulata*)
carp (*Cyprinus carpio*)
fathead minnow (*Pimephales promelas*)
red killifish (*Oryzias latipes*)
golden orfe (*Leuciscus idus*)
zebra fish (*Brachidanio rerio*).

Survival, growth, behavioural and physiological changes are used as quantifiable effects in toxicity tests with fish in the laboratory. Acute toxicity is generally described in terms of the $LC_{50}$ (LC: lethal concentration) – that is, the level of the toxicant (mg/l) that is lethal for 50% of the individuals within a defined exposure period, regardless of how mortality is evoked.

With regard to standard test endpoints, the differences between fish species may be smaller than is sometimes supposed (Sprague, 1970). The various $LC_{50}$ test procedures for fish generally result in a coincident relative ranking of the toxicants; that is, the most toxic compound is always the most toxic, intermediates remain the intermediates and the least-toxic compound is always the least toxic, provided that the physical and chemical test parameters are kept constant. Absolute $LC_{50}$ values are generally more dependent on the specific test set-up than on the inherent sensitivity of the fish species. Various factors may influence the absolute values of the observed toxic concentrations:

Life stage of the species: the toxicity of xenobiotics towards organisms varies with the age relative to lifespan and the sex, size, lipid content, etc. of the species.

Duration of the test: sensitivity to chemicals may increase or decrease during the test period due to alterations in the catabolic status and in the defence mechanisms of the animals.

Feeding regime: sensitivity to chemicals may increase or decrease with the feeding status of the animals, and the food in the test medium may alter the bioavailability of toxicants.

Exposure regime: concentrations of the toxicant will vary according to whether the exposure is static, semi-static or flow-through in the test set-up.

Temperature dependence: sensitivity to chemicals may increase or decrease with changes in the temperature due to alterations in the catabolic status of the animals.

Light intensity: sensitivity to chemicals may increase or decrease with changes in the light intensity due to alterations in the catabolic status of the animals.

pH and buffer capacity of the water phase: pH conditions influence the toxicity of organic acids and bases.

Salinity of the test medium: sensitivity to chemicals may increase or decrease with changes in the salinity and hardness of the water, which may also affect the solubility of the toxicants.

Oxygen content: sensitivity to chemicals may increase or decrease with changes in the oxygen content, which is interrelated with temperature effects and the velocity of respiratory flow, hence the concentrations of contaminants at the gills.

Suspended organic matter: components of soil and sediment (e.g. humic acids) may result in decreased apparent toxicity by serving as a sink compartment for the toxicants due to sorption processes.

Stability of the test compound in water: the substance must not degrade during the experiment; losses in concentration may also occur by evaporation or adsorption to glassware.

Surface-active materials: the presence of solubilizing agents may alter the bioavailable concentration of the test compounds significantly.

Purity of the test compound: pure substances have to be used because even small amounts of soluble impurities can cause large deviations in the measured toxicity.

Cosolute effects: the presence of additional organic solutes, as is the case in 'real' waters, may significantly alter the observed toxicity of the individual compounds.

Analytical method: a specific set-up must be developed for each chemical species.

The difficulties most overlooked in producing reliable toxicity data for fish arise from frequently testing contaminant concentrations above their water solubility. Inventive experimentators add solvents, presumably in inactive amounts, or even coat the test basin with the test compounds dissolved in, say, acetone and let the solvent evaporate before adding the test medium. It has been shown, however, that the use of solvent carriers may result in differences in the acute effects of sparingly soluble chemicals (Calleja and Persoone, 1993). Consequently, the toxicity resulting from the combination of the test substance and a solvent should be evaluated carefully and the magnitude of the interactions taken into account when comparing toxicity data from various sources. Further practical problems in measuring fish $LC_{50}$ values stem from insufficiencies in the chemical analysis. Difficulties arise in the determination of the concentration in the aqueous solution, specially for highly lipophilic chemicals with log $P_{ow} > 6$ and low water solubility $< 10$ $\mu$g/l. Additionally, a major fraction of these compounds will be adsorbed to suspended particles, colloidal organic matter in the water phase (Bruggeman *et al.*, 1984; Schrap and Opperhuizen, 1990) or glassware. The water analyses that are currently applied can hardly distinguish between the bioavailable and the non-bioavailable chemical in the water, thus resulting in uncertainties in the derived $LC_{50}$ data.

Published data on fish toxicants are extremely variable because of the diverse test protocols applied and are hence of limited use for the derivation and validation of sound QSAR models. One of the few reference databases has been generated at the US EPA Environmental Research Laboratory Duluth, Minnesota, on the toxicities of $> 650$ chemicals of 24 classes with 96h $LC_{50}$ values towards the fathead minnow (*Pimephales promelas*) according to a uniform testing scheme (Brooke *et al.*, 1984; Geiger *et al.*, 1985, 1986, 1988, 1990). This valuable resource for QSAR studies, initiated and sustained by G.D. Veith, is superior to other toxicity data compilations because the variation in measured bioactivity is definitively due to differences in the toxic potency of the compounds and not to differences in the test protocol.

For hazard-evaluation purposes, fish toxicants are roughly divided into two categories: the non-specific toxicants (Table 5.2) and a heterogeneous group of specifically acting toxicants.

The non-specific chemicals were recognized to exert toxicity relating to their lipophilicity as expressed by log $P_{ow}$, indicating the predominance of transport phenomena and the relevance of partitioning and hydrophobic membrane interactions for this mechanism of toxic action (Figure 5.4).

A multitude of analogous models has been released since the publication of the Könemann equation (Könemann, 1981b), which describes a linear relationship between guppy (*Poecilia reticulata*) 14(7) d $LC_{50}$ and log $P_{ow}$ for a diverse set of non-polar non-reactive aromatic and aliphatic compounds in the range of log $P_{ow}$ –1–6. For limited series ($n = 5$) of homologous aliphatic alcohols ($C_1$–$C_5$), similar correlations were earlier reported by Hansch and Dunn (1972), when investigating various types of biological activity data relating to hydrophobicity. The respective QSARs (Table 5.3) describe increases in toxicity corresponding to increases in lipophilicity, although the linear dependencies are valid only in the log $P_{ow}$ range < 6.

Limited solubility and/or hindered transport through biological barriers (e.g. membranes) yield a relative decrease in the toxicity of highly lipophilic compounds such that parabolic or bilinear functions describe the observed effects more comprehensively. The fish $LC_{50}$ models for non-polar non-specific toxicants are similar, almost overlapping in the linear section, having congruent slopes (0.8–1.0) and similar intercepts (after conversion to identical units (mmol/l): approximately –2 – –1), the variations being due to differences in the endpoints measured – different species, effects and durations

**Table 5.2** Structural characteristics of non-polar non-specific toxicants towards fish.

Contained elements: C, H, N, O, halogen (excluding Iodine)
log $P_{ow}$ 0–6
MW < 600
Neutral (non-ionized) compound
No interactions between substituents
No activated substituent: e.g. α-Cl-alkene
No epoxide
No peroxide
No primary amine

These structural requirements are not exhaustive: other compounds may also fall into the group of the non-polar non-specific toxicants. However, compounds complying with these rules may be much more toxic than baseline (Verhaar, Leeuen and Hermens, 1992).
Sources: OECD (1992a); Verhaar, Leeuwen and Hermens (1992).

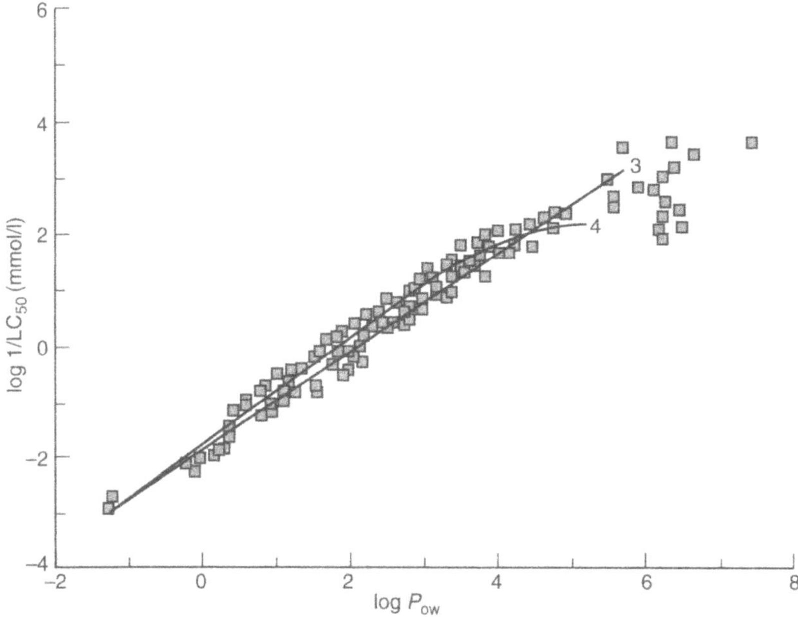

**Figure 5.4**  Comparison of selected QSAR models for non-polar non-specific toxicants relating log l/LC$_{50}$ to log $P_{ow}$ with experimental data on fish lethality (Brooke *et al.*, 1984; Geiger *et al.*, 1985, 1986, 1988, 1990); for identification of the QSAR functions see Table 5.3.

of the experiments. The repeated finding of the same principal relationship with different data sets can be taken as lateral validation, supporting the soundness of the underlying assumptions. Any of these QSARs is applicable to the class of the relatively inert chemicals, which act, at least in acute fish toxicity tests, by hydrophobic bonding to membrane constituents, enzymes, etc. (Franks and Lieb, 1990; Veith and Broderius, 1990; Verhaar, Leeuwen and Hermens, 1992). These interactions occur for all contaminants, also when eventually overruled by further specific effects that result in elevated toxicity. Therefore, these non-specific effects can be regarded as the baseline (minimum) toxicity and any chemical is expected to be at least as toxic as anticipated from its log $P_{ow}$ according to a baseline QSAR model.

Similar relationships have also been derived with other descriptors that are generally collinear with log $P_{ow}$ for non-polar non-specific toxicants; for example, water solubility (Könemann, 1981b; Zaroogian *et al.*, 1985), topological indices (Basak and Magnuson, 1983; Koch, 1983; Sabljic, 1983), or with substructure indicators (Hall, Maynard and Kier, 1989), which may be applied if the log $P_{ow}$ of the test compounds cannot be estimated, and also to cross-check the predictions obtained, especially when there is reasonable doubt about the correctness of the respective log $P_{ow}$ values.

**Table 5.3** Examples of QSAR models for estimating toxicity to fish of non-polar non-specific toxicants (e.g. alkanes, alkenes, saturated and unsaturated halogenated aliphatic hydrocarbons, basic ethers, cyclic ethers, ketones, amides, secondary and tertiary aliphatic and aromatic amines, alkylbenzenes, halogenated benzenes, piperazines, pyrimidines, polychlorinated hydrocarbon pesticides): log $LC_{50}$ correlations with various parameters.

| Model | Fish species | Duration | Units | | $r$ | $n$ | $s$ | Descriptor range |
|---|---|---|---|---|---|---|---|---|
| 1 | Carp | 24 h | mol/l | $\log 1/\mathrm{MLD} = 0.919 \log P_{ow} + 0.967$ | 0.985 | 5 | 0.151 | −0.7–1.4 |
| 2 | Goldfish | 24 h | mol/l | $\log 1/\mathrm{MLD} = 0.881 \log P_{ow} + 0.989$ | 0.958 | 5 | 0.250 | −0.7–1.4 |
| 3 | Guppy | 14(7) d | μmol/l | $\log 1/LC_{50} = 0.871 \log P_{ow} - 4.87$ | 0.988 | 50 | 0.237 | −1.3–5.7 |
| 4 | Fathead | 96 h | mol/l | $\log LC_{50} = -0.94 \log P_{ow}$ $+ 0.94 \log (6.8 \times 10^{-5} P_{ow} + 1) - 1.25$ | 0.97 | 60 | 0.35 | −1.2–5.2 |
| 5 | Goldfish | 24 h | mmol/l | $\log 1/LC_{50} = 1.0 \log P_{ow} - 2.2$ | 0.98 | 8 | 0.18 | 0.6–3.1 |
| 6 | Fathead | 96 h | mmol/l | $\log 1/LC_{50} = 0.79 \log P_{ow} - 1.35$ | 0.92 | 147 | 0.40 | 0–5 |
| 7 | Zebrafish | 28 d | | | | | | |
| | Fathead | 32 d | μmol/l | $\log 1/\mathrm{NOEC} = 0.90 \log P_{ow} - 3.80$ | 0.956 | 30 | 0.33 | 0.5–5.2 |
| 8 | Sheepshead | 96 h | mmol/l | $\log 1/LC_{50} = 0.50 \, {}^{0}\chi - 1.90$ | 0.92 | 19 | 0.37 | 2.7–8.6 |
| 9 | Fathead | 8 d | mol/l | $\log 1/LC_{50} = \Sigma\,(\mathrm{T}) + 3.34$ | 0.921 | 105 | 0.31 | – |

$r$ = regression coefficient; $n$ = number of compounds analysed; $s$ = standard deviation of the residuals; MLD = minimum lethal dose; NOEC = no observed effects concentration, ${}^{0}\chi$ zero-order connectivity index; T = fragment constants for substituents on monoaromatic compounds ($T_{NO_2}$ = 0.58, $T_{Cl}$ = 0.56, $T_{Br}$ = 0.46, $T_{Me}$ = 0.19, $T_{MeO}$ = −0.096, $T_{OH}$ = −0.092, $T_{NH_2}$ = −0.052, $T_{CN}$ = −0.020, $T_{COMe}$ = −0.11, $T_{F}$ = 0.19, $T_{CHO}$ = 0.50.
Sources of models: 1, 2, Hansch and Dunn (1972); 3, Könemann (1981b); 4 Veith, Call and Brooke (1983); 5, Lipnick, Watson and Strausz (1987); 6, Nendza and Russom (1991); 7 Leeuwen, Adema and Hermens (1990); 8 Sabljic (1983); 9 Hall, Maynard and Kier (1989).

Non-reactive chemicals with polar moieties reveal increased toxicity as compared to the baseline model. They interact with the same non-specific sites as non-polar non-reactive toxicants, but presumably by dipolar interactions and hydrogen bonding in addition to hydrophobic bonding (Kamlet *et al.*, 1986; Veith and Broderius, 1990; Veith and Mekenyan, 1993; Verhaar, Ramos and Hermens, 1995). Consequently, because of the change in the mode of interaction between the organisms and the toxicants, the structure–toxicity relationships alter. However, log $P_{ow}$ is satisfactory as the only regressor (Table 5.4), but the intercepts are significantly increased (– 1–0) and the slopes are decreased (approximately 0.6) as compared to models for non-polar toxicants.

The functions are again close, even when different species and test protocols were used to obtain the underlying toxicity data, supporting the aptness of the presumed model. In the case of ionizable compounds, correction of the partition coefficients for the dissociated fraction may be appropriate (section 1.2.3), because the non-specific effects may be attributed solely to the neutral form of the contaminants.

The discrepancy between the toxicity modelling functions for non-polar and polar non-specifically acting compounds, which appears proportional to their difference in hydrogen-bonding capacity, is largest for compounds of low log $P_{ow}$ and diminishes with increasing lipophilicity (Figure 5.5).

For chemicals with log $P_{ow}$ > 3, the differences become negligible. As the log $P_{ow}$ increases, the relative contribution of hydrogen bonding to the chemicals' toxicity decreases in favour of the hydrophobic bonding of the toxicants. Within the two groups of non-specific toxicants, the differences in polar interactions are compensated by the log $P_{ow}$ descriptor, whereas simultaneous modelling would require an additional polarity parameter. Fitting of toxicity data to either of the models may hence be used to classify the interaction between the contaminants and the biosystem under study. If the toxicity data can be described by the QSARs in Table 5.3, a non-polar non-specific mode of action can be assumed. If significant hydrogen bonding occurs with non-reactive compounds, a model for polar non-specific toxicity applies (Table 5.4).

The QSAR functions document identical relative toxicity ranking for non-polar and polar non-reactive chemicals, respectively, due to relying on the same principal property (interphase partitioning). If operated within their parameter domain – that is, the linear functions are always restricted to the log $P_{ow}$ range < 5 so as not to overestimate the toxicity of highly lipophilic chemicals to fish by orders of magnitude – these models can be regarded as principally valid. However, their application for predictive purposes is severely affected by the need to identify beforehand the compounds that have a specific mode of action, which eventually results in a much higher toxicity than anticipated from their $P_{ow}$ alone.

The trivial fact that QSARs can be valid only for homologous classes of compounds with uniform mode(s) of interaction is crucial when allocating

**Table 5.4** Examples of QSAR models for estimating toxicity to fish of polar non-specific toxicants (e.g. alcohols, aldehydes, carboxylic acids, esters, primary aliphatic and aromatic amines ($< 3$ halogen, $< 2$ $NO_2$), phenols ($< 3$ halogen, $< 2$ $NO_2$), mononitroaromatics and halogenated pyridines): log $LC_{50}$ correlations with various parameters.

| Model | Fish species | Duration | Units | | $r$ | $n$ | $s$ | Descriptor range |
|---|---|---|---|---|---|---|---|---|
| 10 | Guppy | 96 h | mmol/l | $\log 1/LC_{50} = 0.59 \log P_{ow} - 0.34$ | 0.976 | 19 | 0.14 | 1.5–4.5 |
| 11 | Bluegill | 24 h | mmol/l | $\log 1/LC_{50} = 0.66 \log P_{ow} - 0.12$ | 0.90 | 7 | – | |
| 12 | Fathead | 96 h | mol/l | $\log LC_{50} = -0.62 \log P_{ow} - 2.55$ | 0.906 | 27 | – | 0.3–6.3 |
| 13 | Fathead | 96 h | mol/l | $\log LC_{50} = -0.65 \log P_{ow} - 2.29$ | 0.95 | 39 | – | 0.9–6.3 |
| 14 | Fathead | 96 h | mmol/l | $\log 1/LC_{50} = 0.62 \log P_{ow} - 0.56$ | 0.91 | 118 | 0.30 | 0–5 |
| 15 | Fathead | 96 h | mol/l | $\log LC_{50}{}^* = -0.72\,{}^1\chi^v - 1.39$ | 0.938 | 15 | 0.25 | 1.3–4.8 |
| 16 | Fathead | 96 h | mol/l | $\log 1/LC_{50} = 0.21\,{}^1\chi + 0.91\,{}^3\chi^v + 1.79$ | 0.934 | 25 | 0.30 | ${}^1\chi$: 3.4–7.8<br>${}^3\chi^v$: 0.8–3.4 |

$r$ = regression coefficient; $n$ = number of compounds analysed; s = standard deviation of the residuals; ${}^1\chi$ = first-order connectivity index; ${}^1\chi^v$ = first-order valence-corrected connectivity index; ${}^3\chi^v$ = third-order valence-corrected connectivity index; * = QSAR specifically for aliphatic and aromatic carboxylic esters.
Sources for models: 10, Saarikoski and Viluksela (1982); 11, Kaiser, Dixon and Hodson. (1984); 12, Schultz, Holcombe and Phipps (1986); 13, Veith and Broderius (1987); 14, Nendza and Russom (1991); 15, Basak, Gieschen and Magnuson (1984); 16, Hall and Kier (1984).

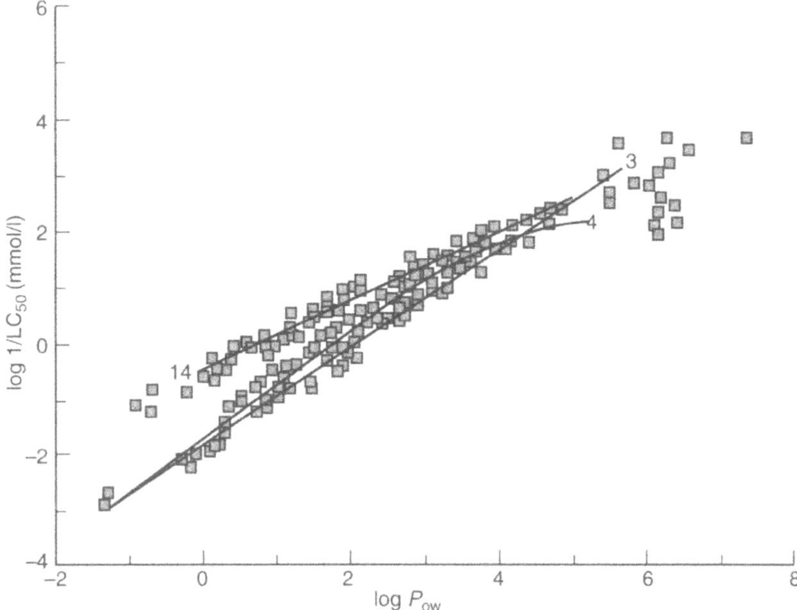

**Figure 5.5**  Comparison of selected QSAR models for non-polar and polar non-specific toxicants relating log $1/LC_{50}$ to log $P_{OW}$ with experimental data on fish lethality (Brooke *et al.*, 1984; Geiger *et al.*, 1985, 1986, 1988, 1990); for identification of the QSAR functions see Tables 5.3 and 5.4.

appropriate models for toxicity predictions, as the required mechanistic knowledge is often scarce. If there is a fundamental understanding of toxicodynamic processes in the context of the various exposure regimes and biological models/endpoints (McKinney, 1985; Bradbury, 1994), the classification of contaminants should be possible. Because mode of action is not a simple single property of the compounds, reliable discrimination criteria encoded in the chemical structures are not yet established and validated for untested substances. Hence the selection of the appropriate prediction models has to be based provisionally on structural moieties characterizing chemical classes of compounds.

Several subgroups of chemicals that fall into the general class of polar, non-specific toxicants were found to be more toxic than calculated from the respective QSARs (Table 5.4) because they exerted additional specific effects. These outliers from the polar non-specific QSARs comprise, for example, phenols and anilines with $pK_a \leq 6.3$, $\geq$ two $NO_2$ substituents and/or $\geq$ three halogen substituents (Bradbury *et al.*, 1989; Nendza and Seydel, 1990; OECD 1992a). They may act as uncouplers of the oxidative phosphorylation with diverging physiological syndromes and hence their QSARs must use different structural descriptors because their toxicity is caused by different chemical properties.

The discrimination between polar non-specific and uncoupling phenols or anilines, respectively, is based on their electrophilicity and proton acidity and may be approached with the respective quantum-chemical parameters (Mekenyan and Veith, 1994). Uncouplers of electron transport and phosphorylation reactions in mitochondria inhibit ATP synthesis without directly affecting the respiratory chain or ATP synthase ($H^+$-ATPase). Most of them are hydrophobic weak acids that have protonophoric activities; that is, they are able to transport $H^+$ through the mitochondrial membranes (Corbett, Wright and Baillie, 1984; Terada, 1990). For induction of uncoupling, an acid dissociable group, a bulky hydrophobic moiety and strongly electron-withdrawing group(s) are required, such as derivatives of these compound classes:

phenols
benzimidazoles
*N*-phenylanthranilates
salicylanilides
phenylhydrazones
salicylic acids
cumarines
aromatic amines.

Their uncoupling effects (BR) on isolated mitochondria were found to be linearly related to their hydrophobicity and their $pK_a$ to quantify the electron-withdrawing properties (Terada, 1990):

$$\log BR = 1.04 \log P_{ow} + 0.27\ pK_a + 2.89$$
$$(n = 25,\ r = 0.90,\ s = 0.37)$$

Corresponding models for uncoupler effects on whole organisms have not been described, but simple linear $\log P_{ow}$-dependent functions were derived to account for fish lethality within these compound classes (Table 5.5).

The elevated specific toxicity results in increased intercepts (approximately 0) again, whereas the slopes are congruent with QSAR models for polar non-specific toxicants, indicating that the amount of the uncouplers' excess toxicity is essentially independent of their $\log P_{ow}$ (Figure 5.6).

The application range of the uncoupler QSARs is limited to chemicals with $\log P_{ow} < 5$; they are not validated for extrapolation to higher lipophilicity.

Excess toxicity of varying extents occurs for acetylcholinesterase (AChE) inhibitors such as organophosphorous compounds (Bruijn and Hermens, 1991a,b). Acetylcholine serves as a neurotransmitter in the central as well as in the peripheral nervous systems of various vertebrates and invertebrates (Corbett, Wright and Baillie, 1984; Fukuto, 1990). When an efferent impulse reaches the nerve ending, acetylcholine is released into the synaptic cleft and diffuses to receptors located in the postsynaptic membrane, causing stimulation of the nerve fibre or the muscle. The quaternary ammonium of the acetylcholine binds by electrostatic interactions to the receptors' anionic centre

**Table 5.5** Examples of QSAR models for estimating toxicity to fish of specific toxicants: log $LC_{50}$ correlations with various parameters.

| Model | Fish species | Duration | Units | | $r$ | $n$ | $s$ | Descriptor range |
|---|---|---|---|---|---|---|---|---|
| Uncouplers | | | | | | | | |
| 17 | Fathead | 96 h | mol/l | $\log LC_{50} = -0.59 \log P_{ow} - 3.22$ | 0.98 | 6 | – | 1–5 |
| 18 | Fathead | 96 h | mmol/l | $\log LC_{50} = -0.67 \log P_{ow} + 0.05$ | 0.91 | 11 | – | 1.5–5.1 |
| AChE inhibitors | | | | | | | | |
| 19 | Guppy | 14 d | µmol/l | $\log 1/LC_{50} = 0.23 \, \Sigma\pi + 0.80 \log k_{NBP} + 2.77$ | 0.92 | 9 | 0.19 | $\Sigma\pi$:−1– 3<br>$\log k_{NBP}$:−4– −5.5 |
| 20 | Guppy | 14 d | µmol/l | $\log 1/LC_{50} = -0.41 \log P_{ow} + 31.4 \, dq(PO) - 48.0$ | 0.93 | 9 | – | $\log P_{ow}$: 2.5–5.5<br>$dq(PO)$:1.44–1.47 |

$r$ = regression coefficient; $n$ = number of compounds analysed; $s$ = standard deviation of the residuals; $\log k_{NBP}$ = first-order rate constant for alkylation of 4-nitrobenzylpyridine; $dq(PO)$ = charge density difference between the phosphorus and the oxygen of the phenoxy leaving group.
Sources of modes: 17, Schultz, Holcombe and Phipps (1986); 18, OECD (1992a); 19, Hermens *et al.* (1987); 20, Schüürmann (1990a).

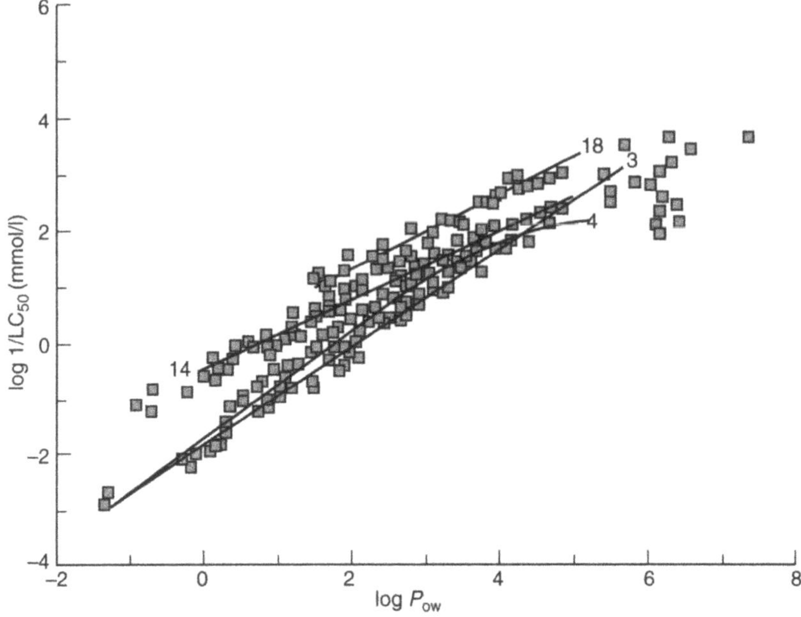

**Figure 5.6** Comparison of selected QSAR models for non-polar and polar non-specific toxicants and uncouplers relating log $1/LC_{50}$ to log $P_{OW}$ with experimental data on fish lethality (Brooke *et al.*, 1984; Geiger *et al.*, 1985, 1986, 1988, 1990); for identification of the QSAR functions see Tables 5.3–5.5.

ω-carboxy-group of aspartic or glutamic acid) and the ester moiety by hydrogen bonding to serine. The excitatory effects are regulated by enzymatic hydrolysis of acetylcholine to produce choline and acetate. The distance of the anionic and the esteratic site in AChE is about 7 Å. Inhibitors of this enzyme (e.g. organophosphorous compounds and carbamates) interact with its active site by a mechanism analogous to the physiological substrate. Like the choline, the respective leaving group is then eliminated, but the rates of the phosphorylated or carbamylated serine hydrolysis are lowered by a factor of at least $10^5$ to $10^6$. As a result of the blocked enzymes, acetylcholine accumulates in the synapses, resulting in continuous stimulation and impairment of proper nerve function. The structural requirements for AChE inhibitors comprise a steric fit to the enzyme pocket and sufficient reactivity to phosphorylate/carbamylate the serine. For the effects of carbamates or organophosphorous compounds on isolated AChE enzyme systems, QSAR models have been described in terms of substituent effects ($\pi$, $\sigma$ and $E_S$), depending on the position in the molecules and indicator variables (Iwamura and Fujita, 1982; Johnson *et al.*, 1985; Licastro *et al.*, 1986; Fukuto, 1990). The lethality to fish of a rather small set of organophosphorous insecticides was modelled on the partitioning and reactivity of the compounds (Table 5.5), but the respective QSARs could not be confirmed when the data set was extended by further derivatives (Bruijn and Hermens, 1991a,b), possibly due to differences in the pharmacokinetic behaviour of the test substances – especially their transformation and bioactivation.

The classification of environmental pollutants as either baseline toxicants (Table 5.2), for which reliable predictive QSARs are available, or as specifically acting (reactive) toxicants, which may be much more toxic, is feasible by experimental studies on mode of action (McKim, Bradbury and Niemi, 1987; Bradbury, Henry and Carlson, 1990; Broderius, Kahl and Hoglund, 1995; Nendza, Wenzel and Wienen, 1995) or from empirically derived structural criteria (Hermens, 1990; Lipnick, 1991; Verhaar, Leeuwen and Hermens, 1992; Jäckel and Nendza, 1994; OECD, 1995; Russom *et al.*, 1997) and has been related to the compounds' reactivity (Veith and Mekenyan, 1993) or to the differences in lipid-phase activity coefficients (Banerjee and Williams, 1993). A comparison of experimental data on fish toxicity with values calculated according to baseline toxicity models (Figure 5.7) identifies outliers with excess toxicity:

$$\text{excess toxicity} = \log 1/LC_{50\ \text{exp.}} - \log 1/LC_{50\ \text{baseline}}$$

Whereas most of the chemicals (around 75%) can be classified as baseline toxicants, about 25% of the compounds are likely to show excess toxicity by more than a factor of five. Analysis of excess toxicants for structural moieties, assumed to be responsible for specific interactions, provides so-called reactive substructures that discriminate them from non-specifically acting baseline toxicants. Chemicals containing functional groups such as:

aldehyde
epoxide
acrylate, pyrethroid
allylhalide
allylic/propargylic compounds with a strong leaving group
allylic/propargylic primary/secondary alcohol
diazo ($>$N–N$<$, –N=N–)
organometallic
haloacetamide
quinone
polynitroaromatic
catechol
thiophosphoric acid ester (P–S)
imidazole
carbamate
$R$–CH–$R'$, with $R$, $R'$ electron-withdrawing substituents

generally reveal excess toxicity. The suspected specific, and accordingly elevated, toxicity of aryl-, alkyl-, and benzylhalides, disulfides, acrylamides,

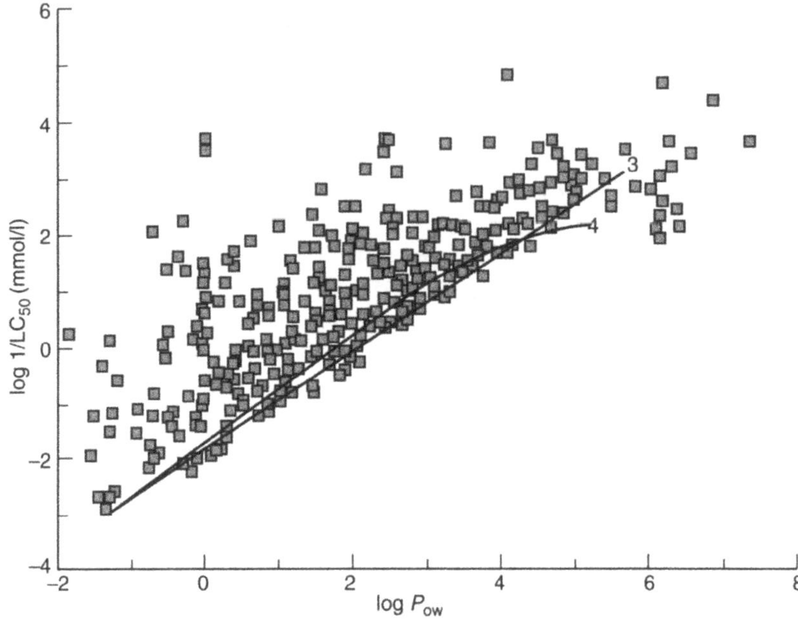

**Figure 5.7** Comparison of baseline QSAR models relating log $1/LC_{50}$ to log $P_{OW}$ with experimental data on fish lethality for a diverse set of chemicals (Brooke *et al.*, 1984; Geiger *et al.*, 1985, 1986, 1988, 1990). About 25% of the compounds are outliers that are severely underestimated in their toxicity on the basis of log $P_{OW}$ alone; for identification of the QSAR functions see Table 5.3. Reproduced from Jäckel and Nendza (1994) with kind permission from Elsevier Science–NL, Amsterdam, The Netherlands.

haloethers, diesters, difunctional ethylenes and oxophosphoric acid esters (P=O) does not systematically occur. Only some of the respective chemicals are excess toxicants, and others do not exceed baseline toxicity. Discrimination of aldehydes with respect to aromatic or aliphatic compounds does not reveal a significant differentiation between these groups, in spite of the generally accepted assumption that aromatic aldehydes are more reactive than aliphatic aldehydes. Some of these substructures can be related to specific modes of action:

Electrophilic reaction with $-NH_2$, $-OH$ and $-SH$ groups of physiological macromolecules such as proteins and DNA bases (e.g. aldehydes, epoxides, allylhalides, catechols). Catechols are easily oxidized to quinones, which are considered to be reactive towards nucleophiles due to an activated C=C bond.

Inhibition of acetylcholinesterase (e.g. carbamates and organophosphorous compounds). With respect to the effects of organophosphorous compounds towards fish, a discrimination between oxo- and thio-derivatives revealed appropriate.

Uncoupling of oxidative phosphorylation (e.g. imidazoles, benzimidazoles and polynitroaromatics).

Formation of stabilized free radicals as reactive intermediates (e.g. malonic acid, malononitriles and $R$–CH–$R'$- containing compounds, with $R$, $R'$ being activating substituents (e.g. esters)).

Application of the respective substructure indicators in combination with a baseline QSAR model allowed the correct prediction of baseline and excess toxicity towards fish for 432 of 500 (86.4%) chemicals analysed (Jäckel and Nendza, 1994). However, 32 outliers (out of 122) were not detected, and 36 of the baseline toxicants (out of 378) were predicted to be more toxic than they really are (mostly aldehydes). Similar improvements were obtained on the basis of data from prolonged testing, where baseline and excess effects were correctly indicated for 93% of the baseline toxicants and for 55% of the outliers, respectively, resulting in an overall correct assignment of 81%. This result is satisfactory for the non-specific compounds, which suggests that the same interactions often predominate under acute and prolonged testing conditions. However, the rate of unrecognized specific toxicants (almost 50%) indicates that a substantial number of acutely non-specific compounds may change their apparent mode of action and affect specific targets at far lower concentrations, when the period of exposure is increased. To characterize the potential of chemicals to cause chronic toxicity by further, acutely latent, interactions with fish, an extended set of relevant substructures has to be identified. Although the identification of potential outliers based on the occurrence of certain structural fragments is an improvement towards reliable QSAR applications, these approaches are principally limited to chemicals with these substructures as well as to the endpoints and organisms for which they were derived.

## 5.2 TOXICITY TO *DAPHNIA*

Daphnids (water fleas) occur ubiquitously in aquatic ecosystems and, as primary consumers, they form an important link in aquatic foodwebs. The standard species *Daphnia magna*, which can be cultured in the laboratory, is used in toxicity tests to record adverse effects on the survival (acute, 48 h tests) or the growth and reproduction (chronic, 21 d tests) of the animals. The toxicity is generally described in terms of the $EC_{50}$ (EC: effective concentration); that is, the level of the toxicant (mg/l) which affects 50% of the individuals exposed, regardless of how the effects are evoked. A variety of factors may considerably influence the absolute values of the observed toxic concentrations:

Life stage of the species: the toxicity of xenobiotics towards organisms varies with the age relative to lifespan.

Duration of the test: sensitivity to chemicals may increase or decrease with changes in the test period due to alterations in the catabolic status and in the defence mechanisms of the animals.

Feeding regime: sensitivity to chemicals may increase or decrease with the feeding status of the animals and the food in the test medium may alter the bioavailability of the toxicants.

Exposure regime: concentrations of the toxicant will vary according to whether the exposure is static, semi-static or flow-through in the test set-up.

Temperature dependence: sensitivity to chemicals may increase or decrease with changes in the temperature due to alterations in the catabolic status of the animals.

Light intensity: sensitivity to chemicals may increase or decrease with changes in the light intensity due to alterations in the catabolic status of the animals.

pH and buffer capacity of the water phase: pH conditions influence the toxicity of organic acids and bases.

Salinity of the test medium: sensitivity to chemicals may increase or decrease with changes in the salinity and hardness of the water, which may also affect the solubility of the toxicants.

Oxygen content: sensitivity to chemicals may increase or decrease with changes in the oxygen content, which is interrelated with temperature effects.

Suspended organic matter: components of soil and sediment (e.g. humic acids) may result in decreased apparent toxicity by serving as a sink compartment for the toxicants due to sorption processes.

Stability of the test compound in water: the substance must not degrade during the experiment; losses in concentration may also occur by evaporation or adsorption to glassware.

Surface-active materials: the presence of solubilizing agents may alter the bioavailable concentration of the test compounds significantly.

Purity of the test compound: pure substances have to be used because even small amounts of soluble impurities can cause large errors in the measured toxicity.

Cosolute effects: the presence of additional organic solutes, as is the case in 'real' waters, may significantly alter the toxicity of the individual compounds.

Analytical method: a specific set-up must be developed for each chemical species individually.

As for the fish toxicity test, the difficulties most overlooked in producing reliable toxicity data for *Daphnia* arise from frequently testing concentrations of contaminants above their water solubility. The uncritical use of solvent carriers may introduce additional effects (Calleja and Persoone, 1993).

Without chemical analysis of the actual contaminant concentrations in the tests, major uncertainties in the experimental $EC_{50}$ data may result.

Published data on *Daphnia* toxicity are highly variable because of the diverse test protocols applied and hence are of limited use for the derivation and validation of sound QSAR models. Validated reference data sets are not available (unlike fish) and if data are retrieved from the literature, therefore, care should be taken that they have been conducted according to standard protocols and evaluated by consistent criteria. The important test parameters should be identical or comparable throughout the data set. If, however, they vary, they should be critically evaluated for their effect on the test results, as such differences may greatly affect the relevance of the QSAR models derived.

Toxicants towards *Daphnia*, as for those against fish, divide into those with non-specific effects and a heterogeneous group with specific effects, although the respective classifications for the different aquatic species do not necessarily coincide. The toxicity of the non-specific contaminants is related to their lipophilicity as expressed by log $P_{ow}$, indicating (as for fish) the predominance of transport phenomena and the relevance of partitioning and hydrophobic membrane interractions for this mechanism of toxic action. As for the fish baseline QSARs for non-polar non-reactive chemicals (section 5.1), log $P_{ow}$-dependent models have been derived to describe *Daphnia* lethality, immobilization of the test animals or inhibition of their growth (Table 5.6).

The slopes and the intercepts are similar to those of the fish baseline models (about 0.8–1 and –2 on the mmol/l scale respectively), indicating a similar sensitivity of fish and daphnids towards non-specific toxicants. The derivation of almost identical models by different investigators with different sets of chemicals supports the reliability of these baseline QSARs for estimating the toxicity of non-polar non-reactive chemicals to *Daphnia*. Any compound is expected to be at least as toxic as predicted from these models.

Similar relationships have been derived using other descriptors that are generally collinear with log $P_{ow}$ for non-polar non-specific toxicants, such as water solubility, molar volume and topological indices (Table 5.6), which may be applied if the log $P_{ow}$ cannot be estimated and may also be used to cross-check the predictions obtained, especially if there is reasonable doubt about the correctness of the respective log $P_{ow}$ values.

Non-reactive chemicals with polar moieties have increased toxicity towards *Daphnia* relative to the baseline model, as for fish (section 5.1). Again, log $P_{ow}$ remains satisfactory as the only regressor and the functions reveal lowered slopes (0.5–0.7) and increased intercepts (Table 5.7).

The difference between the log $P_{ow}$-dependent baseline models for non-polar and polar non-specific contaminants is largest for hydrophilic chemicals and diminishes with increasing lipophilicity. The additional consideration of the $pK_a$ values of the compounds attempts to adjust for ionization effects.

**Table 5.6** Examples of QSAR models for estimating toxicity to *Daphnia* of non-polar, non-specific toxicants: $\log EC_{50}$ correlations with various parameters.

| Model | Endpoint | Duration | Units | | $r$ | $n$ | $s$ | Descriptor range |
|---|---|---|---|---|---|---|---|---|
| 1 | Immobilization | 48 h | µmol/l | $\log 1/EC_{50} = 0.91 \log P_{ow} - 4.72$ | 0.991 | 19 | 0.245 | -1.4–5.7 |
| 2 | Immobilization | 24 h | µmol/l | $\log 1/EC_{50} = 0.873 \log P_{ow} - 5.022$ | 0.98 | 14 | 0.22 | 0.1–4.0 |
| 3 | Reproduction | – | µmol/l | $\log 1/EC_{50} = 1.03 \log P_{ow} - 4.73$ | 0.986 | 10 | 0.38 | -0.3–5.7 |
| 4 | Lethality | 96 h | mol/l | $\log 1/LC_{50} = 0.94 \log P_{ow} + 0.93$ | 0.96 | 18 | 0.33 | 0.9–4.3 |
| 5 | Lethality | 96 h | nmol/l | $\log 1/LC_{50} = 0.535 \, {}^{0}\chi^{v} - 7.00$ | 0.997 | 5 | 0.14 | 3.4–10.9 |
| 6 | Lethality | 48 h | mmol/m³ | $\log LC_{50} = 0.722 \log C_{l} - 0.611$ | 0.931 | 37 | – | – |
| 7 | Immobilization | 48 h | µmol/l | $\log 1/EC_{50} = -4.35 \, V_{i}/100 + 4.86$ | 0.931 | 38 | 0.403 | 61–144 |
| 8 | Growth inhibition | 16 d | µmol/l | $\log 1/NOEC = 0.952 \log P_{ow} - 3.96$ | 0.976 | 6 | 0.46 | 2–5.7 |
| 9 | Growth inhibition | 16 d | µmol/l | $\log 1/NOEC = 0.99 \log P_{ow} - 4.16$ | 0.974 | 10 | 0.50 | -0.3–5.7 |

$r$ = regression coefficient; $n$ = number of compounds analysed; $s$ = standard deviation of the residuals; NOEC = no observed effect concentration; ${}^{0}\chi^{v}$ = zero-order valence-corrected connectivity index; $C_{l}$ = water solubility of subcooled liquid; $V_{i}$ = intrinsic volume ($V_{i}$ = 1.925 $V$ + 3.57; $V$ = molar volume according to Bondi (cm³/mol)).
Sources of models: 1, Hermens *et al.* (1984); 2, Lipnick and Hood (1986); 3, 9, Wolf *et al.* (1988); 4, Ikemoto *et al.* (1992); 5, Govers, Ruepert and Aiking (1984); 6, Abernethy *et al.* (1986); 7, Passino, Hickey and Frank (1988); 8, Hermens *et al.* (1985a).

**Table 5.7** Examples of QSAR models for estimating toxicity to *Daphnia* of polar non-specific toxicants: log $EC_{50}$ correlations with various parameters

| Model | Endpoint | Duration | Units | | $r$ | $n$ | $s$ | Descriptor range |
|---|---|---|---|---|---|---|---|---|
| 10 | Immobilization | 48 h | $\mu$mol/l | $\log 1/EC_{50} = 0.66 \log P_{ow} - 3.53$ | 0.744 | 15 | 0.23 | $1.4\text{--}3.1$ |
| 11 | Immobilization | 24 h | mmol/l | $\log 1/EC_{50} = 0.47 \log P_{ow} + 0.025$ | 0.88 | 17 | 0.23 | $2.1\text{--}5$ |
| 12 | Immobilization | 24 h | mmol/l | $\log 1/EC_{50} = 0.366 \log P_{ow} + 0.025$ $-0.134\ (9.9\text{--}pK_a)$ | 0.91 | 17 | 0.21 | $\log P_{ow}: 2.1\text{--}5$ $pK_a: 4.9\text{--}9.2$ |
| 13 | Lethality | 48 h | mol/l | $\log 1/LC_{50} = -0.062\ \pi^2 + 0.695\ \pi$ $+0.364\ \Sigma\sigma + 3.613$ | 0.965 | 14 | 0.16 | $\pi: 0\text{--}2.9$ $\Sigma\sigma: -0.3\text{--}2$ |
| 14 | Immobilization | 24 h | mmol/l | $\log 1/EC_{50} = 18.49\ V_0 - 4.185\ V_3 - 3.80$ | 0.91 | 57 | 0.57 | $V_0: 0.06\text{--}0.88$ $V_3: 0\text{--}2.2$ |

$r$ = regression coefficient; $n$ = number of compounds analysed; $s$ = standard deviation of the residuals; NOEC = no observed effects concentration; $\pi$ = lipophilicity constants for substituents of phenol derivatives; $\Sigma\sigma$ = sum of $\sigma$ Hammett constants for substituents of phenol derivatives; $V_0$, $V_3$ = autocorrelation vectors.
Sources of models: 10, Deneer *et al.* (1989); 11, 12, Devillers and Chambon (1986); 13, Kopperman, Carlson and Caple (1974); 14, Devillers, Chambon and Zakarya (1987).

However, correction of the partition coefficients for the dissociated fraction may be more appropriate, because the non-specific effects may be attributed solely to the neutral form of the contaminants. The same type of model using substituent constants is furthermore restricted to homologous chemical classes, such as phenols (model 13 in Table 5.7).

The QSARs (Tables 5.6 and 5.7) reveal congruent ranking in the toxicity of non-polar and polar non-reactive chemicals towards *Daphnia*, because they rely on the same principal property (interphase partitioning). If operated within their parameter domain – that is, the linear functions are always restricted to the log $P_{ow}$ range < 5 so as not to overestimate the toxicity of highly lipophilic chemicals by orders of magnitude – these models can be regarded as principally valid. Their application for predictive purposes is severely affected, however, by the need to identify beforehand the compounds that have a specific mode of action, which eventually results in a much higher toxicity than anticipated from their log $P_{ow}$ alone.

Comparison of measured data on acute toxicity with calculated baseline predictions revealed that 21.5% of the compounds (53 out of 246) were at least five times more toxic towards *Daphnia* than estimated, whereas almost 80% had non-specific toxicity. The identification of potential outliers from the baseline QSAR models based on the set of reactive substructures derived from fish tests (section 5.1) is only partly satisfactory: 19 outliers (35.8%) were recognized from their substructures and seven baseline toxicants (3.6%) were incorrectly assumed to exert excess toxicity (Jäckel and Nendza, 1994). Those compounds with reactive substructures actually show excess toxicity towards *Daphnia*, but not all outliers are recognized. Accordingly, the set of indicators devised for fish has to be adjusted to cover different modes of action on specific targets in different organisms.

## 5.3 TOXICITY TO ALGAE

Like daphnids, algae occur ubiquitously in aquatic ecosystems and they are an important link in aquatic foodwebs. The test species are generally fast-growing strains of unicellular green algae, which are easy to culture in the laboratory; for example, *Selenastrum capricornutum*, *Scenedesmus subspicatus* and *Chlorella vulgaris*.

The toxicity tests with algae record adverse effects on cell multiplication within a period of 24–48 h to 7 d (Bringmann and Kühn, 1980; OECD, 1984d). As for *Daphnia* (section 5.2), the toxicity is generally described in terms of the $EC_{50}$; that is, the level of the toxicant (mg/l) that reduces the cell multiplication by 50% as compared to an untreated control culture, regardless of how the effects are evoked. The effects on cell concentration (growth rates) are preferably monitored by obtaining cell counts at defined intervals or by measuring changes in the chlorophyll content (fluorescence) of the cultures. The turbidimetric determination of the cell density may be biased by concurrent bacterial growth. A variety of factors may considerably influence the absolute values of observed toxic concentrations:

Duration of the test: sensitivity to chemicals may increase or decrease with changes in the test period due to alterations in the catabolic status and in the defence mechanisms of the cultures.

Light intensity: sensitivity to chemicals may increase or decrease with changes in the light intensity due to alterations in the catabolic status of the cultures.

Temperature dependence: sensitivity to chemicals may increase or decrease with changes in the temperature due to alterations in the catabolic status of the cultures.

pH and buffer capacity of the water phase: pH conditions influence the toxicity of organic acids and bases.

Salinity of the test medium: sensitivity to chemicals may increase or decrease with changes in the salinity and hardness of the water, which may also affect the solubility of the toxicants.

Oxygen content: sensitivity to chemicals may increase or decrease with changes in the oxygen content, which is interrelated with temperature effects.

Stability of the test compound: the substance must not degrade during the experiment; losses in concentration may also occur by evaporation or adsorption to glassware.

Surface-active materials: the presence of solubilizing agents may alter the bioavailable concentration of the test compounds significantly.

Purity of the test compound: pure substances have to be used because even small amounts of soluble impurities can cause large errors in the measured toxicity.

Cosolute effects: the presence of further organic solutes, as is the case in 'real' waters, may significantly alter the toxicity of the individual compounds.

As for fish and *Daphnia*, the difficulties most overlooked in producing reliable toxicity data for algae arise from frequently testing concentrations of contaminants above their water solubility. The uncritical use of solvent carriers may introduce additional effects (Calleja and Persoone, 1993). The testing of nominal doses without chemical analyses of the actual contaminant concentrations in the tests causes major uncertainties in the experimental $EC_{50}$ data.

Published data on algal toxicity are highly variable because of the diverse test protocols applied and hence are of limited use for the derivation and validation of sound QSAR models. Validated reference data sets are not available for algae (unlike fish) and if data are retrieved from the literature, therefore, care should be taken that they have been conducted according to standard protocols and evaluated by consistent criteria. The important test parameters should be identical or comparable throughout the data set. If, however, they vary they should be critically evaluated for their effect on the test results, as such differences may greatly affect the relevance of the QSAR models derived.

Toxicants against algae, as for those against fish and *Daphnia*, divide into those with non-specific effects and a heterogeneous group of chemicals with specific effects, although the respective classifications for the different aquatic species do not necessarily coincide. The toxicity of non-specific contaminants is related to their lipophilicity as expressed by log $P_{ow}$, indicating (as for fish and *Daphnia*) the predominance of transport phenomena and the relevance of partitioning and hydrophobic membrane interactions for this mechanism of toxic action. Like the fish baseline QSARs for non-polar non-reactive chemicals (section 5.1), log $P_{ow}$-dependent models have been derived to describe the inhibition of algal population growth (Table 5.8).

The slopes (about 0.9) are similar to those of the respective fish baseline models, whereas the intercepts range mostly between –1.5 and –0.5 (on the mmol/l scale), indicating a slightly lower sensitivity of the algae towards non-specific toxicants. The considerable variation in the absolute algal toxicity values (i.e. the sensitivity of the test systems) is attributable to the different species tested and especially the differences in test duration (4 h–8 d), which may severely affect the attainment of steady-state equilibrium.

The derivation of conforming models by different investigators with diverse sets of chemicals in various algal test systems (freshwater and marine, unicellular and macroalgae) supports the reliability of these baseline QSARs for estimating the toxicity of non-polar non-reactive chemicals towards algae. Any compound is expected to be at least as toxic as predicted from these models. Similar relationships have also been derived using other descriptors that are generally collinear with log $P_{ow}$ for non-polar, non-specific toxicants (e.g. water solubility), which may be applied if the log $P_{ow}$ cannot be estimated, and may be used to cross-check the predictions obtained, especially if there is reasonable doubt about the correctness of the respective log $P_{ow}$ values.

Non-reactive chemicals with polar moieties have increased toxicity towards algae relative to the baseline model, as for fish (section 5.1). Again, the QSARs (Table 5.8) reveal a lowered slope of about 0.6 and an increased intercept; log $P_{ow}$ remains satisfactory as the only regressor. The difference between the log $P_{ow}$-dependent baseline models for non-polar and polar non-specific contaminants is largest for hydrophilic chemicals and diminishes with increasing lipophilicity. If operated within their parameter domain – that is, the linear functions are always restricted to the log $P_{ow}$ range < 5 so as not to overestimate the toxicity of highly lipophilic chemicals by orders of magnitude – these models can be regarded as principally valid. Their application for predictive purposes is severely affected, however, by the need to identify beforehand the compounds that have a specific mode of action, which

**Table 5.8** Examples of QSAR models for estimating the toxicity to algae of non-specific toxicants: log $EC_{50}$ correlations with various parameters

| Models | Species | Duration | Units | | $r$ | $n$ | $s$ | Descriptor range |
|---|---|---|---|---|---|---|---|---|
| Non-polar non-specific toxicants | | | | | | | | |
| 1 | *Chlorella vulgaris* | 6 h | mol/l | $\log 1/EC_{50} = 0.99 \log P_{ow} + 0.143$ | 0.97 | 24 | 0.162 | 2–4.8 |
| 2 | *Chlorella vulgaris* | 5 d | mol/l | $\log 1/EC_{50} = 0.94 \log P_{ow} + 0.34$ | 0.97 | 8 | 0.34 | 0.9–5.1 |
| 3 | *Selenastrum capricornutum* | 96 h | mmol/l | $\log 1/EC_{50} = 0.887 \log P_{ow} - 1.545$ | 0.98 | 13 | 0.18 | 1.5–5 |
| 4 | *Selenastrum capricornutum* | 72 h | μmol/l | $\log 1/EC_{50} = 0.91 \log P_{ow} - 4.45$ | 0.95 | 8 | – | 2–4 |
| 5 | *Scenedesmus quadricauda* | 8 d | μmol/l | $\log 1/LOEC = 0.897 \log P_{ow} - 3.40$ | 0.96 | 13 | 0.28 | 0.1–2.9 |
| 6 | *Microcystis aeruginosa* | 8 d | μmol/l | $\log 1/LOEC = 0.909 \log P_{ow} - 3.90$ | 0.95 | 14 | 0.33 | 0.1–2.9 |
| 7 | *Enteromorpha intestinalis* | 24 h | mmol/l | $\log EC_{50} = -0.88 \log P_{ow} + 2.87$ | 0.99 | 8 | – | −0.8–3.0 |
| 8 | *Ankistrodesmus falcatus* | 4 h | mmol/l | $\log 1/EC_{50} = -0.587 \log S + 2.419$ | 0.94 | 12 | 0.29 | $0.02–2.3 \times 10^4$ |
| Polar non-specific toxicants | | | | | | | | |
| 9 | *Scenedesmus quadricauda* | 7 d | mmol/l | $\log 1/EC_{50} = 0.67 \log P_{ow} - 0.86*$ | 0.82 | 16 | 0.55 | 0.5–5 |

$r$ = regression coefficient; $n$ = number of compounds analysed; $s$ = standard deviation of the residuals; LOEC = lowest observed effect concentration; $S$ = water solubility (μmol/l);*QSAR originally reported with chromatographically derived lipophilicity descriptors ($\log k' = 1.03 \log P_{ow} - 0.28$, $r = 0.99$).
Sources of models: 1, Kramer and Trümper (1986); 2, Ikemoto *et al.*(1992); 3, Shigeoka *et al.* (1988); 4, Galassi *et al.* (1988); 5, 6, Lipnick and Hood (1986); 7, Schild *et al.* (1993); 8, Wong *et al.* (1984); 9, Nendza and Seydel (1988a).

eventually results in a much higher toxicity than anticipated from their log $P_{ow}$ alone.

Comparison of measured data on toxicity with calculated baseline predictions revealed that 34.3% of the compounds (34 out of 99) were at least five times more toxic towards algae than estimated, whereas about 65% had non-specific toxicity. The identification of potential outliers from the baseline QSAR models based on the set of reactive substructures from fish tests (section 5.1) is not satisfactory: eight outliers (23.5%) were recognized from their substructures and six baseline toxicants (9.2%) were incorrectly assumed to exert excess toxicity (Jäckel and Nendza, 1994). Those compounds with reactive substructures actually show excess toxicity towards algae, but not all outliers are recognized. The principal sites for specific toxicity in algae and fish thus differ. For example, the targets for AChE inhibitors are not present in unicellular organisms, but green algae are additionally sensitive to contaminants that interact with photosynthesis. Accordingly, the set of indicators devised for fish has to be adjusted to cover different modes of action on specific targets in different organisms.

QSAR models describing the effects of specific toxicants by mode of action are currently not available for algae. However, for several chemical classes confined models have been derived using substituent constants to rank the effects within homologous series of compounds (Table 5.9). Extrapolation to other substances of the same mode of toxic action is not feasible.

## 5.4  TOXICITY TO BACTERIA AND PROTOZOA

Biotest systems using bacteria and protozoa, as well as other unicellular organisms (e.g. yeasts), and *in vitro* cell cultures based on fish or mammalian cell lines (i.e. fibroblasts) have been established to serve as fast, low-cost surrogates for more complex, more expensive aquatic toxicity assessments with higher organisms. Global parameters concerning the viability of the species (e.g. their multiplication rates) are mostly used for measures of toxicity, because these represent the sum of possible effects on the organisms' targets. Because of the variety of test protocols used in different laboratories, generalization of the results on a quantitative basis for the respective taxa is not feasible. However, the data are useful for ranking toxicants and for identifying specifically toxic compounds that need further in-depth testing.

Toxicants against bacteria and protozoa, as for those against other aquatic species, divide into those with non-specific effects and a heterogeneous group of chemicals with specific effects, although the respective classifications for the different species do not necessarily coincide. The toxicity of non-specific contaminants is related to their lipophilicity as expressed by log $P_{ow}$, indicating again the predominance of transport phenomena and the relevance of partitioning and hydrophobic membrane interactions for this mechanism of toxic action. Like the fish baseline QSARs for non-polar non-reactive chemicals

**Table 5.9** Examples of QSAR models for estimating the toxicity to algae of specific toxicants: $\log EC_{50}$ correlations with various parameters.

| Model | Species | Duration | Units | | $r$ | $n$ | $s$ | Descriptor range |
|---|---|---|---|---|---|---|---|---|
| Anilines | | | | | | | | |
| 10 | *Chlorella vulgaris* | 6 h | mol/l | $\log 1/EC_{50} = 0.92\ \pi + 2.41$ | 0.94 | 13 | 0.2 | 0–1.8 |
| Nitrobenzenes | | | | | | | | |
| 11 | *Chlorella pyrenoidosa* | 96 h | μmol/l | $\log 1/EC_{50} = 0.45 \log P_{ow} + 1.15\ \sigma - 3.00$ | 0.91 | 15 | 0.25 | $\sigma^-$:−0.2–0.7 $\log P_{ow}$:1.4–3.1 |
| Thioureas | | | | | | | | |
| 12 | *Chlorella vulgaris* | 6 h | mol/l | $\log 1/EC_{50} = 0.781\ \pi + 0.709$ | 0.99 | 11 | 0.08 | 0.5–3.9 |
| O-alkylcarbamates | | | | | | | | |
| 13 | *Chlorella vulgaris* | 6 h | mol/l | $\log 1/EC_{50} = 0.953\ \pi + 0.252$ | 0.99 | 7 | 0.05 | 1–4 |

$r$ = regression coefficient; $n$ = number of compounds analysed; $s$ = standard deviation of the residuals.
Sources of models: 10, Kramer (1989); 11, Deneer *et al.* (1989); 12, Schelenz and Kramer (1985); 13, Kramer *et al.* (1983).

(section 5.1), log $P_{ow}$-dependent models have been derived to describe the effects on bacteria (Table 5.10) and protozoa (Table 5.11).

The slopes (0.8–1) are similar to those of the respective fish baseline models. However, conclusions from the intercepts of the QSARs about the differences in the test systems' sensitivity may be misleading because of the different test procedures – especially with regard to the individual species tested – and the differences in the test durations. Nevertheless, the derivation of conforming models by different investigators with diverse sets of chemicals in the various test systems supports the principal validity of these baseline QSARs for estimating toxicity of non-polar non-reactive chemicals towards bacteria and protozoa. Any compound is expected to be at least as toxic as predicted from these models. Similar relationships have also been derived using other descriptors that are generally collinear with log $P_{ow}$ for non-polar non-specific toxicants (e.g. connectivity indices), which may be applied if the log $P_{ow}$ cannot be estimated, and may also be used to cross-check the obtained predictions, especially if there is reasonable doubt about the correctness of the respective log $P_{ow}$ values.

Non-reactive chemicals with polar moieties have increased toxicity towards bacteria and protozoa in comparison to the baseline model, as in other aquatic species. Again, the QSARs reveal a lowered slope of about 0.6 and an increased intercept; log $P_{ow}$ remains satisfactory as the only regressor. The difference between the log $P_{ow}$-dependent baseline models for non-polar and polar non-specific contaminants is largest for hydrophilic chemicals and diminishes with increasing lipophilicity. The QSARs (Tables 5.10 and 5.11) confirm the collinear ranking in the toxicity for non-polar and polar non-reactive chemicals, respectively, due to relying on the same principal property (i.e. interphase partitioning). If operated within their parameter domain – linear functions are restricted to the log $P_{ow}$ range < 5 so as not to overestimate the toxicity of highly lipophilic chemicals by orders of magnitude – these models can be regarded as principally valid. Their application for predictive purposes is limited, however, by the need to identify beforehand the compounds that have a specific mode of action, which eventually results in a much higher toxicity than anticipated from their log $P_{ow}$ alone. QSAR models of the effects of specific toxicants by mode of action are currently not available for bacteria and protozoa. However, for some chemical classes confined models using various descriptors to rank the effects within these series of compounds (Table 5.11) have been derived for protozoa.

The apparent disadvantages of the low-complexity test systems have turned into one of their major applications: in connection with QSAR models, they can be used to identify outliers, which are more (or less) toxic than calculated from their physico-chemical properties. Comparisons of measured toxicity data with calculated baseline predictions discriminate non-specific toxicants from those revealing specific effects. The latter group may be further categorized by comparing the actual effects (in a series of assays

**Table 5.10** Examples of QSAR models for estimating the toxicity to bacteria of non-specific toxicants: $\log EC_{50}$ correlations with various parameters.

| Model | Species effects | Duration | Units | | $r$ | $n$ | $s$ | Descriptor range |
|---|---|---|---|---|---|---|---|---|
| Non-polar non-specific toxicants | | | | | | | | |
| 1 | *Bacillus* TL81 dehydrogenase | 30 min | mmol/l | $\log 1/EC_{50} = 0.84 \log P_{ow} - 2.54$ | 0.92 | 24 | 0.36 | 1.5–5 |
| 2 | *Photobacterium phosphoreum* luminescence | 15 min | μmol/l | $\log 1/EC_{50} = 0.995 \log P_{ow} - 5.14$ | 0.952 | 22 | 0.53 | 1–6.1 |
| 3 | *Escherichia coli* multiplication | 8 h | mmol/l | $\log 1/EC_{50} = 0.95 \log P_{ow} - 2.37$ | 0.93 | 11 | 0.26 | –1–2 |
| 4 | *Photobacterium phosphoreum* luminescence | 15 min | μmol/l | $\log 1/EC_{50} = 1.26{}^2\chi^v - 5.87$ | 0.986 | 7 | 0.20 | 0.9–3.3 |
| Polar non-specific toxicants | | | | | | | | |
| 5 | *Bacilus subtilis* multiplication | | mol/l | $\log 1/EC_{50} = 0.617 \log P_{ow} + 1.53$ | 0.976 | 10 | 0.204 | 2–6.3 |
| 6 | *Photobacterium phosphoreum* luminiscence | 30 min | mmol/l | $\log 1/EC_{50} = 0.57 \log P_{ow} - 0.041$ | 0.89 | 15(16?) | 0.35 | 1–4.3 |
| 7 | *Escherichia coli* multiplication | 8 h | mmol/l | $\log 1/EC_{50} = 0.66 \log P_{ow} - 1.58*$ | 0.96 | 40 | 0.21 | 0–4.4 |

$r$ = regression coefficient; $n$ = number of compounds analysed; $s$ = standard deviation of the residuals; ${}^2\chi^v$ = second-order valence-corrected connectivity index;
*QSAR originally reported with chromatographically derived lipophilicity descriptors ($\log k' = 1.03 \log P_{ow} - 0.28$, $r = 0.99$).
Sources of models: 1, Liu, Thomson and Kaiser (1982); 2, Hermens *et al.* (1985b); 3, 7, Nendza and Seydel (1988b); 4, Govers *et al.* (1986); 5, Lien, Hansch and Anderson (1968); 6, Ribo and Kaiser (1984).

**Table 5.11** Examples of QSAR models for estimating the toxicity to protozoa: log $EC_{50}$ correlations with various parameters.

| Models | Species effects | Duration | Units | | $r$ | $n$ | $s$ | Descriptor range |
|---|---|---|---|---|---|---|---|---|
| Non-polar non-specific toxicants | | | | | | | | |
| 1 | *Tetrahymena pyriformis* multiplication | 48 h | mmol/l | $\log 1/EC_{50} = 0.834 \log P_{ow} - 2.069$ | 0.976 | 30 | 0.32 | −0.8–5.7 |
| Polar non-specific toxicants | | | | | | | | |
| 2 | *Tetrahymena pyriformis* multiplication | 48 h | mmol/l | $\log 1/EC_{50} = 0.580 \log P_{ow} - 0.956$ | 0.90 | 95 | 0.32 | 0.9–6.1 |
| 3 | *Uronema parduczi* inhibition | – | μmol/l | $\log 1/LOEC = 0.675 \log P_{ow} - 3.166$ | 0.72 | 12 | 0.57 | – |
| Specific toxicant classes uncoupling phenols and anilines: | | | | | | | | |
| 4 | *Tetrahymena pyriformis* multiplication | 48 h | mmol/l | $\log 1/EC_{50} = 0.425 \log P_{ow} + 0.202$ | 0.95 | 26 | 0.18 | 1.4–5.7 |
| Phenols | | | | | | | | |
| 5 | *Tetrahymena pyriformis* multiplication | 48 h | mmol/l | $\log 1/EC_{50} = 0.567 \log P_{ow} + 0.819 - 0.1885 \ pK_a$ | 0.979 | 21 | 0.225 | $pK_a$ 4.1–10.6 log $P_{ow}$: 0.5–5.7 |
| *N*-heteroaromatic compounds | | | | | | | | |
| 6 | *Tetrahymena pyriformis* multiplication | 72 h | mmol/l | $\log 1/EC_{50} = 0.911^1 \chi^v - 2.969$ | 0.962 | 24 | 0.27 | 1.4–4.7 |

$r$ = regression coefficient; $n$ = number of compounds analysed; $s$ = standard deviation of the residuals; LOEC = lowest observed effect concentration;[1] $\chi^v$ = first-order valence-corrected connectivity index.
Sources of models: 1, 2, 4, Schultz *et al.* (1990); 3, Lipnick and Hood (1986); 5, Schultz (1987); 6, Schultz, Kier and Hall (1982).

representing various endpoints) with models for specific compound classes, producing a measure of similarity of toxic effects. In this way, information is gained about possible modes of action, which may be used for interspecies extrapolations, as well as indications of further effective testing strategies.

## 5.5 PHYTOTOXICITY

Airborne chemicals as well as contaminants in soil may be taken up by terrestrial plants, either directly through stomata or adsorption over leaf surfaces, or with soil water via roots, respectively. Aquatic plants are directly exposed to pollutants in water. The phytotoxicity of chemicals is rarely assessed, except when dealing with agrochemicals. Accordingly, a wide variety of endpoints, such as seedling growth, root elongation, bleaching (reduction in chlorophyll content) and interactions with specific plant components (e.g. chloroplasts) have been investigated.

QSAR models describing phytotoxicity (Table 5.12) are limited to a few chemical classes, mostly those with relatively well-understood modes of action such as herbicidal compounds.

Non-specific toxicity requires descriptors of lipophilicity as the only regressor, although the relatively strong correlations of phytotoxicity to log $P_{ow}$ are restricted to homologous series of chemicals and weaken considerably when applied, for example, to miscellaneous aromatic compounds simultaneously (Hulzebos *et al.*, 1991). If additional interactions (e.g. with the Hill reaction) occur, lipophilicity is used for the transport parameter and electronic and/or steric descriptors are used to model the specific interaction with the photosystem-II electron transport chain (Takemoto *et al.*, 1985). The respective models describe changes in quantitative effects among strictly homologous series of chemicals using substituent constants and hence cannot be extrapolated to other compounds. Because they are intended for use as tools in the design of agrochemicals, the potential of these models for predicting the phytotoxicity of environmental chemicals is limited. Clearly, the effects of specifically phytotoxic compounds towards algae have also to be evaluated according to a phytotoxicity-like model and not to a presumably non-specific aquatic model.

## 5.6 TOXICITY TO MAMMALS

Small laboratory rodents are the established surrogate test species for terrestrial wildlife. For reasons of expediency, the laboratory rat *Rattus norvegicus* is most commonly used in toxicity tests to record adverse effects on mammalian species. The toxicity is generally described in terms of the $LD_{50}$; that is, the single dose of the toxicant (mg/kg) which is acutely lethal after 24–96 h for 50% of the individuals exposed, regardless of how mortality is evoked. The toxicants are administered orally with food to at least six (three

**Table 5.12** Examples of QSAR models for estimating phytotoxicity: correlations with various parameters

| Model | Species effects | Duration | Units | | $r$ | $n$ | $s$ | Descriptor range |
|---|---|---|---|---|---|---|---|---|
| Phenols | | | | | | | | |
| 1 | *Lemna minor* chlorosis | | mol/l | $\log 1/C_{cor} = 1.077\,\pi + 1.531$ | 0.972 | 25 | 0.275 | 0–4.4 |
| Phenylureas | | | | | | | | |
| 2 | Isolated chloroplasts[a] Hill reaction | | mol/l | $pI_{50} = -0.168\,\pi^2 + 1.451\,\pi\; -0.654\,\sigma + 4.491$ | 0.931 | 38 | 0.448 | $\pi$:0–5 $\sigma$:–0.3–0.8 |
| 3 | *Raphanus sativus* chlorophyll concentration | 7 d | mol/l | $\log 1/EC_{50} = 0.523\pi + 0.679\sigma - 0.446\,E_S + 3.186$ | 0.958 | 18 | 0.183 | $\pi$:0–2.6 $\sigma$:0–0.7 $E_S$:–2.4–0 |
| 4 | *Brassica kaber* survival | 14 d | mol/l | $\log 1/EC_{85} = 0.42\,\pi - 5.4\,E_R + 1.3$ | 0.94 | 12 | 0.30 | $\pi$:–1.6–1.3 $E_R$:0–0.4 |
| Triazines | | | | | | | | |
| 5 | Isolated chloroplasts[b] Hill reaction | | mol/l | $pI_{50} = 1.75 \log P_{ow} -0.19\,(\log P_{ow})^2\; + 3.04$ | 0.80 | 28 | 0.38 | 1.4–4.8 |
| Phenoxyacrylates | | | | | | | | |
| 6 | *Echinochlor crus-galli* shoot elongation | | mol/l | $pI_{50} = -0.728 \log P_{ow} + 9.661$ | 0.951 | 21 | 0.308 | 3.5–8.3 |
| 7 | *Oryzae sativa* shoot elongation | | mol/l | $pI_{50} = -0.104\,(\log P_{ow})^2 - 1.865\,\sigma^*\; + 0.80\,E_S + 7.23$ | 0.916 | 17 | 0.508 | $E_S$:–2.5–0.3 $\sigma^*$:–0.3–0.6 $\log P_{ow}$:3.5–6.8 |
| Benzenes, phenols, anilines | | | | | | | | |
| 8 | *Lactuca sativa* growth | 16/21 d | μmol/l | $\log EC_{50} = -0.71 \log P_{ow} + 3.64$ | 0.76 | 54 | – | >1–5.2 |

$r$ = regression coefficient; $n$ = number of compounds analysed; $s$ = standard deviation of the residuals; $C_{cor}$ = equieffective concentration causing chlorosis; $I_{50}$ = molar concentration causing 50% inhibition; $E_R$ = free radical parameter; $\sigma^*$ = Taft $\sigma$ substituent constants; $E_S$= Taft steric substituent constant.
[a] *Raphanus sativus*; [b] spinach.
Sources of models: 1, Fujita (1966); 2, 3, Takemoto *et al.* (1985); 4, Cross *et al.* (1983); 5, Mitsutake *et al.* (1986); 6, 7, Fujinami, Mine and Fujita (1974); 8, Hulzebos *et al.* (1991).

male/three female) animals per dose level. In spite of uncertainties over this exposure route because of varying bioavailability, contaminant levels in the animals (e.g. blood concentrations) are rarely assessed. Comparison of $LD_{50}$ data resulting from different routes of exposure (e.g. i.p. or p.o.) revealed that at least 20% variability in toxicant ranking may be attributed to pharmacokinetic factors (Hodson, 1985); oral $LD_{50}$ values are generally higher owing to apparent losses on this route of exposure. The replicated experimental data for an individual chemical may vary by more than a factor of 10. A variety of factors may influence the absolute values of observed toxic doses in experimental animals:

Life stage of the species: the toxicity of xenobiotics towards organisms varies with age relative to lifespan and their sex, size, lipid content, etc.

Duration of the test: sensitivity to chemicals may increase or decrease with changes in the test period due to alterations in the catabolic status and in the defence mechanisms of the animals.

Feeding regime: sensitivity to chemicals may increase or decrease with the feeding status of an animal and food in the intestines may alter the uptake of toxicants from the gut into the blood circulation.

Exposure regime: the mixing of the toxicants with food may alter the bioavailability of the toxicants.

Stability of the test compound: the substance must not degrade until the complete dose has been consumed by the animal.

Purity of the test compound: pure substances have to be used because even small amounts of impurities can cause large errors in the measured toxicity.

Published data on rat toxicity are highly variable because of the diverse test protocols applied and hence are of limited use for the derivation and validation of sound QSAR models. Validated reference data sets are not available for the laboratory rat (unlike fish) and if data are retrieved from the literature, therefore, care should be taken that they have been conducted according to uniform standard protocols and evaluated by consistent criteria. The important test parameters should be identical or comparable throughout the data set. If, however, they vary they should be critically evaluated for their effect on the test results, as such differences may greatly affect the relevance of the QSAR models derived.

The main problem with applying QSAR techniques to data on acute mammalian toxicity is the generally composite nature of the observed effects of the contaminants; the chemicals may affect various sites in different ways and be influenced by metabolic activation and deactivation. These concurrent processes may differ, even among series of apparently homologous compounds, not only in their relative contributions to the overall effect, but also in their qualitative nature. When this complex situation is rated by a single parameter, such as an $LD_{50}$, this measure will not represent a single well-defined interaction or endpoint as is a prerequisite for sound QSARs.

Accordingly, QSAR modelling of $LD_{50}$ data has been described as notoriously difficult (Adamson, Bawden and Saggers, 1984). Nonetheless, despite some controversy, summary parameters such as the $LD_{50}$ are required by legislative authorities as measures of acute toxicity.

In the rat, the observed effects of oral $LD_{50}$ reflect both the intrinsic toxicity at the ultimate biophase site of action and the factors influencing distribution, membrane transport, protein binding, metabolism and excretion. The manifestation of acute mammalian toxicity is hence a much more complex response than can be described with the use of log $P_{ow}$ alone, although individual processes such as bioavailability and adsorption depend on lipophilicity. The fit to a common QSAR model requires that each of these processes has similar structural dependences, qualitatively and quantitatively, within a given class of compounds. If a different process becomes predominant (i.e. rate limiting), the structure–toxicity relationship must alter; thus the compound will be an outlier even when the principal mechanism of intrinsic toxicity remains the same.

The importance of pharmacokinetics in determining the QSAR of rat oral $LD_{50}$, as opposed to fish $LC_{50}$, constitutes the fundamental difference between these two endpoints. In the fish $LC_{50}$ experiment, the pharmacokinetic factors hardly affect the toxic aquatic concentrations if the duration of the experimental test is sufficient to ensure (pseudo) steady-state partitioning. If such equilibrium conditions are not achieved between the toxicant donor phase and the biophase site of action, as with oral administration, linear QSARs will fail and the effects can instead be described by non-linear models with a maximum response for compounds of optimum pharmacokinetic behaviour (section 3.2.1(c)). These relationships approximate the dependence between the concentration administered and the concentration at the active site after a given period of time under non-equilibrium conditions. They are assumed to reflect in probabilistic terms the movement of the contaminant from the point of introduction into the system to the active sites. Compounds with optimum partitioning properties with regard to the given biosystem will encounter the least resistance to reaching the target and will hence reveal the highest toxicity. The predictive capacity of non-linear pharmacokinetic QSAR models depends on the ability of the respective parameter, generally log $P_{ow}$, to model the rate of transport to the site of action for chemicals with varying hydrogen-bonding capacity, dielectric constant and molecular diameter. The efficacy of certain hypnotic agents that is sometimes ascribed to specific receptor binding may instead reflect increased transport rates (Lipnick, 1989c). With regard to specific toxicants, the non-linear log $P_{ow}$-dependent models of rat oral toxicity evidently represent baseline QSARs, which may underestimate the effects of reactive compounds by orders of magnitude.

For non-reactive aliphatic compounds, such as ethers, saturated alcohols or ketones, log $P_{ow}$-dependent baseline models have been derived from lethality data for rat or mouse (Table 5.13, Figure 5.8).

**Table 5.13** Examples of QSAR models for estimating acute mammalian toxicity after oral administration of toxicants: log $LD_{50}$ correlations with various parameters

| Model | Species | Units | | $r$ | $n$ | $s$ | Descriptor range |
|---|---|---|---|---|---|---|---|
| Ethers | | | | | | | |
| 1 | Mouse | | $\log 1/LD_{50} = 1.12 \log P_{ow} - 0.29 (\log P_{ow})^2 + 1.85$ | 0.86 | 25 | 0.21 | $-0.2$–$2.4$ |
| *n*-alkyl-methyl-ketones | | | | | | | |
| 2 | Mouse | mmol/kg | $\log 1/LD_{50} = 0.417 \log P_{ow} - 0.12 (\log P_{ow})^2 - 1.734$ | 0.946 | 13 | 0.09 | $-0.5$–$4.1$ |
| Saturated monohydric alcohols and saturated monoketones | | | | | | | |
| 3 | Rat | mol/kg | $\log 1/LD_{50} = 0.805 \log P_{ow} + 0.984$ $- 0.971 \log (0.0807 \times 10^{pow} + 1)$ | 0.824 | 54 | 0.21 | $-1$–$8$ |
| Aliphatic amines | | | | | | | |
| 4 | Rat | mol/kg | $\log 1/LD_{50} = 1.02 \log P_{ow} - 0.85 \log (0.199 P_{ow} + 1)$ $- 2.1 (nN_{\%} \log P_{ow}) + 2.06$ | 0.92 | 24 | 0.13 | $\log P_{ow}$:$-1.5$–$4.4$ $nN_{\%}$:$^1/_{12}$–$^1/_2$ |
| Anilines | | | | | | | |
| 5 | Rat | mol/kg | $\log 1/LD_{50} = 1.061 \log P_{ow} - 0.21 (\log P_{ow})^2$ $- 0.305 \, LUMO - 0.038 \, V + 2.46$ | 0.90 | 29 | 0.11 | $\log P_{ow}$: $1.3$–$2.8$ LUMO:$-1.6$–$0.2$ $V$: $33$–$49$ |
| Phenylureas | | | | | | | |
| 6 | Rat | mmol/kg | $\log 1/LD_{50} = 1.39 \log P_{ow} - 0.25(\log P_{ow})^2 + 0.34 \, IP - 5.75$ | 0.81 | 12 | 0.29 | $\log P_{ow}$:$0.9$–$4$ IP:$9.1$–$10.1$ |
| Unsaturated monohydric alcohols | | | | | | | |
| 7 | Rat | mol/kg | $\log 1/LD_{50} = 0.045 (\log P_{ow})^2 + 2.361$ $- 18.7 (\log P_{ow}/MR)$ | 0.92 | 21 | 0.23 | $\log P_{ow}$:$-1.1$–$3.2$ MR: $12$–$51$ |

$r$ = regression coefficient; $n$ = number of compounds analysed; $s$ = standard deviation of the residuals; $nN_{\%}$ = number of N-atoms relative to the total number of atoms in the molecule except H atoms; LUMO = energy of the lowest unoccupied molecular orbital (eV); $V$ = molar volume (cm³/mol); IP = first ionization potential (eV); MR = molar refractivity.
Sources of models: 1, Hansch and Clayton (1973); 2, Tanii, Tsuji and Hashimoto (1986); 3, Lipnick, Pritzker and Bentley (1987); 4, 5, Jäckel and Klein (1991); 6, Nendza (1991a); 7, IUCT (1992).

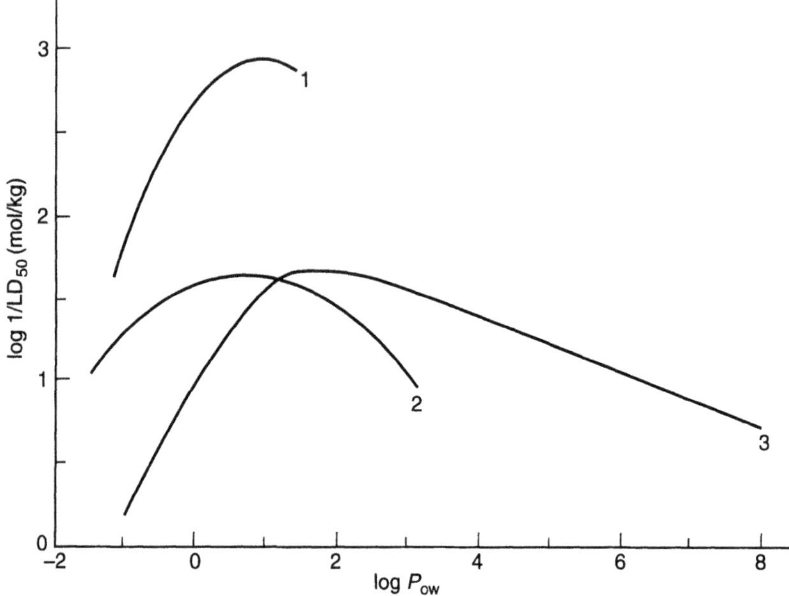


**Figure 5.8**  Comparison of selected QSAR models relating mammalian toxicity (log $1/LD_{50}$ in mol/kg) to log $P_{OW}$; for identification of the QSAR functions see Table 5.13.

Similar bilinear relationships were also obtained with data on toxicity towards mice after intravenous administration (Kubinyi, 1977), but could not be validated due to the lack of further experimental data. For aliphatic or aromatic amines, for example, the respective functions are supplemented by electronic or polarity descriptors or indicator variables. Toxicity predictions for heterogeneous sets of compounds based primarily on substructures (Enslein and Craig, 1978; HDI, 1990) did not yet prove successful, which may be due to different effects of the defined fragments on the various processes involved in lethality. Additionally, the usual constraint associated with the Free–Wilson technique applies: predictions are restricted to the chemical substructures in the training set of data. Even though mammalian toxicity is a multistep process, care has to be taken not to use too many descriptors either for the various interactions or for merely statistical reasons for example, fitting 10 aniline data with seven parameters (Dura *et al.*, 1985). Another frequent mistake in mammalian QSAR studies is the use of the activity data on a mg/kg basis instead of molar units, which must yield chance correlations at best, if chemicals of largely differing molecular weight are compared on the basis of molar descriptors.

Because mammalian oral $LD_{50}$ is not a well-defined endpoint, there must remain considerable doubt about the eventual success of QSARs for modelling this parameter and of the predictive power of these QSARs. QSAR approaches to mammalian toxicity have shown the greatest applicability for structurally related series of compounds and it is unlikely that general models can be derived to explain the toxicity of substances with significantly different structures and physico-chemical properties. Predictions are not yet feasible at any level of accuracy greater than a rough estimate.


## 5.7 MUTAGENICITY

The mutagenic potential of xenobiotics released into the environment constitutes a substantial hazard to any organism, including humans. Mutation may be an important initial step in the carcinogenic process, and hence the identification of mutagens also identifies chemicals with carcinogenic potential. Not all mutagens are carcinogens, however, nor are all carcinogens mutagens, and mutagens may not be active in all genetic assays (Purchase, 1983; Barrett, 1987). Genotoxicity assays indicate the ability of chemicals to interact with DNA (mutational event), which may alter the nucleotide sequence of genes by nucleotide substitution and frameshift mutation (insertion/deletion of nucleotides).

Primary mutagenic reactions, mostly after metabolic activation of the xenobiotics to electrophilic derivatives, with DNA bases comprise:

alkylation
intercalation
cross-linking
incorporation of base analogues
depurination (cleavage of purine bases from the desoxyribose moiety).


Small adducts change the base-pairing specificity of the damaged base and this can lead directly to a mutation on replication; some compounds exert a mutagenic effect by lowering the ability of DNA polymerases to perform accurate DNA synthesis (Silber and Loeb, 1985). In all instances, however, mutagenesis is dependent on DNA replication, even though it may not always be biologically expressed (silent mutations). Mutations may occur at low doses if the exposure time is sufficiently long. DNA damage does not inevitably yield mutagenesis in dividing cells – enzymatic repair systems allow the cell to overcome the challenge of mutagen-induced damage in its DNA. With regard to the permanency of mutations, the frequency of mutagenesis, which is estimated to occur normally once in $\geq 10^9$ DNA replications (Silber and Loeb, 1985), may increase by several orders of magnitude on exposure to mutagens.

According to the various modes of mutagenesis, a wide variety of organic and inorganic compounds is recognized as mutagenic:

nitrosamines
vinylchloride
2-napthylamine
2-aminobiphenyl
PAHs
metals (e.g. As, Cr(VI), Ni and Sb)
asbestos
sawdust.

Regardless of their heterogeneity, most mutagens are, or can be transformed to, alkylating agents (Scribner, 1985).

The primary classification of chemicals as either mutagenic or non-mutagenic is generally based on a battery of short-term *in vitro* assays representing different genetic endpoints (Ashby *et al.*, 1985), such as point mutations and chromosome aberrations, and incorporating some form of metabolic activation system (Neumann, 1986) to identify the potential mutagens acting by different mechanisms. The resulting genotoxicity profiles are then evaluated. The classifications so derived form the basis for any QSAR approach to mutagenicity, which hence attempts simultaneously to model two processes that may or may not be coincident among series of chemicals: metabolic activation and alkylation of the DNA components. Accordingly, any QSAR approach to mutagenicity is mainly restricted by the problem of determining precisely which activity is accounted for:

mutagenesis *by which chemical species*?
mutagenesis *by which reaction*?

QSAR approaches to mutagenic activity are generally based on data from the Ames test, which, though standardized, reveal considerable variability in experimental results attributable only in part to differences in the added metabolic (S9) preparation and the physiological condition of the plated cells (Grafe, Mattern and Green, 1981). The importance of metabolism (i.e. bioactivation) in determining QSARs for mutagenicity, as opposed to the other toxicity parameters, constitutes the fundamental difference between these endpoints. Because of the miscellaneous transformation reactions involved, no uniform QSAR model appears feasible for all kinds of chemicals. Rather, there may be distinctive local models to describe the differences in mutagenic potential within series of closely related compounds.

The derived QSARs on mutagenicity (Table 5.14) account for the composite nature of the endpoint by using several descriptors for quantifying lipophilicity that relate to the transport to the active site, and electronic or polarity parameters that estimate the compounds' reactivity and liability for bioactivating transformations – or a combination of several topological indices.

Because of the specific selection of descriptors for the respective data sets, extrapolations to other structures are inappropriate. The limited predictive

**Table 5.14** Examples of QSAR models for estimating mutagenicity using the Ames test (*Salmonella typhimurium*) for specific toxicants: correlations with various parameters

| Model | Endpoint, units, path-level | | $r$ | $n$ | $s$ | Descriptor range |
|---|---|---|---|---|---|---|
| | **Mono- to tri-nitro polyaromatics** | | | | | |
| 1 | Revertants/nmol <br> > 1: mutagen <br> < 1: non-mutagen | $\log A_{98} = 1.49 \log P_{ow} - 2.83\ \mathrm{LUMO} - 8.63$ | 0.926 | 20 | 0.74 | $\log P_{ow}$: 2.6–5.4 <br> LUMO: −2.9–1.2 |
| | **N-nitroso compounds** | | | | | |
| 2 | Revertants/nmol | $\log A = 2.398\ {}^{0}\chi - 4.095\ {}^{1}\chi^{v} - 5.59$ | 0.964 | 15 | 1.09 | ${}^{0}\chi$:5.1–11.7 <br> ${}^{1}\chi^{v}$:1.6–4.7 |
| | **N-nitroso compounds** | | | | | |
| 3 | Revertants/nmol | $\ln R = -15.3\ (\mathrm{IC_1}) + 3.84\ (\mathrm{IC_1})^2 + 12.0$ | 0.98 | 15 | 0.86 | $(\mathrm{IC_1})$:1.8–3.6 |
| | **Halogenated alkanes and olefins** | | | | | |
| 4 | > 0: mutagen <br> < 0: non-mutagen | $D = 7.865\ \mathrm{MR} - 1.260\ (\mathrm{MR} \log P_{ow}) - 8.503\ {}^{3}\chi^{v} - 128.6$ | – | 51 | – | $\log P_{ow}$: −0.2–4.7 <br> MR: 17–51 <br> ${}^{3}\chi^{v}$:−0.6–4.8 |
| | **Triazines** | | | | | |
| 5 | 30 mutations above background in $10^8$ bacteria (mol/l) | $\log 1/C = 1.04 \log P_{ow} - 1.63\ \sigma^{+} + 3.06$ | 0.974 | 17 | 0.315 | – |

$r$ = regression coefficient; $n$ = number of compounds analysed; $s$ = standard deviation of the residuals; LUMO = energy of the lowest unoccupied molecular orbital (eV); ${}^{0}\chi$ = zero-order connectivity index; ${}^{1}\chi^{v}$ = first-order valence-corrected connectivity index; $\mathrm{IC_1}$ = topological information content; MR = molar refractivity; ${}^{3}\chi^{v}$ third-order valence-corrected connectivity index.

Sources of models: 1, Compadre, Shustermann and Hansch (1988); 2, Kier, Simons and Hall (1978); 3, Basak *et al.* (1986); 4, IUCT (1992); 5, Hansch (1991).

power of a large number of published QSARs on mutagenicity was revealed in validity assessments (ECETOC, 1986; IUCT, 1992), which disclosed various methodological shortcomings in the analyses, such as inconsistent experimental data from various sources or highly intercorrelated descriptors (as, for example, in Koch, Strobel and Nagel, 1985; Flora *et al.*, 1985). Discriminant models based on substructure identification mostly revealed high specificity for the structures underlying the model and the presumably discriminating fragments occurred with mutagens as well as non-mutagens when analysed with further compounds (as, for example, in Klopman, Frierson and Rosenkranz, 1990). Thus, the application of the models to additional chemicals of known mutagenicity often resulted in insufficient identification of the actual mutagens that were not in the training set. These findings emphasize the limited predictive capabilities of the available models on mutagenesis. However, these models may provide valuable information concerning the structural requirements of mutagens and the interpretation of the mode of action within homologous series of compounds.

There are currently no validated predictive models for assessing the carcinogenicity of chemicals directly from their structure. Carcinogenic properties have been related to hydrophobic and electronic factors: The carcinogens tend to be more hydrophobic and more electrophilic than the non-carcinogens. These average differences, however, are not sufficient for an effective discrimination between negatives and positives (Benigni *et al.*, 1995). Neither QSARs based on physico-chemical parameters (e.g. Singer, Taylor and Lijinsky, 1977; Wishnok *et al.*, 1978) nor those based on substructure/connectivity/geometry (e.g. Jurs, Chon and Yuan, 1979; Yuan and Jurs, 1980; Klopman, 1984) could be successfully applied to compounds that were not contained in the training sets of the models (IUCT, 1992).

# Validation status of QSAR models for exposure- and effects-related parameters

The fact that QSARs are local models (unless the respective endpoint can be traced back to a single uniform rate-limiting process) inevitably causes varying degrees of validity and reliability for their predictive applications. Depending on the chemical class and the endpoint concerned, the estimates may range between satisfactory to very rough approximations. Corresponding to the variability in the underlying experimental data, the estimates may be exact by ± 10% or may scatter by a factor of 10 or more. Furthermore, the accuracy of the QSAR predictions depends on the selection of the appropriate model on a case-by-case basis. A traditional illusion is that physico-chemical measurements are more exact than biological tests, so it is not surprising that severe problems remain for predicting such apparently simple properties as water solubility or melting points, whilst complex endpoints such as aquatic toxicity can be calculated with a rather high degree of confidence. The apparent success of QSARs for selected endpoints strongly depends on the (experimental) efforts devoted to the respective parameters, as also influenced by the demands of the regulating authorities for certain summarizing parameters.

The exposure-related parameters of contaminants may be estimated with varying degrees of reliability (Table 6.1).

Besides the $pK_a$ values, which can be calculated for aromatic compounds according to the Hammett equation, the log $P_{ow}$ estimates are in general reasonably accurate. However, the application of the currently available fragment/factor methods for log $P_{ow}$ calculations is precluded for certain classes of compounds, such as dissociated compounds, charged compounds, surfactants, chelating compounds, organometallics, organophosphorus compounds, compounds with unknown fragment values and mixtures of unknown composition (including impurities). If the restrictions on the methods are considered, estimated log $P_{ow}$ values between 0 and 5 can be generally regarded as reliable, which is the indispensable prerequisite for the application of further QSAR models based on log $P_{ow}$ as the input parameter. However, substantial limitations apply to the predictive tools for all other compound-specific

**Table 6.1** Methods and validation status for estimating exposure-related parameters

| Parameter | Method | Status[a] |
|---|---|---|
| $\log P_{ow}$ | Fragment/factor method | ++ |
| $pK_a$ | Aromatic compounds: Hammett | ++ |
| | Aliphatic compounds: Taft | + |
| $S_w$ | Liquid compounds: $\log P_{ow}$ | + |
| | Solid compounds: $\log P_{ow}$, $T_m$ | – |
| $T_m$ | | –– |
| $T_b$ | Meissner's method | – |
| $p_v$ | Liquid, gaseous compounds: $T_b$, $K_f$ | + |
| | Solid compounds: $T_b$, $T_m$, $K_f$ | + |
| | (Halogenated) hydrocarbons: $T_b$, $T_m$ | + |
| $H$ | $p_v/S_w$ (compounds of low $S_w$) | + |
| | Fragment method | + |
| $K_{oc}$ | Non-polar compounds: $\log P_{ow}$ | + |
| | Polar compounds: diverse descriptors | –– |
| $K_{(hyd)}$ | Only few compound classes: Hammett, Taft | ++ |
| $k_{(photo)}$ | AOP[b] | + |
| Biodegradation | Only some compound classes: | |
| | substructures diverse descriptors | + |
| BCF | $\log P_{ow}$ | + |

++ = valid procedure; + = procedure in general valid, but with limitations; – = procedure of limited applicability; –– = procedure not recommended.
AOP = Atmospheric Oxidation Program.
Sources: IUCT (1992); Nendza *et al.* (1993);OECD (1993a, 1994).

properties. Although the water solubility of liquid compounds can be estimated with a fair degree of accuracy, the models tend to fail for solids, especially those with low solubility (i.e. < 1 mg/l) due to the complexity of factors influencing water solubility. The procedures for calculating melting points are so far insufficient, and those for boiling points are of only limited applicability, therefore measured values of these properties should be acquired. The methods for estimating vapour pressure are in general valid, but can be applied only with limitations: even then, the predictions tend to underestimate the vapour pressure, which may affect the estimated exposure levels and thus contribute to a misjudgement of the risks associated with a chemical (OECD, 1994). The extrapolation of the Henry's law constants from the water solubility and vapour pressure of a contaminant may be biased by the inaccuracy in the input parameters. The QSARs concerning soil sorption can generally not be validated, due to the inhomogeneity of the experimental material; only for non-polar compounds are they of some efficacy (they cannot be recommended for polar compounds). The abiotic degradation by hydrolysis is relevant only for a few compound classes, and QSARs are not available for all of them. If appropriate models exist, however, their predictive power is adequate. The assessment of photodegradation in the atmosphere is principally a worst-case approach, because only degradable substructures are considered. The method is generally accepted, however, although an extensive external validation has not been feasible because almost all available experimental data were included in the model when it was derived. Biodegradability predictions may be used with some confidence if their result is 'non-readily biodegradable' (i.e. there is a high probability that compounds rated non-degradable are really persistent) but the identification of compounds that readily degrade is rather unreliable. The bioconcentration potential in aquatic species may be estimated with sufficient reliability from the log $P_{ow}$ values, which has to be considered a worst-case approach, because it neglects any factor reducing the accumulation of chemicals in exposed organisms.

Estimations of effects-related parameters are subject to varying degrees of uncertainty (Table 6.2) that depend on the inhomogeneity of the underlying experimental material.

The acute aquatic toxicity of non-reactive chemicals can be predicted from their log $P_{ow}$ with quite a high level of confidence. Problems remain for specific toxicants, for which the appropriate QSAR has to be selected prospectively: while straightforward criteria for the identification of the relevant modes of action are still lacking, merely empirical expert knowledge has to be applied for deciding on a suitable model. If the wrong QSAR is used, the predictions may be wrong by several orders of magnitude. Principally the same restrictions concern (sub)chronic aquatic toxicity, for which the rate of outliers from baseline QSARs is even larger. The available QSARs for phytotoxicity concern only a few selected classes of compounds; for most environmental chemicals no reliable predictions are feasible. The complex nature

**Table 6.2** Methods and validation status for estimating effects-related parameters.

| Parameter | Method | Status |
|---|---|---|
| Fish | Non-specific toxicants: log $P_{ow}$ | ++ |
| | Specific toxicants: log $P_{ow}$, substructures | + |
| *Daphnia* | Non-specific toxicants: log $P_{ow}$ | + |
| | Non-specific toxicants: geometric descriptors | + |
| Algae | Non-specific toxicants: log $P_{ow}$ | + |
| | Only a few classes of specific toxicants: diverse descriptors | + |
| Bacteria | Non-specific toxicants: log $P_{ow}$ | + |
| Protozoa | Non-specific toxicants: log $P_{ow}$ | + |
| | Only a few classes of specific toxicants: diverse descriptors | + |
| Plants | Only a few classes of specific toxicants: diverse descriptors | − |
| Rodents | Only a few classes of specific toxicants: log $P_{ow}$, diverse descriptors | − |
| Mutagenicity | Only a few classes of specific toxicants: diverse descriptors | − |

++ Valid procedure: +, Procedure in general valid, but with limitations; − Procedure of limited applicability.
Sources: IUCT (1992); Nendza *et al.* (1993); OECD (1994).

of the toxicity measures for mammalian species has meant that there are no global QSARs for general application, and the efficacy of the few class-specific models is obscured by the lack of validated QSARs for the major proportion of compounds. Largely the same limitations apply to the modelling of mutagenic potential; for only a few selected compound classes can fair predictions be made.

In conclusion, validity assessments of the currently available QSARs in environmental sciences indicate that there are models for predicting some of the properties of potential contaminants reliably. If operated in accordance with their limitations, they can be valuable tools to fill data gaps. However, a complete coverage of all parameters and endpoints relevant to the hazard and risk assessment of chemicals has not yet been achieved. To improve and extend the techniques for estimating the intrinsic properties of environmental chemicals, experimental as well as theoretical aspects of these parameters need further elaboration.

# PART THREE

# *Application of QSARs in Environmental Hazard and Risk Assessments*

The considerable lack of experimental data for the fate and effects assessment of environmental contaminants requires the provisional use of substitutes. For this purpose, QSAR estimates are superior to the alternative of using uniform default values, which do not allow differentiation and ranking of the potential hazards of diverse chemicals. QSAR data-quality requirements for assessing environmental hazards and risks have to be evaluated relative to the quality of the available experimental data and the relevance of the accuracy of the individual data for subsequent hazard and risk classifications. Within these assessment schemes, several factors are usually ignored for the sake of scientific precision and standardization. The simplifications of the real world apply, regardless of whether experimental or calculated input data are used:

Environmental contaminations are unlikely to be single chemicals, but rather mixtures of compounds are present concurrently; however, evaluations are conducted for each chemical individually and mutual interactions are ignored.

Assessments of environmental effects rely on extrapolating no-effect levels from a multitude of standard endpoints for single species, mostly lethality. The toxicity threshold obtained is strongly dependent on the presumed abundance of target and non-target species and the availability of the respective data sets, as well as on the selected exposure scenario (i.e. ecosystem sensitivity). The variance in determination of the toxicity thresholds depends on the sensitivity and variance of the parameters (i.e. if the most sensitive species and effects actually have been identified and the exactness of the measurement or estimation of them). The ambiguity of many endpoints may introduce further uncertainty.

In assessments of exposure, the environmental (e.g. climatic and hydrogeological) conditions and the release routes and rates play a major role besides the contaminants' physico-chemical properties.

Evaluations of the risk of fatal impacts on environmental systems have to take into account that the effects may often be caused by impurities and not by the declared chemical.

Environmental contaminants are often encountered as mixtures and the behaviour of a chemical in a mixture may not correspond to that predicted from data on the pure compound. Interactions of components in a mixture can cause complex and substantial changes in the apparent properties of its constituents. The cosolutes in a mixture may induce either increased (synergistic) or decreased (antagonistic) effects as compared to the ideal behaviour (independently additive effects). No satisfactory approaches are yet available to account systematically for mutual interactions with regard to different properties and endpoints.

The classification of responses of chemical combinations as synergistic, antagonistic or additive has been essentially based on two postulates of combined drug action (Garrett, 1971): (a) additivity, where the combined response is additive with respect to the separate responses of the components, and (b) equivalence, in which the components act in the same manner with the same (parallel) dose–response curves, separately or in combination, except for a difference in the weight of the effective concentration. Additivity is expected on the basis that the components of the mixture act independently and do not affect each other's mode, degree or efficacy of action. Equivalence is expected on the basis that different amounts of the same or equivalent compounds or potency factors are combined. The use of this criterion for classifications of synergism or antagonism takes account of the fact that most chemicals have non-linear dose–response relationships, which otherwise might be classified as synergistic or antagonistic with themselves. Experimental methods for mixture assessment have been established in pharmacological research since the beginning of the twentieth century (e.g. Frei, 1913; Loewe and Muischnek, 1926), mostly relying on the graphic evaluation of equipotent combinations (Figure 7.1).

With regard to toxicity, interactions with co-contaminants concern not only the actual toxicant/target interaction, but may also influence the adsorption, distribution and excretion (i.e. kinetics), the biotransformation (i.e. metabolism) or the bioavailability (Tichy, Cikrt and Roth, 1994). The primary impacts may be due to alterations in transport and distribution, changes in membrane properties, effects on metabolizing enzymes or competition for conjugation reactions. The resulting activities of a chemical mixture may be

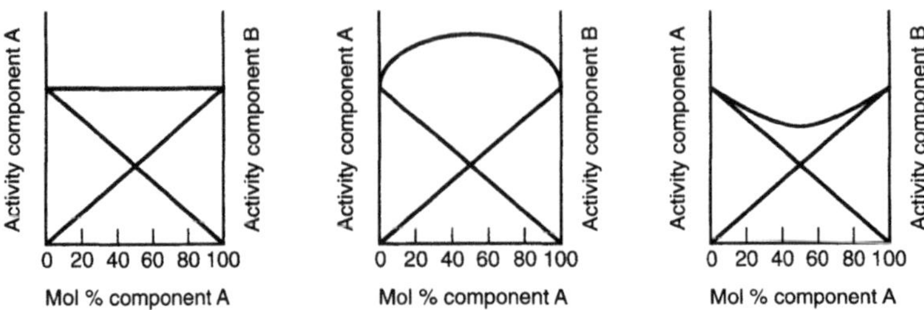

**Figure 7.1** Joint action of equipotent mixtures of components A and B: the diagonal lines represent the activity of the individual components A and B at concentrations corresponding to the respective mol %, the horizontal curves sum the joint action in the mixture with varying fractions of A and B in the case of additive (left), synergistic (middle) or antagonistic (right) effects.

additive; more often they are likely to be not. Additive effects occur only if the principal activity (e.g. acute toxicity) is not changed if one compound is partially replaced by an equipotent amount of another.

Assuming that non-specific effects are the minimum baseline toxicity of any chemical, all components of a mixture are equivalent with regard to this one mode of action. As a consequence, the fractional minimal acute toxicity from hydrophobicity of the individual components must be concentration additive, and any mixture must hence be at least as toxic as corresponds to the cumulative sum of the fractional baseline toxic concentrations of the components (Könemann, 1981a; Hermens, Leeuwangh and Musch, 1984; Hermens *et al.*, 1984; Hermens *et al.*, 1985a–c; Broderius and Kahl, 1985). In a mixture of chemicals all acting by different mechanisms, the specific effects may not be exposed due to the very low concentrations of the individual components, but their contributions to the baseline toxicity will persist and may markedly add to the overall observed response. The synergism observed between specific toxicants at higher concentration levels may be suspected to be not yet triggered, and the additivity of the same components observed at much lower concentrations may be attributed to the non-specific joint toxic action (Broderius and Kahl, 1985; Verhaar, Busser and Hermens, 1995).

Mixtures prepared with equitoxic concentrations (e.g. equal fractions of the $LC_{50}$ of each component) are concentration additive in their acute as well as their sublethal effects (Könemann, 1981a; Hermens, Leeuwangh and Musch, 1984; Hermens *et al.*, 1984; Broderius and Kahl, 1985; Hermens *et al.*, 1985a–c; Deneer *et al.*, 1988; Wolf *et al.*, 1988; Nendza and Seydel, 1990). For example, the concentrations of the individual chemicals in a mixture of 50 compounds, producing 50% mortality, were 0.02 times their $LC_{50}$.

Accordingly, 1000 chemicals at $1/1000$ $LC_{50}$ concentration each will together kill 50% of the exposed organisms. The toxicity of a mixture can thus be expressed as the sum of the toxic units (TU: dimensionless ratio of the exposure and the effective concentrations) contributed by each component (McCarty *et al.*, 1992). For QSAR modelling, these findings imply that the baseline toxicity equations can be extended for mixture assessments:

$$\log 1/LC_{50(mix)} = a \log (\Sigma (x_i (P_{ow})_i)) + b$$

This general expression assumes that all mixture components (i) shall contribute to the predictive descriptor value, and hence to the overall activity of the mixture, according to their molar fraction $(x_i)$ in the mixture. The application of this model is limited, though, to mixtures with precisely known composition. Only then is it possible to estimate a theoretical $\log P_{ow}$ value for the mixture based on the lipophilicity of the individual compounds and their relative abundance in the sample: $P_{ow}$ can be calculated for each component, each $P_{ow}$ value is multiplied by the mole fraction of the corresponding component, the products are summed and the logarithm is taken to give the weighted average $\log P_{ow}$ of the mixture (Roberts, 1991). For most environmental samples (e.g. effluents or chemical preparations including byproducts) the exact composition of the mixture is not known, and then there is no way of computing QSAR parameter values, such as a hypothetical mixture $\log P_{ow}$. Because of the trivial fact that structural descriptors cannot be calculated when the individual structures are unknown, the assessment of mixtures of undefined composition has to rely on some experimental determinations. Neither complete chemical analyses, which are practically impossible, nor extensive ecotoxicity testing, which is very costly, can be the methods of choice for any screening-level assessments. The use of HPLC-MS techniques for the detection of lipophilic contaminants and the estimation of their $\log P_{ow}$ values in complex mixtures (Burkhard, Kuehl and Veith, 1985) may be biased – for example, by the efficacy of the extraction method applied to the samples (Hendriks *et al.*, 1994). As an alternative, a surrogate parameter for the total lipophilicity content of a mixture has been proposed, which avoids the need for identifying and quantifying individual compounds (Hermens *et al.*, 1994; Verbruggen, Loon and Hermens, 1994; Verhaar, Busser and Hermens, 1995). The procedure is based on a separation of the mixture on an rp-HPLC C18 column into fractions of increasing hydrophobicity, determination of the total molar concentration in each fraction using, for example, vapour pressure osmometry and consecutive extrapolation of a mean $\log P_{ow}$ from the retention time and the molar concentration for each fraction. Using a different partitioning system, this approach was applied to assess the total baseline toxicity content of water samples; thus this can be a surrogate parameter for the detection of (sub)acute levels of contaminations in mixtures of unknown composition (Verhaar, Busser and Hermens, 1995).

The ultimate application of mixture QSARs concerns the evaluation of environmental samples. In the field, contaminants always occur as mixtures, generally at very low levels of the individual components, and an analysis of single chemicals is likely to be misleading. The principal difficulties in dealing with mixtures, however, limit the quantitative application of QSARs in environmental field research. The restrictions mostly relate to the severe deficiencies in the characterization of the physico-chemical properties, which are the basic input parameters for any QSAR. The current approaches refer to selected individual contaminants, such as PAHs, benzene derivatives or phthalates (Matthiessen *et al.*, 1993; Boxall *et al.*, 1994), and allow at best a qualitative analysis of major contaminants or groups of contaminants. For exploratory data analyses and the interpretation of environmental field observations, however, descriptive QSARs are extremely useful for deriving a basic understanding of relevant interactions and molecular mechanisms. On this basis, QSARs can provide a framework for designing and interpreting monitoring programmes to link biological effects *in situ* with chemical analyses.

Common to the currently available practical approaches for the hazard assessment of mixtures in the environment is the principal assumption of additive joint activity. Although evidently more likely than evaluating multi-chemical exposures for each contaminant separately (Logan and Wilson, 1995), and generally supported by empirical observations for a long time (Brown, 1968; Sprague, 1970), the additivity model is merely a working

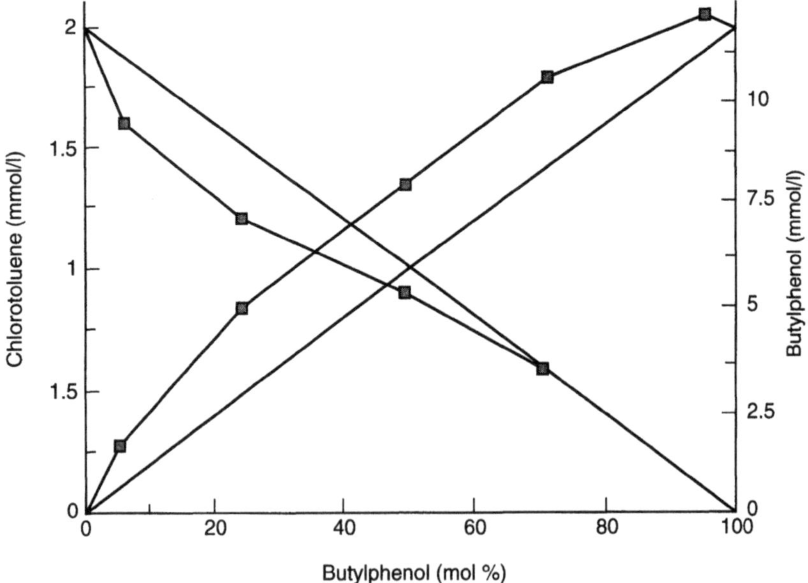


**Figure 7.2** Mutual interactions of cosolutes on the components' water solubility: comparison of experimental data with ideal behaviour according to Raoult's law (Fredenslund, Nendza and Herbst, 1995).

concept; it does not necessarily reflect the reality. Its invalidity for effects such as irritation, corrosion, sensitization, mutagenesis, teratogenesis and carcinogenesis is obvious due to the particular mechanisms of action as compared to direct (sub)acute toxicity (Martens *et al.*, 1984). But even for such apparently simple properties as water solubility, deviations from the additive behaviour of ideal mixtures as described by Raoult's law occur in binary mixtures of rather simple compounds, such as substituted benzenes (Fredenslund, Nendza and Herbst, 1995). The solubility of the same compound may be either enhanced or decreased depending on the respective cosolute (Figure 7.2).

Mutual interactions are hence likely to influence the apparent properties of chemicals substantially, and assessments based on data from pure compounds, either from experimental or QSAR sources, may not realistically reflect the hazard posed by environmental contaminants. However, because no systematic approaches are yet available to account quantitatively for interferences caused by multichemical exposures, the additivity concept remains a practical option for the time being.

# 8

# *Interspecies correlations*

For chemicals acting by a uniform mode in different organisms (e.g. non-specific toxicants), interspecies extrapolations constitute a further means of obtaining toxicity data not available from experiments. Several statistical tools – the same as used to derive QSARs – may be used to establish taxonomic correlations. Regression analysis provides a relation on a one-by-one species basis (e.g. Holcombe, Phipps and Fiandt, 1983; Suter, Vaughan and Gardner, 1983; Janardan, Olson and Schaeffer, 1984; LeBlanc, 1984; Thurston *et al.*, 1985; Suter and Rosen, 1986; Yoshioka, Ose and Sato, 1986; Blum and Speece, 1991). For a two-species comparison this can be displayed by a regression function, for $n$ species by a correlation matrix. Correlations of acute effects data between related aquatic or terrestrial species may predict toxicity within one order of magnitude: Suter, Vaughan and Gardner (1983) calculated the $r^2$ coefficient of determination for intercorrelations of aquatic toxicity data for congeneric species $r^2 = 0.90$, for genera $r^2 = 0.89$, for families $r^2 = 0.82$ and for orders $r^2 = 0.74$.

The identical relative ranking of chemicals in different test systems depends on the similarity of the targets attacked in the different organisms. The toxicity for a variety of species will be collinear as long as the individual compounds have the same mode of action in each of the different species (i.e. the toxicological profiles are homologous). Therefore the extrapolation of toxic concentrations by interspecies correlations is restricted to those toxicants with a uniform mode of action with regard to the different species. A comprehensive presentation of intercorrelations between the toxicity data for several species is accessible using multivariate statistical procedures, such as PCA and FA (section 3.2.2). The advantage of these methods over simple regression analysis is that they take into account also the inherent multiple correlations. Furthermore, the results can be presented in informative plots. The vectors represent the endpoints investigated, and the angles between the vectors are related to the correlation coefficients of the corresponding test systems. Small angles between vectors indicate the collinearity of the test systems. The relative ranking in the toxicity of, for example, selected substituted aromatics is essentially identical in different aquatic test systems using fish, *Daphnia*, algae, protozoa or bacteria (Figure 8.1).

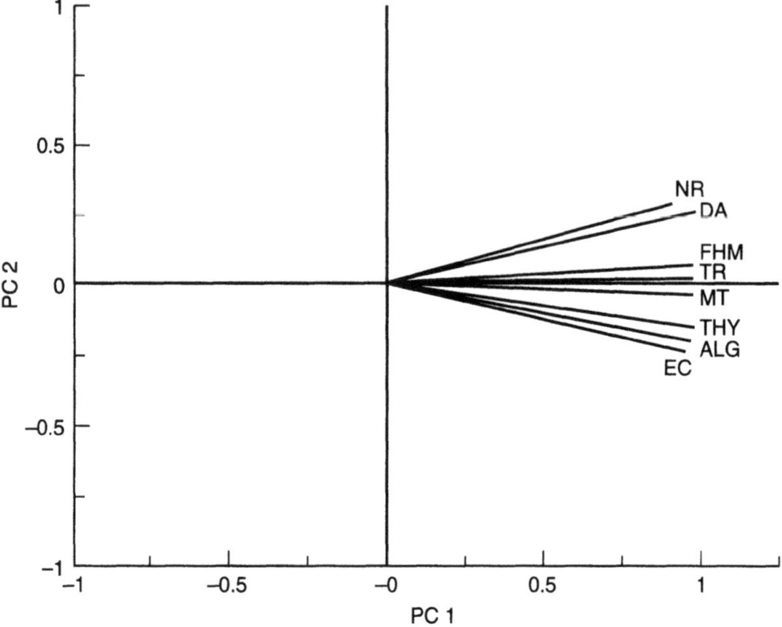



**Figure 8.1** Principal component analysis of toxicity data obtained using a variety of aquatic species and *in vitro* assays. The effects by selected toxicants are ranked collinear by fish (FHM = fathead minnow; TR = trout); crustaceans (DA = *Daphnia*); protozoa (THY = *Tetrahymena*); algae (ALG = *Selenastrum capricornutum*); bacteria (EC = *Escherichia coli*; MT = *Photobacterium phosphoreum*) and cell cultures (NR = neutralred assay).

The interactions with similar sites in the different species may be determined by the same principal processes and thus by the same properties of the xenobiotics; that is, analogous QSARs (e.g. Hansch and Dunn, 1972; Ribo and Kaiser, 1983; Slooff, Canton and Hermens, 1983; Lipnick 1985c; Moulton and Schultz, 1986; Nendza and Seydel, 1988a,b; Nendza and Klein, 1990). In collinear test systems, the parallelism in response, including the pattern of outliers from lipophilicity-dependent baseline toxicity, can be recognized either from the comparison of the QSAR scatter plots for identical test compound sets on each of the endpoints or, even more conclusively, from the relationship between $\log P_{ow}$ and the principal component scores for the chemicals extracted from the multivariate data set on toxicity towards several species (Figure 8.2).

Multivariate statistics are extremely useful for assessing collinearity among the effects on various species and endpoints (e.g. Dunn *et al.*, 1984; Coats *et al.*, 1985; Schaper and Seydel, 1985; Szydlo *et al.*, 1985; Buckley *et al.*, 1987; McKim, Bradbury and Niemi, 1987; Nendza and Seydel, 1988a; Bradbury, Henry and Carlson, 1990; Nendza and Klein, 1990; Breukelen and Brock, 1993; Nendza and Wenzel, 1993). The identical ranking of chemicals in a multitude of biotests constitutes the basis for interspecies extrapolations.

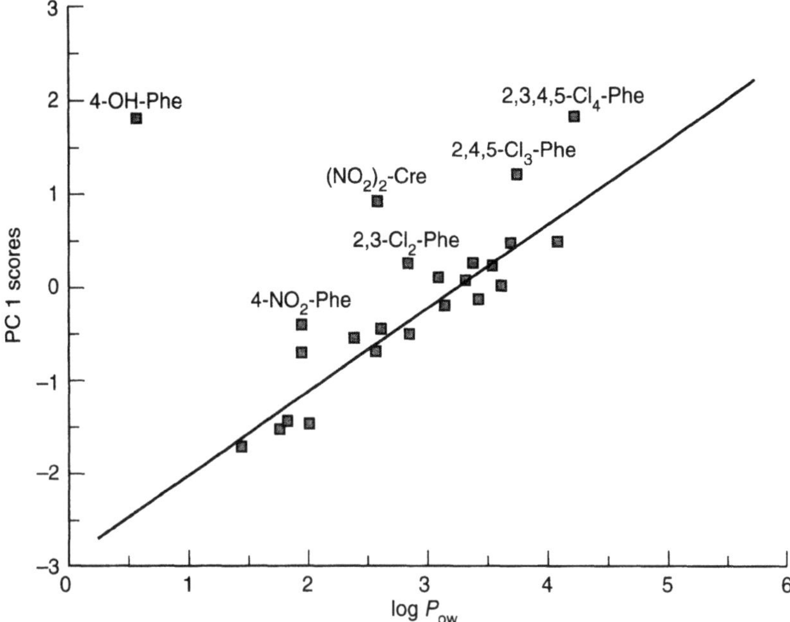



**Figure 8.2**  Relationship between the principal component scores for selected toxicants and log $P_{OW}$. The non-specific toxicants are identically ranked according to their lipophilicity; the excess toxicity of specifically acting compounds is equally detected by the different tests (Figure 8.1).

The absolute effects concentrations may, however, differ considerably due to differences in the sensitivity of different species in the respective test conditions. The particular level of susceptibility of each of these biotest systems becomes evident by comparing the respective baseline QSARs, each of which pertained to the same set of toxicants (Figure 8.3).

The classic acute tests using fish or *Daphnia* are generally more sensitive towards low concentrations of toxic components than the simpler assays with single cells and suborganismic tests *in vitro*. The comparative analysis of toxicity data obtained *in vivo* and *in vitro* demonstrated that the low-complexity assays are of far lesser sensitivity than the traditional tests on organisms (Wenzel *et al.*, 1997), disproving the common assumption of increasing sensitivity with the reduction in complexity of the test systems. Readily substantiated by evaluation of the published data, the rather low susceptibility of the (sub)organismic low-complexity biotests may be due to several reasons. The relatively high concentrations of the test compounds may be related to the short test durations (e.g. kinetic factors). Cytotoxic chemicals may be more effective in whole animals because of the presence of especially sensitive target tissues (e.g. the central nervous system). Furthermore, the inequivalence of the endpoints in organisms and in enzymatic assays *in vitro* has to be

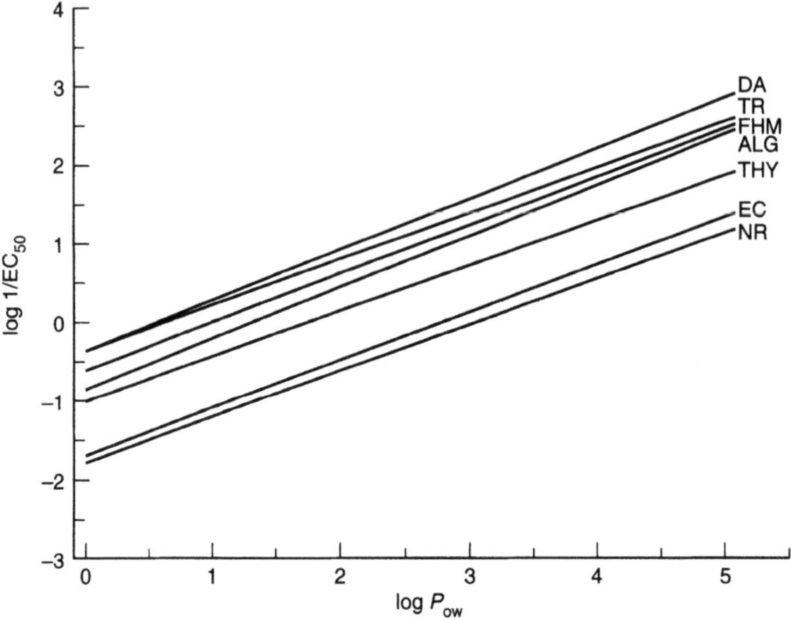



**Figure 8.3** Comparison of baseline QSAR models for polar non-specific toxicants indicating the differences in sensitivity of various aquatic species and test systems: The tests with fish (FHM = fathead; TR = trout) and crustaceans (DA = *Daphnia*) are generally more sensitive than the low-complexity assays with protozoa (THY = *Tetrahymena*), algae (ALG = *Selenastrum capriconutum*), bacteria (EC = *Escherichia coli*) and cell cultures (NR = neutralred assay).

considered. A fish LC$_{50}$ of an AChE inhibitor is not produced by blocking 50% of the respective enzymes; the animals' death may have already occurred with much lesser impairment.

The differences in sensitivity of organisms and test systems, although a possible complication for interspecies extrapolations, do not affect the reliability of these correlations. Much more critical are the actual limitations of their validity stemming from the different modes of toxic action of environmental contaminants in various species. Chemicals (e.g. herbicides and AChE inhibitors) that have different toxicity mechanisms in fish and algae, respectively (Figure 8.4), reveal distinct QSARs; the interspecies correlations thus break down (e.g. Gehring and Rao, 1977; Smissaert and Jansen, 1984; Wallace and Niemi, 1988; Nendza and Wenzel, 1993).

Evidently, the toxicity ranking by fish and algae cannot correlate for chemicals such as non-specific toxicants and photosynthesis-inhibitors together. These compounds may all be baseline toxicants with regard to fish but some (i.e. the herbicides) are definitely excess toxicants towards algae. The lack of parity, resulting from the non-uniformity of the chemicals' toxicological pro-

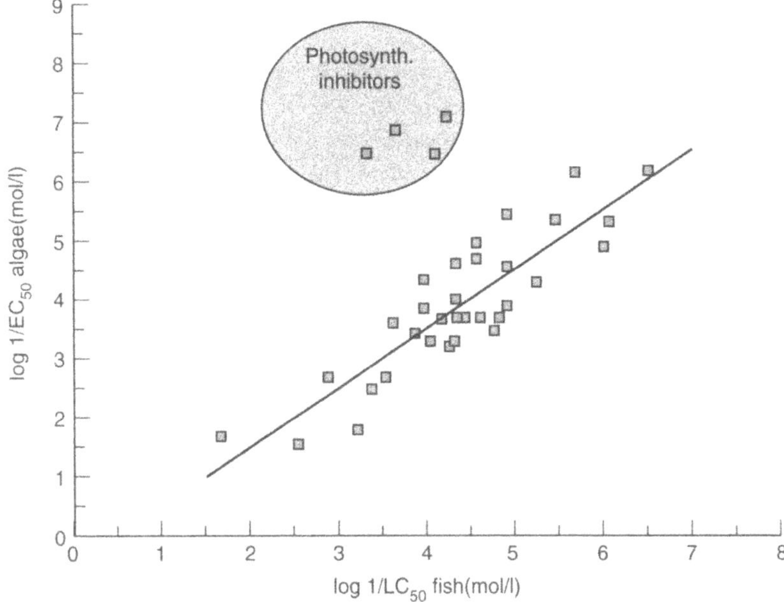



**Figure 8.4** Relationship between the toxicity to fish and algae for compounds with different modes of action in the respective species.

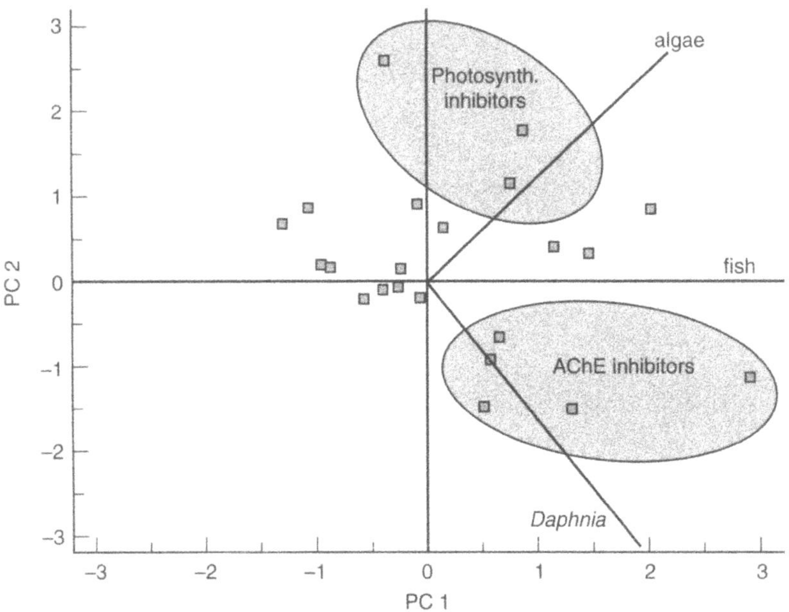



**Figure 8.5** Principal component analysis of toxicity data obtained with non-specific and specific toxicants in three aquatic assays. The collinearity in toxic responses breaks down for chemicals with non-homologous modes of action with regard to the different test species and interspecies correlations are no longer valid.

files, depends on the varying modes of action of the toxicants in different species, so the interspecies correlations are exempted (Figure 8.5).

For the reliable application of taxonomic correlations, therefore, the same conceptions about the compounds' potential modes of action are required, as for QSARs. Subjected to similar limitations, QSARs and interspecies correlations use similar principles and methods and may complement each other when used to obtain toxicity estimates if experimental results are unavailable.

# 9

# Application of QSAR estimates in hazard evaluation and risk assessments

Environmental risk assessments for the preliminary priority setting and the identification of testing needs are one practical field for the application of QSARs. Based on the integration of compound-specific data, hazard assessments are concerned with the quantification of adverse effects, and risk assessments try to evaluate the probability of hazards occurring (e.g. Barnthouse and Suter, 1986; Nendza, Volmer and Klein, 1990; Cowan *et al.*, 1995; Feijtel *et al.*, 1995; Leeuwen and Hermens, 1995; Logan and Wilson, 1995).

Risk may arise from any chemicals released into the environment, intentionally or by accident, during manufacture, use or dissipation of the products. The expected concentration time profile for chemicals at specific locations in the various environmental media – water, air, soil and biota – may result in hazards to humans and the environment. Hence, ecological risk assessment must evaluate the probability whether the exposure level of potential contaminants may exceed effective (i.e. toxic) concentrations in the environmental compartment of concern. The basic principle is the comparison of the environmental and the toxic concentrations of the contaminants:

$$\text{risk} = \text{probability (environmental concentration} \geq \text{toxic concentration)}$$

The scientific process of risk assessment may be defined as assigning magnitudes and probabilities to the occurrence of undesired impacts. This operation is descriptive and analytical, it does not include the people's perception and valuation of a known or potential risk (Morgan, 1985). Because of its probabilistic nature, the risk can never be zero if only a single contaminant molecule is present. Judging the acceptability of risk from a particular hazard is always a political question, dependent on public sensibility about the hazard, and influenced by the degree of information, current publicity and habituation. The pragmatic concept of avoiding unreasonable adverse effects on the environment takes into account the economic, social and environmental costs and benefits of the use of a compound, such as a pesticide.

The information necessary for the assessment of environmental hazard and risk, resulting from the use of chemicals, is preferably obtained from field monitoring and laboratory studies, but often it has to be supplemented by estimates based on various modelling techniques. The input data set required for risk assessments comprises data on the persistence, accumulation, mobility and ecotoxicity of chemicals as well as detailed information on the particular environment, to estimate in stepwise fashion the exposure concentration and the toxic levels for a regional scenario (Figure 9.1).

For exposure assessments, the sources and the release rates of the compounds have to be determined and the target environment, in which the contaminants will be distributed, has to be described. From data on the distribution and persistence of the compounds estimates of the environmental concentrations can be made. If the experimental data are insufficient for the exposure assessment, QSARs and transport models may be applied to obtain approximations of the contaminant concentrations for the defined environmental scenario. For the assessment of effects, the targets of possible adverse effects have to be identified and suitable endpoints to quantify these effects selected (e.g. the toxicity to organisms regarded representative for the respective environment). The effects on the target organisms or communities may be quantified by laboratory and field measurements and/or estimation methods, including QSARs, interspecies correlations, and analysis of sensitivity distributions. The actual risk assessments then combine the information obtained by comparing the measured or estimated environmental concentration and the toxic concentrations. The two methods for doing this (the quotient method and probabilistic procedures) are based on different principles. The quotient method is the direct arithmetic comparison of benchmark concentrations from toxicity testing or from QSARs with expected environmental concentrations:

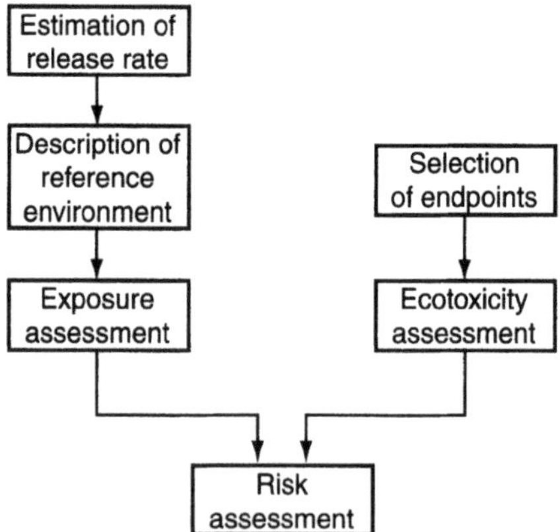


**Figure 9.1** Principal elements of environmental risk assessment (modified from Barnthouse and Suter, 1986). Reproduced from Nendza, Volmer and Klein (1990) with kind permission from Kluwer Academic Publishers, Dordrecht.

$$Q = \frac{\text{environmental concentration}}{\text{toxic concentration}}$$

The US EPA proposed the quotient method for use in ecological risk assessment of pesticides (Urban and Cook, 1986). No risk is presumed if $Q < 0.1$ with $LC_{50}$ for aquatic species taken as the effective toxic concentrations. Risk is regarded unacceptable if $Q > 0.5$, whereas for $Q = 0.1$–$0.5$ regulatory action may be taken (e.g. restrictions imposed on the use of the respective pesticide). The quotient method may be used for screening purposes in tiered risk assessment schemes, but it is unsuitable for quantifying risk. Its major disadvantage is the neglect of uncertainty resulting from the variability in the determination of the environmental and the effective concentrations.

The uncertainties in ecological risk assessments chiefly result from natural variabilities (e.g. the climate, the (hydro)geology and the composition of the community with respect to the sensitivity of the species in the habitat). Any deficiencies in precise knowledge about the composition, structure and functioning of the particular receiving biotic and abiotic environmental systems increase the variance, and hence the uncertainty, associated with the determination of the environmental and the toxic concentrations. Model error, natural variability and parameter error contribute to the analytical uncertainty. Obviously, an ecological risk assessment can only be as good as the data on which it is based (Landis *et al.*, 1994). The sources of error concern the ambiguity in the definition of the specific input parameters, interspecies differences in sensitivity between taxonomic groups, variable relationships between acute and chronic effects of chemicals used for the extrapolation of toxicity endpoints and, last but not least, the variability in the measurements or extrapolations used for the quantification of the properties and toxicity of the potential contaminants.

In accounting explicitly for the variance – the uncertainty associated with the exposure levels and the toxic concentrations – Barnthouse and Suter (1986) described a probabilistic method for quantifying risk to aquatic species, the analysis of extrapolation error (AEE). Based on the assumption that environmental and effective concentrations are log-normally distributed (Figure 9.2), the overlap of the corresponding probability density functions provides the measure of risk.

Risk will be greater the smaller the distance between the mean concentrations and the larger their variance (i.e. the larger the width of the curves). In this way it is possible to evaluate the risk for selected single species; differences in sensitivity among species are not accounted for. An extension is necessary to consider the varying sensitivity distributions (Kooijman, 1987) for different species in aquatic communities (Figure 9.3).

From a log-logistic distribution function of a battery of toxicity data (e.g. $LC_{50}$ values obtained with different species), the effects concentrations can

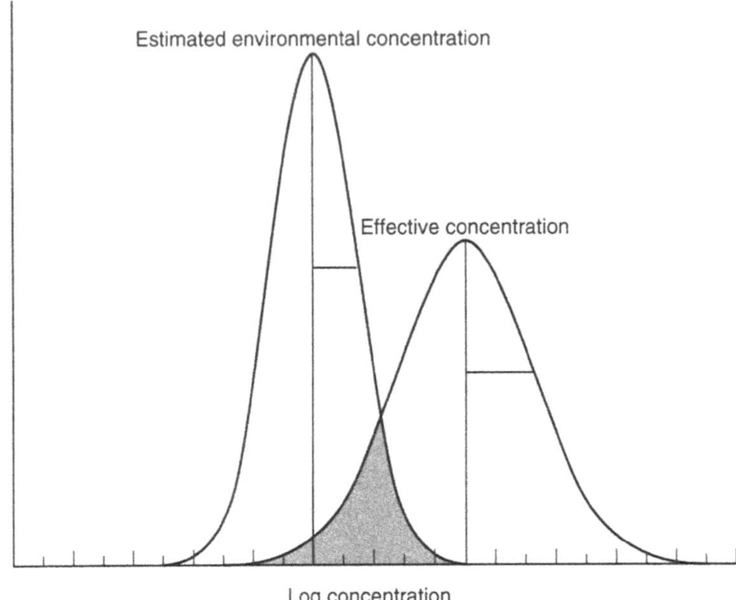


**Figure 9.2** Quantification of risk for environmental contaminants using probability density functions. The probability of impacts corresponds to the extent of overlap of the estimated environmental concentration (left curve) and the effective toxic concentration (right curve). The risk can be quantified from the distance between the mean values of the concentrations and the variance in the respective concentrations (due to measurement variability and/or extrapolation errors) represented by the width of the distributions. Reproduced from Nendza, Volmer and Klein (1990) with kind permission from Kluwer Academic Publishers, Dordrecht.

be determined for the most sensitive species of a community. The experimental error is neglected, because it is assumed to be small with respect to variation in toxicity among species. If the number of species is known, the concentration level can be calculated at which the probability is less than $p\%$ (frequently $p = 5$) that the most sensitive species is exposed to toxic concentrations. These 'most sensitive species' are not real species, such as *Daphnia* or rainbow trout, which are generally considered very sensitive, but refer to hypothetical species based on statistical considerations. This abstraction is supposed to compensate for the varying sensitivity of species to different chemicals (e.g. *Daphnia* may be highly sensitive to one compound but resistant to another). The uncertainty of the estimated $LC_{50}$ values for the most sensitive species is again characterized by a log-normal distribution, which is then the basis for quantifying risk for the entire habitat (OECD, 1992b).

The concept of risk assessment presented in Figure 9.4 is used here to demonstrate an application of QSAR estimates in risk assessment.

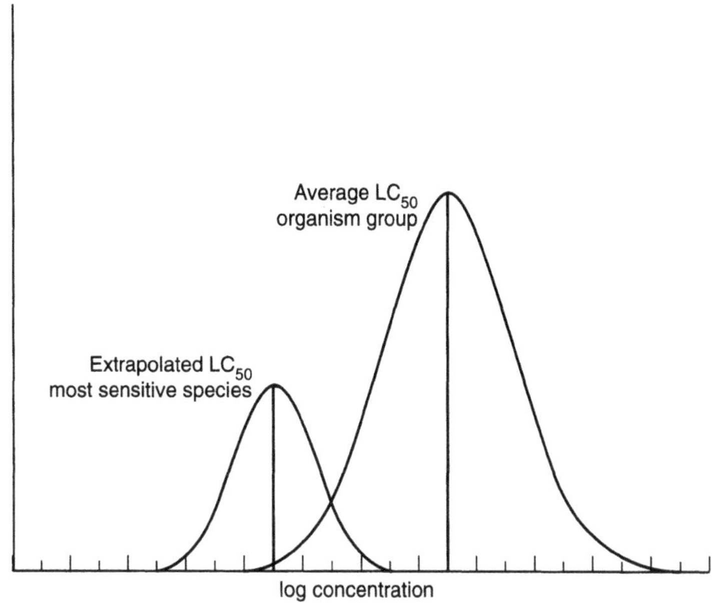



**Figure 9.3**  Estimation of toxicity for the hypothetical most sensitive species. Based on the assumption of log-normally distributed effect concentrations causing a definite toxic response (e.g. LC$_{50}$) with the different species of the habitat (right curve), the respective impact level for the most sensitive species can be extrapolated provided that the total number of species in the community is known (left curve). Reproduced from Nendza, Volmer and Klein (1990) with kind permission from Kluwer Academic Publishers, Dordrecht.

An ubiquitous contaminant (phenol) is taken as an example (Nendza, Volmer and Klein, 1990). The comparison of the computed results with those obtained from experimental data illustrates the possibilities and limitations of modelling techniques. Before the actual risk quantification, it has to be realized clearly that the crucial point is not the mathematical calculation but the ecological relevance of the assessment. The numerical result has to be interpreted in the context of the underlying data – the scientific knowledge about the fate and effects of the chemicals in the given environment. The example of phenol in a dynamic river system obviously cannot be transferred simply to a marine environment, where the contaminants cannot be washed out after the release has ceased. As a result, persistent chemicals with very long half-lives (e.g. decades in the case of organochlorine pesticides) remain a threat for marine ecosystems years after regulation because of re-mobilization from sediment reservoirs. Furthermore, the presence of co-contaminants has to be taken into account because exposure to only one chemical at a time is unrealistic. Rather, the compounds' contribution to the total risk has to be evaluated. The limitations and constraints of environmental hazard and risk

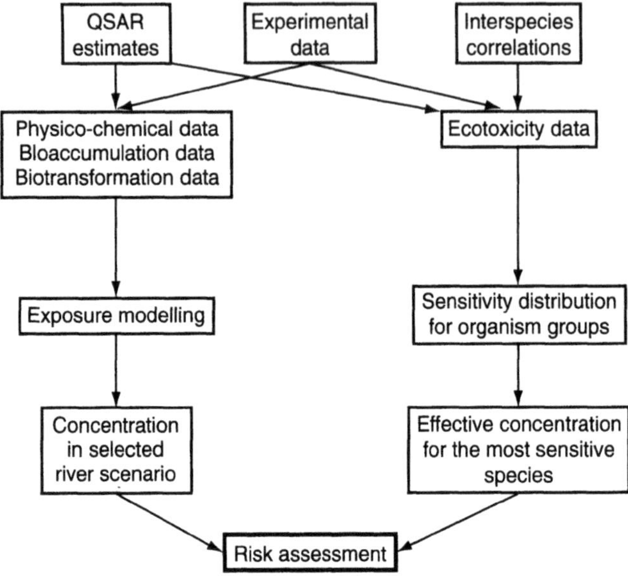


**Figure 9.4** Risk assessment for an aquatic environment based on a probabilistic procedure into which the concept of varying sensitivity in multispecies communities is incor-
porated (Nendza, Volmer and Klein, 1990). Exposure and effects are determined separately from experimental or, if not available, QSAR data. Physico-chemical data and
information on bioaccumulation and biotransformation are the input for computer simulations of transport and distribution processes that estimate the concentrations of a potential contaminant in a selected river scenario, using, for example, the EXAMS model
(Burns, Cline and Lassiter, 1982). For the effects assessment, the log-normal sensitivity
distribution is calculated from ecotoxicological data and the effective concentrations for
the most sensitive species are determined. The exposure concentrations and toxicity data
are then compared by analysis of variance to give a measure of risk for the environment.
Modified from Nendza, Volmer and Klein (1990) with kind permission from Kluwer
Academic Publishers, Dordrecht.

assessment, however, are not specifically related to the use of QSARs and
modelling techniques; they apply equally to experimental data from the laboratory and field. The only promising device for overcoming the difficulties is
the application of comprehensive scientific knowledge.

## 9.1 EXPOSURE ASSESSMENTS

Exposure may be defined as the 'concentration of toxic materials in space and
time at the interface with target populations' (Travis *et al.*, 1983).
Descriptions of exposure scenarios must include the biota exposed to the contaminants as well as the hydrological, topographical, geological and meteorological characteristics of the environment, which may strongly affect the
transport and the transformation of the chemicals. If measurements of ambi-

ent contaminant concentrations are unavailable, the key step in exposure assessments is the use of transport and transformation models to quantify the movement of the contaminants from the source through the environment to the target populations (i.e. the contaminants' bioavailability).

Model representations of ecosystems have been established to study the interactions between the receiving compartments and the pollutants. Ideally, these simulations should give a 1:1 representation of the reality. In practice, no model can be as complex as the real scenarios. To handle these multivariate situations, any evaluation has to be restricted to a well-defined segment of the environment for which the number of variables can be fixed, hence reduced. These variables may be altered systematically for individual parameters or sets of influential factors. Controlled input parameters are assumed invariant for the scenario under study, and they have to be determined separately for each release of contaminants. These fixed variables generally comprise:

The climate conditions based on the average of observations obtained during several years.
The geometry and hydrology of water bodies, including the ground water system.
The type of soil and sediment.
The species composition, the population density and the sensitivity of the members of a community.

The use of fixed variables reduces the number of parameters in the system and corresponds to an artificial restriction of the possible interactions between the pollutants and the ecosphere. This approach results in environmental models that are not designed for general application but have to be adapted for the particular local conditions. Accordingly, a model established for an alpine creek is not transposable to a lowlands river of similar size in a glacial valley. The (hydro)geological factors determining the distribution of the contaminants may vary to a considerable degree, such as the cross-section of the river, the flow rate, and the composition of the soil and sediment, resulting in significantly different concentrations in, for example, the water body. Also, interactions with other chemical loads may be of influence. Consequently, any hazard and risk assessment scheme can be valid only on a regional scale.

Properties of both the contaminants and the target environmental components determine the effective exposure levels, which in turn depend on several processes, such as propagation, distribution, accumulation, soil/sediment sorption and abiotic (e.g. hydrolysis and photolysis) and biotic (e.g. microbial) transformation. Several models of varying complexity have been developed to calculate and predict the distribution of chemicals in the environment (OECD, 1989a, 1993c). Most of them are derived from the Mackay model (Mackay, 1979; Mackay and Paterson, 1981, 1982, 1990, 1991; Mackay, Paterson and Shiu, 1992) to estimate the environmental compartment (e.g. air, soil, water or biota), in which the chemicals are most likely to be found. Based on the concept of fugacity, models were described for four levels of increasing complexity. Level I assumes equilibrium between all compartments, level II includes transformation and steady-state input of the contaminants, level III is extended to slow transport processes between the compartments, and level IV allows variation of fugacity, input rate and concentration with time. The EXAMS (exposure analysis modelling system) of Burns, Cline and Lassiter (1982) can be used to obtain estimates of the fate and persistence of chemicals in aquatic environments for non-steady-state conditions. Some features of the EXAMS model correspond to Mackay level IV, with flow data of the compartments additionally included. Volumes and compositions of water and sediment can be specified and transformations by photolysis, hydrolysis and biodegradation can be taken into account. For the purpose of an error analysis, the input parameters of the compartment model must include generalized site descriptions specifying the variability in, for example, the hydrological and meteorological conditions, and the first-order rates for the relevant transport and transformation processes. With EXAMS, several general scenarios are available, such as rivers, ponds, and oligotrophic and eutrophic lakes. Its principal applicability for modelling phenol exposure was demonstrated in a field study (Pollard and Hern, 1985) that compared the predicted concentrations with ambient phenol concentrations resulting from the effluents of a steel plant into an American river. The results of the exposure simulation agreed satisfactorily with the phenol concentrations measured in the river.

In other simulations with phenol (Nendza, Volmer and Klein, 1990), EXAMS was adapted for two river scenarios regarded to be representative for Germany: the Rhine was selected to represent the large rivers, and the Leine was chosen to represent the smaller, shorter rivers with lower water volume. Besides features of the particular environmental scenario, various compound-specific data are included in the model:

The release rate into the environment estimated from manufacturing data and the use pattern of the chemical.

The persistence and distribution of the chemical estimated from its physico-chemical properties:

vapour pressure ($p_v$)

water solubility ($S_w$)

1-octanol/water partition coefficient ($P_{ow}$)

hydrolysis half-life time ($Ht_{1/2}$)

photolysis half-life time ($Pt_{1/2}$)

soil sorption coefficient ($K_{oc}$)

bioconcentration factor (BCF)

biodegradability.

The kinetics of the transport and transformation processes, such as:

volatilization

direct photolysis

hydrolysis

oxidation

chemical behaviour, including the equilibrium partitioning of the compound into ionic species, each of which may occur in a dissolved, sedimentsorbed or biosorbed state.

The exposure-related physico-chemical properties of phenol were estimated according to the methods in Chapter 4. The estimates were mostly obtained from general models, except for soil sorption and biotransformation for which QSARs for polar non-reactive aromatics were applied. Overall, there is good agreement between the calculated data and the experimental data taken from databases (Table 9.1).

The exposure levels in the two different river scenarios (Rhine and Leine) were estimated from either calculated or experimental data. For reasons of comparison, the respective scenario data were kept constant and a realistic release rate of phenol into the river ecosystems of 10 kg/h was also uniformly assumed. Environmental levels of around 70 $\mu$g/l in the Rhine and around 7 $\mu$g/l in the Leine were estimated (Table 9.2).

The differences in the calculated phenol concentrations in each of the rivers based on either QSAR or experimental data are negligible, which reflects the good agreement between the two sets of input data. Additionally, the features of EXAMS, which strongly depend on the assumed hydrological properties (e.g. flow rate and discharge, and river geometry), may contribute to the similarity of the results. The substantial differences between the two

**Table 9.1** Comparison of experimental and calculated exposure-related data for phenol. Reproduced from Nendza, Volmer and Klein (1990) with kind permission from Kluwer Academic Publishers, Dordrecht.

| Parameter | Experimental data[a] | Calculated data | QSAR Model reference[b] |
|---|---|---|---|
| Melting point (K) | 312–314 | 312.3 | 1 |
| Boiling point (K) | 453 | 453.6 | 2 |
| Vapour pressure (25 °C) | 40.7 hPa | 125.6 Pa | 3 |
| Water solubility (g/l) | 66.7 | 11.8 | 4 |
| log $P_{ow}$ | 1.46 | 1.48 | 5 |
| Hydrolysis $t\frac{1}{2}$ | – | – | |
| Photolysis $t\frac{1}{2}$ (d) | 0.8 | 0.45 | 6 |
| Soil sorption $K_{oc}$ | 39–91 | 18.6 | 7 |
| Bioconcentration BCF | 2–200 | 7.8 | 8 |
| Biotransformation rate (1 organism$^{-1}$ h$^{-1}$) | $7.1 \times 10^{-12}$ | $6.9 \times 10^{-12}$ | 9 |

[a] Taken from the AQUIRE and ECDIN databases.
[b] 1, EPA (1986a); 2, Rechsteiner (1990); 3, Mackay *et al.* (1982); 4, Müller and Klein (1992); 5, MedChem (1989); 6, Atkinson (1987); 7, Karickhoff, Brown and Scott (1979); 8, Veith *et al.* (1980); 9, Paris, Wolfe and Steen (1982).

**Table 9.2** Estimated average phenol concentrations (geometric mean) from October to March in two different river ecosystems using QSAR estimates or experimental input data. Reproduced from Nendza, Volmer and Klein (1990) with kind permission from Kluwer Academic Publishers, Dordrecht.

| River | Section | Phenol ($\mu$g/l) | $s$ | Source of input data |
|---|---|---|---|---|
| Rhine | 0.1 × 10 km segment | 69.8 | 0.01 | Experiments |
|  | at Mannheim | 70.6 | 0.01 | QSARs |
| Leine | 20 km segment | 6.9 | 0.018 | Experiments |
|  | at Hannover | 6.9 | 0.018 | QSARs |

$s$ = standard deviation (log) of the estimates.

rivers result primarily from the differences in the geometry of the modelled river sections, which suggests that environmental factors predominate over chemical properties in the exposure levels of this contaminant.

## 9.2 EFFECTS ASSESSMENTS

The toxicity endpoints for environmental risk assessments are selected according to what is considered to be biologically relevant and important to human society. Adverse effects are generally quantified for single species, presumed to represent a group of organisms or a compartment of the ecosystem. Standard toxicity data are not directly usable for risk assessments, because the objective can hardly be to evaluate the probability for 50% mortality of some species at a particular exposure level in the environment. Much lesser effects also have to be avoided. NOEC values may be biased because only selected effects are monitored in the particular experimental set-up; other impairments may still occur at this concentration level and affect the interaction between the ecosystem components. In addition to direct toxicity, indirect effects (e.g. via the foodchain) and effects on the structure and functioning of the target ecosystem must be considered. Extrapolations from data on acute, subacute and chronic exposure are thus inevitable for determining toxicity threshold levels. Evidently, the uncertainties related to these unavoidable extrapolations apply to QSAR estimates as well as to experimental input data for environmental risk assessments.

These uncertainties depend on parameter sensitivity (i.e. the susceptibility of the selected endpoints) and parameter variance (i.e. how exact the measurement or estimation is). The ambiguity of many endpoints may introduce further uncertainty. Substantial inaccuracies may result from insufficient knowledge of the abundance and sensitivity of the species in the ecosystem community. Variations in the sensitivity of different taxonomic groups may cause major imprecision. The crucial point in effects assessments for environmental risk evaluations is hence precise knowledge of the species that might be affected and the identification of the appropriate endpoints.

Estimation of the chemical concentrations evoking toxic effects towards the (hypothetical) most sensitive species of an aquatic or terrestrial community requires data on the sensitivity distribution of the exposed group of organisms. This information is integrated based on toxicity data on individual species in the habitat:

The relevant toxicity endpoints are identified, taking into account the various modes of direct and indirect impacts on the species and the viability of the populations.

Toxicity data for the different species in the target environment are obtained. The effects data should be homogeneous with regard to the toxic endpoint and the exposure duration, because the simultaneous use of, for example, acute and chronic $LC_{50}$ and $EC_{50}$ data may result in unpredictable variance of the estimated toxicity values for the most sensitive species.

A frequency distribution of the toxicity data, which depicts at which contaminant levels toxicity has been observed, is calculated. The most important features of these frequency-distribution curves are the average toxicity, as reflected by the geometric mean of all toxicity values, and the width of the curve, reflected by the standard deviation.

The distribution curves reflect the sensitivity of members of a group of organisms and from these the toxicity values for the (hypothetical) most sensitive species can be extracted.

For the example of a risk assessment of phenol exposures in river environments, toxicity estimates (Table 9.3) were obtained for several aquatic species (fish, crustaceae, algae and bacteria). Two criteria are used in choosing the appropriate prediction models:

relevance of the particular endpoints for the risk assessment scenario;
applicability and validity of the available QSARs for the chemical under investigation.

The first point especially addresses ecological considerations about the abundance and the distribution of sensitive species in the environmental compartment of concern; the second aspect relates to the correct operation of QSAR methods. In the cases of the model contaminant phenol, which was chosen because of the availability of ample experimental data for the comparative exercise, sufficient knowledge is at hand to guide the retrieval of suitable QSARs. For toxicity to fish, the available models vary regarding (a) the endpoints measured – different fish species, different effects (NOEC, $LC_{50}$, $LC_{100}$, etc.) and different durations of the experiments (4 h–30 d), and (b) the various classes of chemicals studied. Because phenol is known to be a polar non-specific toxicant, suitable QSARs for fish (section 5.1) can be selected in a straightforward manner. These models are generally log $P_{ow}$-dependent,

**Table 9.3** Comparison of experimental and calculated ecotoxicity data for phenol. Reproduced from Nendza, Volmer and Klein (1990) with kind permission from Kluwer Academic Publishers, Dordrecht.

| Parameter | Experimental data[a] | Calculated data | QSAR model reference[b] |
|---|---|---|---|
| Fish $LC_{50}$ (mg/l) | | | |
| *Pimephales promelas* | 28.8 | 64.1 | 1 |
| | 32.4 | 41.3 | 2 |
| | 49.7 | 86.6 | 3 |
| *Poecilia reticulata* | 32.9 | 27.6 | 4 |
| *Lepomis macrochirus* | 17.4 | 13.1 | 5 |
| | | 37.2 | 3 |
| *Carassius auratus* | 44.5 | 57.7 | 6 |
| *Salmo gairdneri* | 8.9 | 17.5 | 3 |
| *Cyprinodon variegatus* | – | 40.8 | 3 |
| Shrimp: $LC_{50}$ (mg/l) | | | |
| *Crangon septemspinosa* | 18.2 | 9.1 | 3 |
| *Mysidopsis bahia* | – | 31.5 | 3 |
| Daphnids: $EC_{50}$ (mg/l) immobilization | | | |
| *Daphnia magna* | 36 | 18.8 | 7 |
| | | 15.6 | 3 |
| Algae: $EC_{50}$ (mg/l) | | | |
| *Scenedesmus subspicatus* | 600 | 69.5 | 8 |
| *Selenastrum capricornutum* | 100 | 83.0 | 3 |
| *Skeletonema costatum* | 50 | 40.9 | 3 |
| Bacteria: $EC_{50}$ (mg/l) | | | |
| *Escherichia coli* | 659 | 377 | 8 |
| *Photobacterium phosphoreum* | 35.8 | 53.6 | 5 |
| (Microtox assay) | | 66.0 | 3 |

[a] Taken from the AQUIRE and ECDIN databases.
[b] 1, Hall and Kier (1984); 2, Nendza and Russom (1991); 3, Nendza and Klein (1990); 4, Saarikoski and Viluksela (1982); 5, Kaiser, Dixon and Hodson (1984); 6, Nendza, Volmer and Klein (1990); 7, Devillers and Chambon (1986); 8, Nendza and Seydel (1988).

indicating the assumed predominance of transport phenomena and the relevance of membrane perturbation for the mechanism of toxic action. Because of the rather low lipophilicity of phenol (log $P_{ow}$ 1.48), it is not necessary explicitly to consider non-linear equations, which account for the comparatively reduced toxicity of highly lipophilic compounds (log $P_{ow}$ > 5) due to their limited solubility and/or hindered transport through biological barriers such as membranes. QSARs similar to those for fish have also been derived for crustaceans (section 5.2). Baseline toxicity models for non-polar and polar compounds relate, for example, the immobilization of *Daphnia magna* to the lipophilicity of the contaminants. Assuming also a non-specific mode of action, the QSARs suitable for estimating phenol effects on different shrimp and cladoceran species can be readily identified. Various species of algae have been used for toxicity testing and derivation of QSAR models (section 5.3). The inhibition of growth of algal cell populations by non-reactive toxicants has been related to partitioning features. The respective models for polar compounds apply to phenol because there is no evidence that the compound interacts with the algal photosystem, and hence is more toxic than is assumed for non-specific activity. Diverse endpoints on microorganism toxicity have been modelled by QSAR techniques (section 5.4), such as the inhibition of cell population growth and the Microtox test. QSAR analyses using log $P_{ow}$ or connectivity indices as the only regressor yielded significant correlations, which are appropriate for phenol. Although the toxicity is ranked identically by the various bacterial test systems (i.e. they are collinear), the considerable differences in their sensitivity yield large differences in the absolute toxicity estimates.

Comparison of the QSAR estimates for six fish species, three crustaceans, three algal species and two bacterial strains with experimental data retrieved from the literature (Table 9.2) reveals generally good agreement. The QSAR estimates do not differ much from the experimental data – in most cases the divergence is within a factor of two. Furthermore, these deviations are not large relative to the within-species variations in experimental data. For some endpoints and properties, QSARs appear even more precise than individual tests because they summarize large amounts of information; for example, the QSAR models for non-specific toxicants show less variance than measurements of toxicity to fish. Nonetheless, although experimental data often show high variability, they are always considered to be true. In contrast, the disadvantages and limitations of computed data are judged more strictly. Major sources of erroneous predictions are the utilization of unsuitable QSAR functions with regard either to the compound classes or to the parameter range covered. Errors in descriptor estimates may introduce additional uncertainties. It always has to be remembered that the reliable application of QSARs is restricted to homologous classes acting by a uniform mode of action. Only if the applicability and validity of the model for the chemical under study are substantiated can reliable estimates be calculated.

For chemicals acting by a uniform mode in different organisms (e.g. non-specific toxicants), interspecies extrapolations (Chapter 8) are a further means of obtaining toxicity data for environmental risk assessments. When experimental results are unavailable, they can be used with QSARs to fill data gaps. Biologists may prefer interspecies extrapolations, as they proceed from real (i.e. experimental) data. For mathematicians, chemists and statisticians, QSARs may be more reliable, as their input parameters, chemical structures and physico-chemical properties are usually considered to be subject to much less variability as compared to biological data. In practice, there mostly remains only the pragmatic point of view: all available information should be collected, scrutinized expertly, and then used (or rejected) for hazard and risk assessments.

For the phenol example, toxicity data for individual species can be combined into frequency distributions, which characterize the average effective phenol concentrations in the assumed community (mean values) and the range of sensitivity of the exposed species (width of the curves). The functions (Figure 9.5) derived from either experimental or calculated 96 h $LC_{50}$ data for five fish species (fathead minnow, rainbow trout, goldfish, bluegill sunfish and guppy) are in excellent agreement, because the coinciding input data result in almost identical mean values and standard deviations.

Based on the distribution characteristics summarized in Table 9.4, the $LC_{50}$ values for the most sensitive species can be estimated according to the method of Kooijman (1987), provided that the number of species in the community is known. Assuming an aquatic community of 200 species, $LC_{50}$ values for the most sensitive species have been extracted (Table 9.5).

The similarity in the sensitivity distributions derived from QSAR and experimental acute toxicity data results in only minor differences in the extrapolated phenol toxicity values for the most sensitive species: about 4 mg/l from the experimental data and 4.6 mg/l from the QSAR estimates.

The extracted toxicity value for the most sensitive species is strongly dependent on the model assumptions about the species composition of the

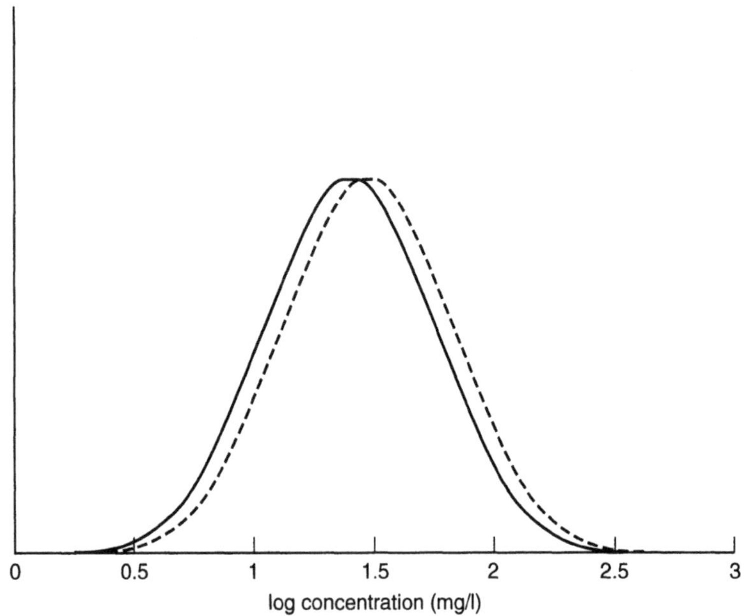



**Figure 9.5** Sensitivity distribution functions estimated from experimental (left curve) and QSAR data (right curve). Acute toxicity data for five fish species were used for the estimate resulting in almost identical distribution functions. Reproduced from Nendza, Volmer and Klein (1990) with kind permission from Kluwer Academic Publishers, Dordrecht.

**Table 9.4** Characteristics of log-normal sensitivity distribution derived from fish and aquatic macroinvertebrate data on acute toxicity (geometric mean) of phenol. Reproduced from Nendza, Volmer and Klein (1990) with kind permission from Kluwer Academic Publishers, Dordrecht.

| Organism group | Number of species | Toxicity (mg/l) | $s$ | Source of input data |
|---|---|---|---|---|
| Fish | 5 | 24.5 | 0.288 | Experiments |
| | 5 | 29.8 | 0.306 | QSARs |
| | 17 | 17.6 | 0.353 | Experiments |
| Macroinvertebrates | 12 | 41.3 | 0.615 | Experiments |
| Fish and macroinvertebrates | 29 | 25.0 | 0.495 | Experiments |

$s$ = Standard deviation (log) of the estimates.

**Table 9.5** Estimated toxicity of phenol for the hypothetical most sensitive of 200 species in the River Rhine. Modified from Nendza, Volmer and Klein (1990) with kind permission from Kluwer Academic Publishers, Dordrecht.

| Organism group | Average toxicity (mg/l), ($s$) | Minimum toxicity (mg/l), ($s$) | Source of input data |
|---|---|---|---|
| Fish ($n = 5$) | 24.5 (0.288) | 4.0 (0.443) | Experiments |
| | 29.8 (0.306) | 4.6 (0.447) | QSARs |
| Fish ($n = 17$) | 17.6 (0.353) | 1.7 (0.272) | Experiments |
| Fish and macroinvertebrates ($n = 29$) | 25.0 (0.495) | 0.9 (0.310) | Experiments |

$s$ = Standard deviation (log) of the estimates.

community: a more diverse community exhibits a wider range of sensitivity, resulting in a larger width of the distribution function and hence in a lower toxicity threshold. Extension of the data set on fish acute toxicity to 17 species reveals the standard deviation as a measure of the distribution width to be considerably higher than in smaller data sets (Table 9.4). Fish are only a small group within aquatic communities. Consideration of acute toxicity data for macroinvertebrates as well results in a further increase in the width of the distribution compared with that for solely fish populations (Figure 9.6).

Although the sensitivity calculated for the more diverse, and hence more realistic, group of test organisms, is similar to the average toxicity obtained for the small, fish-only data set, the differences in variability result in a significantly lower toxicity value for the postulated most sensitive species (Table 9.5). These findings demonstrate again the importance of the assumed local scenario characteristics, such as the abundance of sensitive organism groups in a community, over compound-specific properties in environmental risk assessments. The latter, however, play a major part in comparisons of the impact of several chemicals with different persistence and toxicity in exactly the same ecosystem.

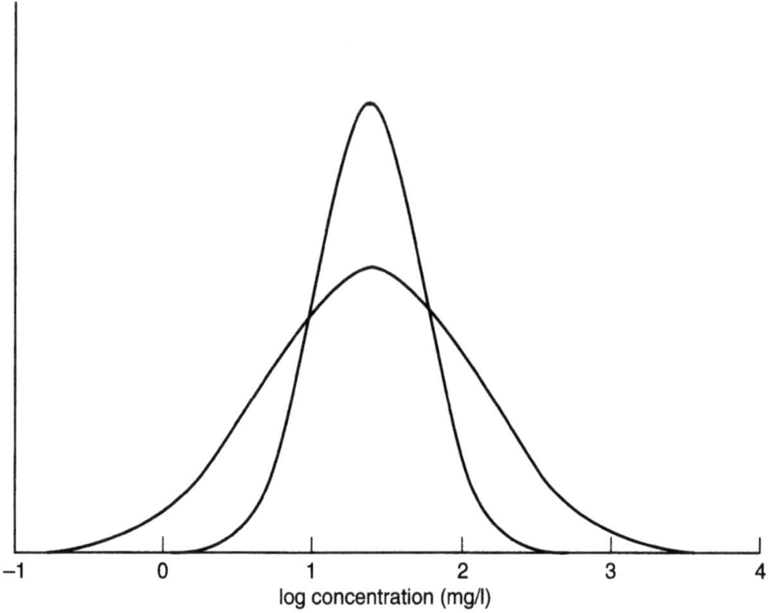

**Figure 9.6** Comparison of sensitivity distribution functions based on five fish species (upper curve) or a more diverse set of 29 fish and macroinvertebrate species (lower curve). Reproduced from Nendza, Volmer and Klein (1990) with kind permission from Kluwer Academic Publishers, Dordrecht.

## 9.3 COMPARATIVE RISK ASSESSMENT

Relating toxicity thresholds (section 9.2) to the estimated environmental concentrations of phenol (section 9.1) makes it possible to determine the likelihood of adverse effects in the target populations, and, thus, to make an ecological risk assessment. For the example of phenol release into the River Rhine, the risk for the most sensitive species was evaluated using the quotient method and the quantitative probabilistic procedure (Table 9.6) for four different communities.

No differences occur if for the same scenario either experimental data or QSAR predictions of the compound-specific parameters are used. Possible influences of parameter inaccuracies are by far outweighed by the predominance of the environmental factors in this example. The realistic assumption of a phenol release rate of 10 kg/h, resulting in an exposure level of about 70 µg/l, is unlikely to conflict with the effective concentrations (> 1 mg/l). Compared with the values based on the small data set ($n = 5$ species), the increased width of the sensitivity distribution for the fish and macroinvertebrate community ($n = 29$ species) yields considerably lower $LC_{50}$ estimates for the most sensitive species. Consequently, the ratio of environmental and

**Table 9.6** Estimation of risk from phenol exposure for the most sensitive of 200 species in the River Rhine. Modified from Nendza, Volmer and Klein (1990) with kind permission from Kluwer Academic Publishers, Dordrecht.

| Organism group | Exposure (mg/1) | Minimum toxicity (mg/1) | Risk quotient[a] | Calculated risk (%)[b] | Source of input data |
|---|---|---|---|---|---|
| Fish ($n = 5$) | 0.07 | 4.0 | 0.017 | 0.004 | Experiments |
| | 0.07 | 4.6 | 0.015 | 0.002 | QSARs |
| Fish ($n = 17$) | 0.07 | 1.7 | 0.04 | <0.001 | Experiments |
| Fish and macro-invertebrates ($n = 29$) | 0.07 | 0.9 | 0.076 | 0.024 | Experiments |

[a] Quotient of the environmental concentration and the estimated $LC_{50}$ value (minimum toxicity).
[b] Risk estimated according to Barnthouse and Suter (1986).

effective concentrations is decreased, and the risk increases. However, for all assumed populations the calculated risk is small ($< 0.03\%$) and no discriminations are meaningful for the different data sets.

To illustrate the possible influence of sensitivity distribution characteristics on risk estimates at significant risk levels, risk analyses were re-examined at higher exposure levels (Table 9.7).

Hypothetical release rates were defined, resulting in 320 μg/l and 630 μg/l phenol in the respective segment of the Rhine. At 320 μg/l, risk for the most sensitive species of a 200-species community is minor for the fish communities, but if macroinvertebrates are included, the risk estimate increases to 8%. Negligible differences occur if the risk assessment is based on the same set of fish species using either experimental data or QSAR estimates. Further increase in the environmental concentration (630 μg/l) constitutes general risk. Comparatively high risk values obtained for the more diverse communities are evident from the increased width of the respective sensitivity distribution, resulting in a lower toxicity threshold for the most sensitive species. These findings clearly demonstrate that risk estimates can be affected by the

**Table 9.7** Estimation of risk for the hypothetical most sensitive of 200 species at different levels of exposure to phenol in the River Rhine. Modified from Nendza, Volmer and Klein (1990) with kind permission from Kluwer Academic Publishers, Dordrecht.

| Organism group | Minimum toxicity (mg/l) | Risk at exposure level (%) 320 μg/l | Risk at exposure level (%) 630 μg/l | Source of input data |
|---|---|---|---|---|
| Fish ($n = 5$) | 4.0 | 0.9 | 3.8 | Experiments |
| | 4.6 | 0.9 | 3.1 | QSARs |
| Fish ($n = 17$) | 1.7 | 0.9 | 7.5 | Experiments |
| Fish and macro-invertebrates ($n = 29$) | 0.9 | 8 | 31 | Experiments |

characteristics of the respective sensitivity distribution, which, in turn, is determined by the composition of the data set (i.e. selection of toxicity parameters) used for calculating the distribution parameters.

The example of phenol release into river ecosystems clearly illustrates that the risk assessment procedures currently used are far from perfect. The major limitations concern the issues of

relevant species and modes of toxic action
indirect exposure
joint toxicity of multiple releases of chemicals.

These constraints, however, apply independently of the source of the input data, either QSARs or experiments. Therefore, within the framework available, the comparison of risk assessments for the different scenarios demonstrate positively that QSAR estimates and exposure modelling can be useful for environmental risk assessments. If the established validation criteria and limitations are considered, the reliability of the QSAR data and the accuracy of the modelling mostly correspond to the variability in the underlying experimental data. Special care has to be taken to obtain representative data sets, such as accounting for the toxicity to all relevant species in an aquatic community. As long as the wide variety of data required for quantitative risk assessments are not available from experimental sources, QSARs remain a tool for providing estimates of the exposure-relevant properties of chemicals and the toxicity of the compounds towards various species, thus allowing a more reliable quantification of the potential hazards and risks from environmental contaminants.

> QSARs can be good tools for generating data, but
> their interpretation remains the responsibility of the scientist.

## *Appendix: QSAR-related computer programs and software*

QSAR development and applications depend on computer programs and software as well as on databases for the necessary input data. Basic software for data storage, spreadsheet calculations, statistics and graphic representations are a necessity and the actual selection from the large variety on the market depends on the particular hardware facilities and on the preferences and ambitions of the user. Detailed descriptions of exposure-assessment models and overviews of databases are given by, for example, OECD (1989a, 1993c) and McCutcheon *et al.* (1990), respectively. The following compilation of QSAR-related programs is by no means complete and presents only some of the available systems, which, due to their diversity in features, vary considerably in hardware requirements and in price. Most programs are available for personal computers and/or workstations. The rapid development of improved versions and new software systems precludes any recommendations concerning the strengths and weaknesses of individual programs. Rather, future users should consult with experienced colleagues, and attempt to work with different programs, to identify the software most suitable for their needs.

The **MedChem** (1989) software provides estimates of physico-chemical properties and access to the Thor database (partition coefficients, $pK_a$).

*PC properties*: log $P_{ow}$ and MR.

The **ProLogP** (1992) software estimates partition coefficients based on the Rekker method.

*PC properties*: log $P_{ow}$.

The **LOGKOW** (1993) software calculates log $P_{ow}$ values.

*PC properties*: log $P_{ow}$.

**GAMMA** (DMU, 1992) is a computer program for calculating liquid-phase activity coefficients by means of UNIFAC.

*PC properties:* $\gamma$.

The **AOP** (1990) program calculates transformation rate constants for the gas-phase reactions between hydroxyl radicals and organic chemicals in the atmosphere using the fragment method of Atkinson and rate constants for the reaction between ozone and olefins/acetylenes.

*Degradation*: atmospheric degradation.

**CHEMEST** (EPA, 1986) is an interactive computer system for estimating environmentally important properties of organic chemicals, featuring the methods compiled by Lyman, Reehl and Rosenblatt (1990).

*PC properties*: $p_v$, H, $T_m$, $T_b$, $pK_a$, $K_{oc}$ and $S_w$.
*Ecotoxicity*: BCF.

The **ASTER-System** (EPA, 1991) provides estimates of chemical properties, environmental fate and toxicity and access to the AQUIRE database.

*Descriptors*: heat of vaporization, molar volume, surface area, molar refractivity, parachor, connectivity indices and functional group indices.
*PC properties*: log $P_{ow}$, $p_v$, H, $T_m$, $T_b$, $pK_a$, $K_{oc}$ and $S_w$.
*Degradation*: hydrolysis half-life, atmospheric degradation and biodegradation.
*Ecotoxicity*: BCF, aquatic toxicity, mammalian toxicity, phytotoxicity and genetic/mutagenic assessment.

The **UBA(FHG)-SAR-System** (IUCT, 1992) provides tools for estimating physico-chemical, toxicological and degradation properties. The procedures are based on molecular structure and correlations with measured properties of chemicals with similar structure as well as on property–property relationships.

*Descriptors*: molar volume, surface area, Hammett $\sigma$ constants, molar refractivity, quantum-chemical descriptors (HOMO, LUMO, hardness, electronegativity, heat of formation, ionization potential, dipole moment) and connectivity indices.
*PC properties*: log $P_{ow}$, $p_v$, H, $T_b$, $pK_a$, $K_{oc}$ and $S_w$.
*Degradation*: atmospheric degradation and biodegradation.
*Ecotoxicity*: BCF, aquatic toxicity, mammalian toxicity and mutagenicity.
*Exposure*: Mackay level I.

**ECOSAR** (EPA, 1994) is a computer program for estimating the ecotoxicity of industrial chemicals.

*Ecotoxicity*: aquatic toxicity.

# *References and Bibliography*


Abernethy, S. G., Mackay, D. and McCarty, L. S. (1988) 'Volume fraction' correlation for narcosis in aquatic organisms: the key role of partitioning. *Environ. Toxicol. Chem.*, **7**, 469–81.

Abernethy, S. G., Bobra, A. M., Shiu, W. Y. Wells, P. G. and Mackay, D. (1986) Acute lethal toxicity of hydrocarbons and chlorinated hydrocarbons to two planktonic crustaceans: the key role of organism–water partitioning. *Aquat. Toxicol.*, **8**, 163–74.

Abraham, M. H., Chadha, H. S., Whiting, G. S. and Mitchell, R. C. (1994) Hydrogen bonding. 32. An analysis of water–octanol and water–alkane partitioning and the $\Delta \log P$ parameter of Seiler. *J. Pharm. Sci.*, **83**, 1084–100.

Abraham, M. H., Grellier, P. L., Prior, D. V., Duce, P. P., Morris, J. J. and Taylor, P. J. (1989) Hydrogen bonding. Part 7. A scale of solute hydrogen-bond acidity based on values for complexation in tetrachloromethane. *J. Chem. Soc. Perkin. Trans.*, **II**, 699–711.

Adamson, G. W., Bawden, D. and Saggers, D. T. (1984) Quantitative structure–activity relationship studies of acute toxicity ($LD_{50}$) in large series of herbicidal benzimidazoles. *Pestic. Sci.*, **15**, 31–9.

Albert, A. and Serjeant, E. P. (1962) *Ionisation Constants of Acids and Bases*, Methuen, London.

Altomare, C., Tsai, R. S., ElTayar, N., Testa, B., Carotti, A., Cellamare, S. and Benedetti, P. G. de (1991) Determination of lipophilicity and hydrogen-bond donor acidity of bioactive sulphonyl-containing compounds by reversed-phase HPLC and centrifugal partition chromatography and their application to structure–activity relations. *J. Pharm. Pharmacol.*, **43**, 191–7.

Altschuh, J., Brüggemann, R. and Karcher, W. (1993) Attempts to classify QSARs with respect to their validity: vapour pressure estimation as an example. *Sci. Total Environ.*, Suppl., **1993**, 1409–19.

Anderson, E., Veith, G. D. and Weininger, D. (1987) *SMILES: A Line Notation and Computerized Interpreter for Chemical Structures*, Environmental Research Brief, EPA-600/M-87-021.

Anliker, R., Moser, P. and Poppinger, D. (1988) Bioaccumulation of dyestuffs and organic pigments in fish. Relationships to hydrophobicity and steric factors. *Chemosphere*, 17, 1631–44.

AOP (1990) *Atmospheric Oxidation Program Version 1.31*, Syracuse Research Corporation, Syracuse, NY.

Aoyama, T. and Ichikawa, H. (1991) Basic operating characteristics of neural networks when applied to structure–activity studies. *Chem. Pharm. Bull.*, **39**, 358–66.

AQUIRE (Aquatic Toxicity Information Retrieval Data Base), Chemical Information Systems, Baltimore, MD.

Arbuckle, W. B. (1983) Estimating activity coefficients for use in calculating environmental parameters. *Environ. Sci. Technol.*, **17**, 537–42.

Ariens, E. J. (1984) Domestication of chemistry by design of safer chemicals: structure–activity relationships. *Drug Metab. Rev.*, **15**, 425–504.

Ariens, E. J., Mutschler, E. and Simonis, A. M. (1978) *Allgemeine Toxikologie*, Thieme, Stuttgart.

Ashby, J., Serres, F. J. de, Draper, M., Ishidate Jr, M., Margolin, B. H., Matter, B. and Shelby, M. D. (1985) Overview and conclusions of the IPCS collaborative study on *in vitro* assay systems. *Progr. Mutat. Res.*, **5**, 117–74.

ASTM (1984) *Book of ASTM Standards*, Vol. 11.04, American Society for Testing and Materials, Philadelphia, PA.

Atkinson, R. (1987) Structure–activity relationship for the estimation of rate constants for the gas-phase reactions of hydroxyl radicals with organic compounds. *Int. J. Chem. Kinet.*, **19**, 799–828.

Atkinson, R. (1988) Estimation of gas–phase hydroxyl radical rate constants for organic chemicals. *Environ. Toxicol. Chem.*, **7**, 435–42.

Babeu, L. and Vaishnav, D. D. (1987) Prediction of biodegradability for selected organic chemicals. *J. Industr. Microbiol.*, **2**, 107-15.

Babich, H. and Borenfreund, E. (1987) Structure–activity relationship (SAR) models established *in-vitro* with the neutral red cytotoxicity assay. *Toxicol. in Vitro*, **1**, 3–9.

Babich, H. and Davis, D. L. (1981) Phenol: a review of environmental and health risks. *Rec. Toxicol. Pharmacol.*, **1**, 90–109.

Bahner, L. H., Wilson, A. J., Sheppard, J. M., Patrick, L. R., Goodman, L. R. and Walsh, G. E. (1977) Kepone[R] bioconcentration, accumulation, loss and transfer through estuarine food chains. *Chesapeake Sci.*, **18**, 299–308.

Bahnick, D. A. and Doucette, W. J. (1988) Use of molecular connectivity indices to estimate soil sorption coefficients for organic chemicals. *Chemosphere*, 17, 1703–15.

Balasubramanian, K. (1994) Integration of graph theory and quantum chemistry for structure–activity relationships. *SAR QSAR Environ. Res.*, **2**, 59–77.

Banerjee, S. and Baughman, G. L. (1991) Bioconcentration factors and lipid solubility. *Environ. Sci. Technol.*, **25**, 536–9.

Banerjee, S. and Williams, C. L. (1993) Compatibility of functional groups in $K_{ow}$-based QSARs: application to nitro compounds. *Environ. Toxicol. Chem.*, **12**, 1847–9.

Barnthouse, L. W. and Suter II, G. W. (eds) (1986) *User's Manual for Ecological Risk Assessment*, ORNL-6251, Oak Ridge National Laboratory, Oak Ridge, TN.

Barrett, J. C. (1987) *Mechanisms of Environmental Carcinogenesis*, CRC Press, Boca Raton, FL.

Basak, S. C. (1990) A nonempirical approach to predicting molecular properties using graph-theoretic invariants, in *Practical Applications of Quantitative Structure–Activity Relationships (QSAR) in Environmental Chemistry and Toxicology* (eds W. Karcher and J. Devillers), Kluwer, Dordrecht.

Basak, S. C. and Grunwald, G. D. (1994) Molecular similarity and risk assessment: analog selection and property estimation using graph invariants. *SAR QSAR Environ. Res.*, **2**, 289–307.

Basak, S. C. and Grunwald, G. D. (1995) Tolerance space and molecular similarity. *SAR QSAR Environ. Res.*, **3**, 265–77.

Basak, S. C. and Magnuson, V. R. (1983) Molecular topology and narcosis. *Drug Res.*, **33**, 501-3.

Basak, S. C., Bertelsen, S. and Grunwald, G. D. (1995) Use of graph theoretic parameters in risk assessment of chemicals. *Toxicol. Lett.*, **79**, 239–50.

Basak, S. C., Gieschen, D. P. and Magnuson, V. R. (1984). A quantitative correlation of the $LC_{50}$ values of esters in *Pimephales promelas* using physicochemical and topological parameters. *Environ. Toxicol. Chem.*, **3**, 191–9.

Basak, S. C., Frane, C. M., Rosen, M. E. and Magnuson, V. R. (1986) Molecular topology and mutagenicity: a QSAR study of nitrosamines. *IRCS Med. Sci.*, **14**, 848–9.

Baughman, G. L. and Paris, D. F. (1981) Microbial bioconcentration of organic pollutants from aquatic systems – a critical review. *CRC Crit. Rev. Microbiol.*, **8**, 205–28.

Benigni, R., Andreoli, C., Cotta-Ramusino, M., Giorgi, F. and Gallo, G. (1995) The electronic properties of carcinogens, and their role in SAR studies of noncongeneric chemicals. *Toxicity Modeling*, **1**, 157–67.

Berg, M. van den, Blank, F., Heeremans, C., Wagenaar, H. and Olie, K. (1987) Presence of polychlorinated dibenzo-*p*-dioxins and polychlorinated dibenzofurans in fish-eating birds and fish from the Netherlands. *Arch. Environ. Contam. Toxicol.*, **16**, 149–58.

Berg, M. van den, Meent, D. van de, Peijnenburg, W. J. G. M., Sijm, D. T. H. M., Struijs, J. and Tas, J. W. (1995) Transport, accumulation and transformation processes, in *Risk Assessment of Chemicals* (eds C. J. Leeuwen and J. L. M. Hermens), Kluwer, Dordrecht.

Biagi, G. L., Gandolfi, O., Guerra, M. C., Barbaro, A. M. and Cantelli-Forti, G. (1975) $R_m$ values of phenols. Their relationship with log *P* values and activity. *J. Med. Chem.*, **18**, 868–73.

Biddinger, G. R. and Gloss, S. P. (1984) The importance of trophic transfer in the bioaccumulation of chemical contaminants in aquatic ecosystems. *Resid. Rev.*, **91**, 103-45.

Blum, D. J. W. and Speece, R. E. (1991) A database of chemical toxicity to environmental bacteria and its use in interspecies comparisons and correlations. *Research Journal WPCF*, **63**, 198–207.

BMU (1992) *10 Jahre Chemikaliengesetz–Bilanz und Perspektiven*, Bundesministerium für Umwelt, Naturschutz und Reaktorsicherheit, Bonn.

Boecklen, W. J. and Niemi, G. J. (1994) Multivariate association of graph theoretic variables and physicochemical properties. *SAR QSAR Environ. Res.*, **2**, 79–87.

Boethling, R. S. (1986) Application of molecular topology to quantitative structure–biodegradability relationships. *Environ. Toxicol. Chem.*, **5**, 797–806.

Boethling, R. S. and Sabljic, A. (1989) Screening-level models for aerobic biodegradability based on a survey of expert knowledge. *Environ. Sci. Technol.*, **23**, 672–9.

Bonazountas, M. and Wagner, J. M. (1984) *SESOIL: A Seasonal Soil Compartment Model* (prepared for the Office of Toxic Substances, US EPA), Arthur D. Little, Cambridge, MA.

Bondi, A. (1964) Van der Waals volumes and radii. *J. Phys. Chem.*, **68**, 441.

Boxall, A. B. A., Maltby, L. L., Forrow, D. M. and Betton, C. (1994) The application of QSARs to predict the toxicity of road runoff to *Gammarus pulex*. Abstracts, 6th QSAR Workshop, Belgirate, Italy.

Bradbury, S. P. (1994) Predicting modes of toxic action from chemical structure: an overview. *SAR QSAR Environ. Res.*, **2**, 89–104.

Bradbury, S. P. and Broderius, S. J. (1994) QSAR applications for mixture assessments: issues in predicting the toxicity of reactive chemicals. Abstracts, 6th QSAR Workshop, Belgirate, Italy.

Bradbury, S. P., Carlson, R. W. and Henry, T. R. (1989) Polar narcosis in aquatic organisms, in *Aquatic Toxicology and Hazard Assessment* (eds U. M. Cowgill and L. R. Williams), STP 1027, ASTM, Philadelphia, PA.

Bradbury, S. P., Henry, T. R. and Carlson, R. W. (1990) Fish acute toxicity syndromes in the development of mechanism specific QSARs, in *Practical Applications of Quantitative Structure–Activity Relationships (QSAR) in Environmental Chemistry and Toxicology* (eds W. Karcher and J. Devillers), Kluwer, Dordrecht.

Breukelen, S. W. F. van and Brock, T. C. M. (1993) Response of a macro-invertebrate community to insecticide application in replicated freshwater microcosms with emphasis on the use of principal component analysis. *Sci. Total Environ.*, Suppl., **1993**, 1047–58.

Bridie, A. L., Wolff, C. J. M. and Winter, M. (1979) BOD and COD of some petrochemicals. *Water Res.*, **13**, 627–30.

Briggs, G. G. (1981) Theoretical and experimental relationships between soil adsorption, octanol–water partition coefficients, water solubilities, bioconcentration factors, and the parachor. *J. Agric. Food Chem.*, **29**, 1050–9.

Bringmann, G. and Kühn, R. (1975) Die beginnende Hemmwirkung der Zellvermehrung von *Pseudomonas* als wassertoxikologischer Test. *Vom Wasser*, 44, 119–29.

Bringmann, G. and Kühn, R. (1980) Comparison of the toxicity thresholds of water pollutants to bacteria, algae, and protozoa in the cell multiplication inhibition test. *Water Res.*, **4**, 231–41.

Broderius, S. and Kahl, M. (1985) Acute toxicity of organic chemical mixtures to the fathead minnow. *Aquat. Toxicol.* **6**, 307–22.

Broderius, S. J., Kahl, M. D. and Hoglund, M. D. (1995) Use of joint toxic response to define the primary mode of action for diverse industrial organic chemicals. *Environ. Toxicol. Chem.*, **14**, 1591–605.

Brooke, D. N., Dobbs, A. J. and Williams, N. (1986) Octanol:water partition coefficients (*P*) measurement, estimation and interpretation, particularly for chemicals with $P > 10^5$. *Ecotoxicol. Environ. Saf.*, **11**, 251–60.

Brooke, D., Nielsen, I., Bruijn, J. de and Hermens, J. (1990) An interlaboratory evaluation of the stir-flask method for the determination of octanol/water partition coefficients (log $K_{ow}$). *Chemosphere*, **21**, 119–33.

Brooke, L. T., Call, D. J., Geiger, D. L. and Northcott, C. E. (1984). *Acute toxicities of Organic Pollutants to Fathead Minnow (Pimephales promelas)*, Vol. I, Center for Lake Superior Environmental Studies, Superior, WI.

Brown, D. S. and Flagg, E. W. (1981) Empirical prediction of organic pollutant sorption in natural sediments. *J. Environ. Qual.*, **10**, 382–6.

Brown, V. M. (1968) The calculation of the acute toxicity of mixtures of poisons to rainbow trout. *Water Res.*, **2**, 723–33.

Bruggeman, W. A., Opperhuizen, A., Wijbenga, A. and Hutzinger, O. (1984) Bioaccumulation of superlipophilic chemicals in fish. *Toxicol. Environ. Chem.*, **7**, 173–89.

Brüggemann, R. and Altschuh, J. (1991) A validation study for the estimation of aqueous solubility from *n*-octanol/water partition coefficients. *Sci. Tot. Environ.*, **109/110**, 41–57.

Bruijn, J. de and Hermens, J. (1991a) Inhibition of acetylcholinesterase and acute toxicity of organophosphorous compounds to fish: a preliminary structure–activity analysis; in Hydrophobicity, *Biokinetics and Toxicity of Environmental Pollutants: a Structure–Activity Approach* (by J. de Bruijn). Thesis, Utrecht University, Utrecht.

Bruijn, J. de and Hermens, J. (1991b) Qualitative and quantitative modelling of toxic effects of organophosphorous compounds to fish. *Sci. Tot. Environ.*, **109/110**, 441–55.

Bruijn, J. de, Busser, F., Seinen, W., and Hermens, J. L. M. (1989) Determination of octanol/water partition coefficients for hydrophobic organic chemicals with the 'slow-stirring' method. *Environ. Toxicol. Chem.*, **8**, 499–512.

Buckley, S., Ford, M. G., Leake, L. D., Salt, D. W., Burt, P. E., Moss, M. D. V., Brealey, C. J. and Livingstone, D. J. (1987) A neurotoxicological investigation of the action of synthetic pyrethroid insecticides against adult houseflies *Musca domestica* L., in *QSAR in Drug Design and Toxicology* (eds D. Hadzi and B. Jerman-Blazic), Elsevier, Amsterdam.

Buikema, A. L., McGinniss, M. J. and Cairns, J. (1979) Phenolics in aquatic ecosystems: a selected review of recent literature. *Marine Environ. Res.*, **2**, 87–181.

Burkhard, L. P., Kuehl, D. W. and Veith, G. D. (1985) Evaluation of reverse phase liquid chromatography/mass spectrometry for estimation of *n*-octanol/water partition coefficients for organic chemicals. *Chemosphere*, **14**, 1551–60.

Burns, L. A., Cline, D. M. and Lassiter, R. R. (1982) *Exposure Analysis Modeling System (EXAMS): User Manual and System Documentation*, EPA-600/3-82/023, Athens, GA.

Butte, W., Willig, A. and Zauke, G. P. (1987) Bioaccumulation of phenols in zebrafish determined by a dynamic flow through test, in *QSAR in Environmental Toxicology – II* (ed. K. L. E. Kaiser), D. Reidel, Dordrecht.

Bysshe, S. E. (1990) Bioconcentration factor in aquatic organisms, in *Handbook of Chemical Property Estimation Methods* (eds W. J. Lyman, W. F. Reehl and D. H. Rosenblatt), American Chemical Society, Washington, DC.

Calamari, D. and Vighi, M. (1990) Quantitative structure activity relationships in ecotoxicology: value and limitations. *Rev. Environ. Toxicol.*, **4**, 1–112.

Calleja, M. C. and Persoone, G. (1993) The influence of solvents on the acute toxicity of some lipophilic chemicals to aquatic invertebrates. *Chemosphere*, **26**, 2007–22.

Chen, F., Holten-Andersen, J. and Tyle, H. (1993) New developments of the UNIFAC model for environmental application. *Chemosphere*, **26**, 1325–54.

Chen Fredenslund, F., Nendza, M. and Herbst, T. (1995) *Mutual Interactions of Chemicals with Regard to their Water Solubility*, Working Report, National Environmental Research Institute, Department of Policy Analysis, Roskilde, Denmark.

Chiou, C. T. (1985) Partition coefficients of organic compounds in lipid–water systems and correlations with fish bioconcentration factors. *Environ. Sci. Technol.*, **19**, 57–62.

Chiou, C. T., Peters, L. J. and Freed, V. H. (1979) A physical concept of soil–water equilibria for nonionic organic compounds. *Science*, **206**, 831–2.

Chiou, C. T., Freed, V. H., Schmedding, D. W. and Kohnert, R. L. (1977) Partition coefficients and bioaccumulation of selected organic chemicals. *Environ. Sci. Technol.*, **11**, 475–8.

Clayton, J. R., Pavlou, S. P. and Breitner, N. F. (1977) Polychlorinated biphenyls in coastal marine zooplankton: bioaccumulation by equilibrium partitioning. *Environ. Sci. Technol.*, **11**, 676–82.

Clements, R. G., Nabholz, J. V., Zeeman, M. and Auer, C. M. (1995) The application of structure–activity relationships (SARs) in the aquatic toxicity evaluation of discrete organic chemicals. *SAR QSAR Environ. Res.*, **3**, 203–15.

Coats, E. A., Cordes, H. P., Kulkarni, V. M., Richter, M., Schaper, K. J., Wiese, M. and Seydel, J. K. (1985) Multiple regression and principal component analysis of

**232**  *References and bibliography*

antibacterial activities of sulfones and sulfonamides in whole cell and cell-free systems of various DDS sensitive and resistant bacterial strains. *Quant. Struct.–Act. Relat.*, **4**, 99–109.

Collander, R. (1951) The partition of organic compounds between higher alcohols and water. *Acta Chem. Scand.*, **5**, 774–80.

Compadre, R. L. L. de, Shusterman, A. J. and Hansch, C. (1988) The role of hydrophobicity in the Ames test. The correlation of mutagenicity of nitropolycyclic hydrocarbons with partition coefficients and molecular orbital indices. *Int. J. Quant. Chem.*, **34**, 91–101.

Connell, D. W. (1988) Bioaccumulation behaviour of persistent organic chemicals with aquatic organisms. *Rev. Environ. Contam. Toxicol.*, **101**, 117–54.

Connell, D. (1989) Biomagnification by aquatic organisms – a proposal. *Chemosphere*, **19**, 1573–84.

Connell, D. W. and Hawker, D. W. (1988) Use of polynominal expressions to describe the bioconcentration of hydrophobic chemicals by fish. *Ecotoxicol. Environ. Saf.*, **16**, 242–57.

Connell, D. W., Braddock, R. D. and Mani, S. V. (1993) Prediction of the partition coefficient of lipophilic compounds in the air–mammal tissue system. *Sci. Total Environ.*, Suppl., **1993**, 1383–96.

Corbett, J. R., Wright, K. and Baillie, A. C. (1984) *The Biochemical Mode of Action of Pesticides*, Academic Press, London.

Cowan, C. E., Versteeg, D. J., Larson, R. J. and Kloepper-Sams, P. J. (1995) Integrated approach for environmental assessment of new and existing substances. *Reg. Toxicol. Pharmacol.*, **21**, 3–31.

Craig, P. N. (1971) Interdependence between physical parameters and selection of substituent groups for correlation studies. *J. Med. Chem.*, **14**, 680–4.

Cramer III, R. D., Bunce, J. D. and Patterson, D. E. (1988) Crossvalidation, bootstrapping, and partial least squares compared with multiple regression in conventional QSAR studies. *Quant. Struct.–Act. Relat.*, **7**, 18–25.

Crick, F. (1989) The recent excitement about neural networks. *Nature*, **337**, 129–32.

Cronin, M. T. D., Dearden, J. C. and Dobbs, A. J. (1991) QSAR studies of comparative toxicity in aquatic organisms. *Sci. Total Environ.*, **109/110**, 431–9.

Cross, B., Hoffman, P. P., Santora, G. T., Spatz, D. M. and Templeton, A. R. (1983) The design of postemergence phenylurea herbicides using physicochemical parameters and structure–activity analyses. *J. Agric. Food Chem.*, **31**, 260–4.

Dannenfelser, R. M., Paric, M., White, M. and Yalkowsky, S. H. (1991) A compilation of some physico-chemical properties for chlorobenzenes. *Chemosphere*, **23**, 141–65.

Davies, R. P. and Dobbs, A. J. (1984) The prediction of bioconcentration in fish. *Water Res.*, **18**, 1253–62.

Dearden, J. C. (1990) Physico-chemical descriptors, in *Practical Applications of Quantitative Structure–Activity Relationships (QSAR) in Environmental Chemistry and Toxicology* (eds W. Karcher and J. Devillers), Kluwer, Dordrecht.

Dearden, J. C. (1991) The QSAR prediction of melting point, a property of environmental relevance. *Sci. Total Environ.*, **109/110**, 59–68.

Degner, P. (1991) *Abschätzung der biologischen Abbaubarkeit mittels SARs*. Dissertation, Universität Duisburg.

Degner, P., Nendza, M. and Klein, W. (1991) Predictive QSAR models for estimating biodegradation for aromatic compounds. *Sci. Total Environ.*, **109/110**, 253–9.

Deneer, J. W., Sinnige, T. L., Seinen, W. and Hermens, J. L. M. (1987) Quantitative structure–activity relationships for the toxicity and bioconcentration factor of nitrobenzene derivatives towards the guppy (*Poecilia reticulata*). *Aquat. Toxicol.*, **10**, 115–29.

Deneer, J. W., Sinnige, T. L., Seinen, W. and Hermens, J. L. M. (1988) The joint acute toxicity to *Daphnia magna* of industrial organic chemicals at low concentrations. *Aquat. Toxicol.*, **12**, 33–8.

Deneer, J. W., Leeuwen, C. J. van, Maas-Diepeveen, J. L. and Hermens, J. L. M. (1989) QSAR study of the toxicity of nitrobenzene derivatives towards *Daphnia magna*, *Chlorella pyrenoidosa* and *Photobacterium phosphoreum*. *Aquat. Toxicol.*, **15**, 83–98.

Derde, M. P. and Massart, D. L. (1982) Extraction of information from large data sets by pattern recognition. *Fresenius Z. Anal. Chem.*, **313**, 484–95.

Desai, S. and Govind, R. (1990) Development of quantitative structure activity relationships for predicting biodegradation kinetics. *Environ. Toxicol. Chem.*, **9**, 473–7.

DEV (1987) Normenausschuß Wasserwesen (NAW) im DIN Deutsches Institut für Normung e.V.: *German Standard Methods for the Examination of Water, Waste Water and Sludge; General Measures of Effects and Substances (Group H); Determination of the Biological Oxygen Demand after n Days, Dilution Methods (Dilution BOD$_n$) (H51)*, DIN 38409, Teil 51, Beuth Verlag, Berlin.

Devillers, J. and Chambon, P. (1986) Toxicity and QSAR of chlorophenols on *Daphnia magna*. *Bull. Environ. Contam. Toxicol.*, **37**, 599–605.

Devillers, J., Chambon, P. and Zakarya, D. (1987) A predictive structure–toxicity model with *Daphnia magna*. *Chemosphere*, **16**, 1149–63.

Devillers, J., Chambon, P., Zakarya, D. and Chastrette, M. (1986) A new approach to ecotoxicological QSAR studies. *Chemosphere*, **15**, 993–1002.

Dewar, M. J. S. and Thiel, W. (1977) Ground states of molecules. 38. The MNDO method. Approximations and Parameters. *J. Am. Chem. Soc.*, **99**, 4899–907.

Dewar, M. J. S., Zoebisch, E. G., Healy, E. F. and Stewart, J. J. P. (1985) AM1: A new general purpose quantum mechanical molecular model. *J. Am. Chem. Soc.*, **107**, 3902–9.

DMU (1992) *Manual for Program GAMMA: Calculation of Activity Coefficients by Means of UNIFAC*, Afdeling for Systemanalyse, Danmarks Miljoundersogelser, Roskilde.

Donkin, P. (1994) Quantitative structure–activity relationships, in *Handbook of Ecotoxicology* (ed. P. Calow), Blackwell, Oxford.

Draper, N. R. and Smith, H (1981) *Applied Regression Analysis*, Wiley, New York, NY.

Drossman, H., Johnson, H. and Mill, T. (1988) Structure activity relationships for environmental processes 1: hydrolysis of esters and carbamates. *Chemosphere*, **17**, 1509–30.

Dunn III, W. J. and Wold, S. (1990) Pattern recognition techniques in drug design, in *Quantitative Drug Design* (ed. C. A. Ramsden), Pergamon Press, Oxford.

Dunn III, W. J., Grigoras, S. and Johansson, E. (1986) Principal component analysis of partition coefficient data: an approach to a new method of partition coefficient estimation, in *Partition Coefficient Determination and Estimation* (eds W. J. Dunn III, J. H. Block, R. S. Pearlman), Pergamon Press, New York, NY.

Dunn III, W. J., Wold, S., Edlund, U., Hellberg, S. and Gasteiger, J. (1984) Multivariate structure–activity relationships between data from a battery of biological tests and an ensemble of structure descriptors: the PLS method. *Quant. Struct.-Act. Relat.*, **3**, 131–7.

Dura, Gy., Krasovski, G. N., Zholdakova, Z. I., Mayer, G. (1985) Prediction of toxicity using quantitative structure–activity relationships. *Arch. Toxicol.*, Suppl., **8**, 481–7.

ECDIN, *Environmental Chemicals Data and Information Network*, Ispra Establishment of the EU Joint Research Centre, Ispra.

ECETOC (1986) *Structure–Activity Relationships in Toxicology and Ecotoxicology: An Assessment*, Monograph No. 8, Ecetoc, Brussels.

Edward, J. T., Johanson, E. and Wold, S. (1988) Use of molecular connectivity indices in socio-historical and business management studies, in *Graph Theory Notes of New York XV* (eds J. W. Kennedy and L. V. Quintas), New York Academy of Sciences, New York, NY.

EEC (1990) *Proposal for a Council Regulation (EEC) on the Evaluation and the Control of the Environmental Risks of Existing Substances*, Commission of the European Communities, Brussels.

Ehrlich, P. (1910) *Studies in Immunity*, Wiley, New York, NY.

ElTayar, N., Testa, B. and Carrupt, P. A. (1992) Polar intermolecular interactions encoded in partition coefficients: an indirect estimation of hydrogen-bond parameters of polyfunctional solutes. *J. Phys. Chem.*, **96**, 1455–9.

ElTayar, N., Tsai, R. S., Testa, B., Carrupt, P. A. and Leo, A. (1991) Partitioning of solutes in different solvent systems: the contribution of hydrogen-bonding capacity and polarity. *J. Pharm. Sci.*, **20**, 590–8.

Enslein, K. and Craig, P. N. (1978) A toxicity estimation model. *J. Environ. Pathol. Toxicol.*, **2**, 115–21.

EPA (1982) *Environmental Effects Guide-lines*, EPA-560/6-82-002. US Environmental Protection Agency, Washington, DC.

EPA (1986a) *CHEMEST* Office of Toxic Substances, Washington, DC.

EPA (1986b) *Environmental Partitioning User's Guide*, Office of Toxic Substances, Washington, DC.

EPA (1991) *ASTER* Office of Toxic Substances, Washington, DC.

EPA (1994) *ECOSAR*, Office of Pollution Prevention and Toxics, Washington, DC.

Eriksson, L. and Hermens, J. L. M. (1995) Multivariate approach to quantitative structure–activity and structure–property relationships, in *Handbook of Environmental Chemistry*, Vol. 2H (ed. J. Einax), Springer Verlag, Berlin.

Eriksson, L., Jonsson, J. and Tysklind, M. (1995) Multivariate QSAR modeling of biodehalogenation half-lives of halogenated aliphatic hydrocarbons. *Environ. Toxicol. Chem.*, **14**, 209–17.

Eriksson, L., Verhaar, H. J. M. and Hermens, J. L. M. (1994) Multivariate characterization and modeling of the chemical reactivity of epoxides. *Environ. Toxicol. Chem.* **13**, 683–91.

Esser, H. O. (1986) A review of the correlation between physicochemical properties and bioaccumulation. *Pestic. Sci.*, **17**, 265–76.

European Commission (1996) Technical guidance document in support of Commission Directive 93/67/EEC on risk assessment for new notified substances and Commission Regulation (EC) No 1488/94 on risk assessment for existing substances. Part III, Brussels.

Exon, J. H. and Koller, L. D. (1984) A review of chlorinated phenols. *Vet. Hum. Tox.*, **26**, 508–20.

Fan, W., ElTayar, N., Testa, B. and Kier, L. B. (1990) Water-dragging effect: a new experimental hydration parameter related to hydrogen bond donor acidity. *J. Phys. Chem.*, **94**, 4764–6.

Feijtel, T. C. J. (1995) Evaluation of the use of QSARs for priority setting and risk assessment. *SAR QSAR Environ. Res.*, **3**, 237–45.

Feijtel, T., Veerkamp, W., Koch, V. and Niessen, H. (1995) Use of environmental fate models in the risk assessment of substances. *Toxicol. Model.*, **1**, 5–19.

Ferguson, J. (1939) The use of chemical potentials as indices of toxicity. *Proc. R. Soc. Lond.*, **B127**, 387–404.

Fini, O., Brusa, F. and Chiesa, L. (1981) Bestimmung der Totzeit in der 'Reversed-Phase' Hochleistungsflüssigkeitschromatographie. *J. Chromatogr.*, **210**, 326–30.

Flora, S. de, Koch, R. Strobel, K. and Nagel, M. (1985) A model based on molecular structure descriptors for predicting mutagenicity of organic compounds. *Toxicol. Environ. Chem.*, **10**, 157–70.

Franke, C., Studinger, G., Berger, G., Böhling, S., Bruckmann, U., Cohors-Fresenborg, D. and Jöhncke, U. (1994) The assessment of bioaccumulation. *Chemosphere*, **29**, 1501–14.

Franke, R. (1979) On the interpretability of quantitative structure–activity relationships (QSAR). *Il Farmaco Ed. Sci.*, **34**, 545–70.

Franke, R. (1980) *Optimierungsmethoden in der Wirkstoff-Forschung*, Akademie Verlag, Berlin.

Franke, R. (1984) *Theoretical Drug Design Methods*, Elsevier, Amsterdam.

Franke, R. (1985) QSAR parameters, in *QSAR and Strategies in the Design of Bioactive Compounds* (ed. J. K. Seydel), VCH, Weinheim.

Franke, R. and Schmidt, W. (1973) Parabelförmige Zusammenhänge zwischen biologischer Aktivität und Hydrophobizität in kongeneren Serien: Transport oder Eiweißbindung? *Acta Biol. Med. Germ.*, **31**, 273–87.

Franks, N. P. and Lieb, W. R. (1990) Mechanisms of general anaesthesia. *Environ. Health Persp.*, **87**, 199–205.

Fredenslund, Aa., Gmehling, J. and Rasmussen, P. (1977) *Vapour–Liquid Equilibria Using UNIFAC*, Elsevier, Amsterdam.

Fredenslund, F. C., Nendza, M. and Herbst, T. (1995) Cosolute interactions on aqueous contaminant concentrations. *SAR QSAR Environ. Res.*, **3**, 293–300.

Frei, W. (1913) Versuche über Kombination von Desinfektionsmitteln. *Z. Hygiene Infekt.*, **75**, 433–96.

Fujinami, A., Mine, A. and Fujita, T. (1974) Relationship between chemical structure and selectivity in herbicidal activity of trans-β-(2,4-dichlorophenoxy)-acrylates against rice plant and barnyard-grass. *Agr. Biol. Chem.*, **38**, 1399–403.

Fujita, T. (1966) The analysis of physiological activity of phenols with substituent constants. *J. Med. Chem.*, **9**, 797–803.

Fujita, T., Iwasa, J. and Hansch, C. (1964) A new substituent constant $\pi$ derived from partition coefficient. *J. Am. Chem. Soc.*, **86**, 5175.

Fukuto, T. R. (1990) Mechanism of action of organophosphorus and carbamate insecticides. *Environ. Health Persp.*, **87**, 245–54.

Galassi, S., Mingazzini, M., Vigano, L., Cesareo, D. and Tosato, M. L. (1988) Approaches to modeling toxic responses of aquatic organisms to aromatic hydrocarbons. *Ecotoxicol. Environ. Saf.*, **16**, 158–69.

Garrett, E. R. (1966) The use of microbial kinetics in the quantification of antibiotic action. *Drug Res.*, **16**, 1364–9.

Garrett, E. R. (1971) Drug action and assay by microbial kinetics. *Prog. Drug Res.*, **15**, 271–352.

Geating, J. (1981) *Literature Study of the Biodegradability of Chemicals in Water*, Vols 1–2, EPA-600/2-81-175 and EPA-600/2-81-176, Municipal Environmental Research Laboratory, Cincinnati/Office of Research and Development, US Environmental Protection Agency, Cincinnati, OH.

Gehring, P. J. and Rao, K. S. (1977) Toxicologic data extrapolation, in *Industrial Hygiene and Toxicology*, Vol. III (ed. F. A. Patty), Wiley, New York, NY.

Geiger, D. L., Brooke, L. T. and Call, D. J. (1990) *Acute Toxicities of Organic Pollutants to Fathead Minnow (Pimephales promelas)*, Vol. V, Center for Lake Superior Environmental Studies, Superior, WI.

Geiger, D. L., Call, D. J. and Brooke, L. T. (1988) *Acute Toxicities of Organic Pollutants to Fathead Minnow (Pimephales promelas)*, Vol. IV, Center for Lake Superior Environmental Studies, Superior, WI.

Geiger, D. L., Northcott, C. E., Call, D. J. and Brooke, L. T. (1985) *Acute Toxicities of Organic Pollutants to Fathead Minnow (Pimephales promelas)*, Vol. II, Center for Lake Superior Environmental Studies, Superior, WI.

Geiger, D. L., Poirier, S. H., Brooke, L. T. and Call, D. J. (1986) *Acute Toxicities of Organic Pollutants to Fathead Minnow (Pimephales promelas)*, Vol. III, Center for Lake Superior Environmental Studies, Superior, WI.

Geladi, P. and Kowalski, B. R. (1986) Partial least squares regression: a tutorial. *Anal. Chim. Acta*, **185**, 1–17.

Geladi, P. and Tosato, M. L. (1990) Multivariate latent variable projection methods: SIMCA and PLS, in *Practical Applications of Quantitative Structure–Activity Relationships (QSAR) in Environmental Chemistry and Toxicology* (eds W. Karcher and J. Devillers), Kluwer, Dordrecht.

Gestel, C. A. M. van, Ma, W. and Smit, C. E. (1991) Development of QSARs in terrestrial ecotoxicology: earthworm toxicity and soil sorption of chlorophenols, chlorobenzenes and dichloroaniline. *Sci. Total Environ.*, **109/110**, 589–604.

Geyer, H., Politzki, G. and Freitag, D. (1984) Prediction of ecotoxicological behaviour of chemicals: relationship between n-octanol/water partition coefficient and bioaccumulation of organic chemicals by algae *Chlorella*. *Chemosphere*, **13**, 269–84.

Geyer, H. J., Steinberg, C. E. W. and Kettrup, A. (1992) Biokonzentration von persistenten superlipophilen Stoffen in aquatischen Organismen. *UWSF-Z. Umweltchem. Ökotox.*, **4**, 74–7.

Geyer, H., Sheehan, D., Kotzias, D., Freitag, D. and Korte, F. (1982) Prediction of ecotoxicological behaviour of chemicals: relationship between physicochemical properties and bioaccumulation of organic chemicals in the mussel. *Chemosphere*, **11**, 1121–34.

Gobas, F. A. P. C. (1991) A thermodynamic analysis of quantitative structure–activity relationships for hydrophobic organic chemicals. *Sci. Total Environ.*, **109/110**, 89–104.

Gobas, F. A. P. C. and Schrap, S. M. (1990) Bioaccumulation of some polychlorinated dibenzo-*p*-dioxins and octachlorodibenzofuran in the guppy (*Poecilia reticulata*). *Chemosphere*, **20**, 495–512.

Gobas, F. A. P. C., Shiu, W. Y. and Mackay, D. (1987) Factors determining partitioning of hydrophobic organic chemicals in aquatic organisms, in *QSAR in Environmental Toxicology – II* (ed. K. L. E. Kaiser), D. Reidel, Dordrecht.

Gobas, F. A. P. C., McNeil, E. J., Lovett-Doust, L. and Haffner, G. D. (1991) Bioconcentration of chlorinated aromatic hydrocarbons in aquatic macrophytes. *Environ. Sci. Technol.*, **25**, 924–9.

Gombar, V. K. and Enslein, K. (1989) Topological shape and electronic descriptors and their correlation with toxicity to *Photobacterium phosphoreum*. *In Vitro Toxicol.*, **2**, 117–27.

Govers, H., Ruepert, C. and Aiking, H. (1984) Quantitative structure–activity relationships for aromatic hydrocarbons: correlation between molecular connectivity, physico-chemical properties, bioconcentration and toxicity in *Daphnia pulex*. *Chemosphere*, **13**, 227–36.

Govers, H., Ruepert, C., Stevens, T. and Leeuwen, C. J. van (1986) Experimental determination and prediction of partition coefficients of thioureas and their toxicity to *Photobacterium phosphoreum*. *Chemosphere*, **15**, 383–93.

Govers, H. A. J., Parsons, J. R., Krop, H. B. and Cheung, C. L. (1995) Thermodynamic descriptors for (bio-)degradation, in *Proceedings of the Workshop 'Quantitative Structure Activity Relationships for Biodegradation'*, [September 1994, Belgirate, Italy], Report No. 719101021, National Institute of Public Health and Environmental Protection, Bilthoven.

Grafe, A., Mattern, I. E. and Green, M. (1981) A European collaborative study of the Ames assay I. Results and general interpretation. *Mutat. Res.*, **85**, 391–410.

Grain, C. F. (1990) Vapor pressure, in *Handbook of Chemical Property Estimation Methods* (eds W. J. Lyman, W. F. Reehl and D. H. Rosenblatt), American Chemical Society, Washington, DC.

Güsten, H., Klasinc, L. and Maric, D. (1984) Prediction of the abiotic degradability of organic compounds in the troposphere. *J. Atmos. Chem.*, **2**, 83–93.

Guthrie, F. E. (1984) Adsorption and distribution, in *Introduction to Biochemical Toxicology* (eds E. Hodgson and F. E. Guthrie), Elsevier, New York, NY.

Hall, L. H. and Kier, L. B. (1984) Molecular connectivity of phenols and their toxicity to fish. *Bull. Environ. Contam. Toxicol.*, **32**, 354–62.

Hall, L. H., Kier, L. B. and Phipps, G. (1984) Structure–activity relationship studies on the toxicities of benzene derivatives: I. An additivity model. *Environ. Toxicol. Chem.*, **3**, 355–65.

Hall, L. M., Maynard, E. L. and Kier, L. B. (1989) Structure–activity relationship studies on the toxicity of benzene derivatives: III. Predictions and extension to new substituents. *Environ. Toxicol. Chem.*, **8**, 431–6.

Hamaker, J. W. and Thompson, J. M. (1972) Adsorption, in *Organic Chemicals in the Soil Environment* (eds C. A. I. Goring and J. W. Hamaker), Marcel Dekker, New York, NY.

Hamelink, J. L. and Spacie, A. (1977) Fish and chemicals: the process of accumulation. *Ann. Rev. Pharmacol. Toxicol.*, **17**, 167–77.

Hamelink, J. L., Waybrant, R. C. and Ball, C. (1971) A proposal: exchange equilibria control the degree chlorinated hydrocarbons are biologically magnified in lentic environments. *Trans. Am. Fish. Soc.*, **100**, 207–14.

Hammett, L. P. (1940) *Physical Organic Chemistry*, McGraw Hill, New York, NY.

Hansch, C. (1968a) Physicochemical parameters in drug design. *Ann. Rep. Med. Chem.*, 248–357.

Hansch, C. (1968b) The use of substituent constants in drug modification. *Il Farmaco Ed. Sci.*, 23, 293–320.

Hansch, C. (1969) A quantitative approach to biochemical structure–activity relationships. *Acc. Chem. Res.*, **2**, 232–40.

Hansch, C. (1975) Les relations structure–activité quantitatives en chimie thérapeutique, in *Relations Structure–Activité*, Societé de Chimie Thérapeutique, Paris.

Hansch, C. (1977) On the predictive value of QSAR, in *Biological Activity and Chemical Structure* (ed. J. A. Keverling Buisman), Elsevier, Amsterdam.

Hansch, C. (1980) The role of the partition coefficient in environmental toxicology, in *Dynamics, Exposure and Hazard Assessment of Toxic Substances* (ed. R. Hague), Ann Arbor Science, Ann Arbor, MI.

Hansch, C. (1984) The QSAR paradigm in the design of less toxic molecules. *Drug Metab. Rev.*, **15**, 1279–94.

Hansch, C. (1991) Structure–activity relationships of chemical mutagens and carcinogens. *Sci. Total Environ.*, **109/110**, 17–29.

Hansch, C. (1993) Quantitative structure–activity relationships and the unnamed science. *Acc. Chem. Res.*, **26**, 147–53.

Hansch, C. and Clayton, J. M. (1973) Lipophilic character and biological activity of drugs II: the parabolic case. *J. Pharm. Sci.*, **62**, 1–21.

Hansch, C. and Dunn III, W. J. (1972) Linear relationships between lipophilic character and biological activity of drugs. *J. Pharm. Sci.*, **61**, 1–19.

Hansch, C. and Fujita, T. (1964) $\rho$-$\sigma$-$\pi$ Analysis. A method for the correlation of biological activity and chemical structure. *J. Am. Chem. Soc.*, **86**, 1616–26.

Hansch, C. and Leo, A. J. (1979) *Substituent Constants for Correlation Analysis in Chemistry and Biology*, Wiley, New York, NY.

Hansch, C., Kiehs, K. and Lawrence, G. L. (1965) The role of substituents in the hydrophobic bonding of phenols by serum and mitochondrial proteins. *J. Am. Chem. Soc.*, **87**, 5770–3.

Hansch, C., Leo, A. and Taft, R. W. (1991) A survey of Hammett substituent constants and resonance and field parameters. *Chem. Rev.* **91**, 165–95.

Hansch, C., Quinlan, J. E. and Lawrence, G. L. (1968) The linear free-energy relationship between partition coefficients and the aqueous solubility of organic liquids. *J. Org. Chem.*, **33**, 347–50.

Hansch, C., Muir, R. M., Fujita, T., Malone, P. P., Geiger, F. and Streich, M. (1963) The correlation of biological activity of plant growth regulators and chloromycetin derivatives with Hammett constants and partition coefficients. *J. Am. Chem. Soc.*, **85**, 2817.

Hansen, H. K., Rasmussen, P., Fredenslund, Aa., Schiller, M. and Gmehling, J. (1991) Vapour–liquid equilibria by UNIFAC group contribution. 5. Revision and extension. *Ind. Eng. Chem. Res.*, **30**, 2352–5.

Harnisch, M., Möckel, H. J. and Schulze, G. (1983) Relationship between log $P_{ow}$ shake-flask values and capacity factors derived from reversed-phase high-performance liquid chromatography for n-alkylbenzenes and some OECD reference substances. *J. Chromatogr.*, **282**, 315–32.

Harris, J. C. (1990) Rate of hydrolysis, in *Handbook of Chemical Property Estimation Methods* (eds. W. J. Lyman, W. F. Reehl and D. H. Rosenblatt), American Chemical Society, Washington, DC.

Hauk, A., Richartz, H., Schramm, K. W. and Fiedler, H. (1990) Reduction of nitrated phenols: a method to predict half-wave potentials of nitrated phenols with molecular modelling. *Chemosphere*, **20**, 717–28.

Hawker, D. W. (1990) Description of fish bioconcentration factors in terms of solvatochromic parameters. *Chemosphere*, **20**, 467–77.

Hawker, D. W. and Connell, D. W. (1986) Bioconcentration of lipophilic compounds by some aquatic organisms. *Ecotoxicol. Environ. Saf.*, **11**, 184–97.

Hawker, D. W. and Connell, D. W. (1989) A simple water/octanol partition system for bioconcentration investigations. *Environ. Sci. Technol.*, **23**, 961–5.

Hawker, D. and Connell, D. (1991) An evaluation of the relationship between bioconcentration factor and aqueous solubility. *Chemosphere*, **23**, 231–41.

HDI (1990) *Topkat Predictions of Toxicity from Chemistry*, Health Designs, Inc., New York, NY.

Hebert, C. E. and Keenleyside, K. A. (1995) To normalize or not to normalize? Fat is the question. *Environ. Toxicol. Chem.*, **14**, 801–7.

Heinen, W. (1972) Inhibitors of electron transport and oxidative phosphorylation, in *Methods in Microbiology* (eds J. R. Norris and D. W. Ribbons), Academic Press, London.

Hellberg, S. (1986) *A Multivariate Approach to QSAR*. Thesis, Umea University, Umea.

Hendriks, A. J. (1995) Modelling equilibrium concentrations of microcontaminants in organisms of the Rhine Delta: can average field residues in the aquatic foodchain be predicted from laboratory accumulation? *Aquat. Toxicol.*, **31**, 1–25.

Hendriks, A. J., Maas-Diepeveen, J. L., Noordsij, A. and Gaag, M. A. van der (1994) Monitoring response of XAD-concentrated water in the Rhine Delta: a major part of the toxic compounds remains unidentified. *Water Res.*, **23**, 581–98.

Herington, E. F. G. and Kynaston, W. (1952) The effect of solvent on the ultra-violet absorption spectra of aromatic hydrocarbons with special reference to the mechanism of salting-out. Part II: aqueous salt solutions. *J. Chem. Soc. (Lond.)*, 3143–9.

Hermens, J. L. M. (1986) Quantitative structure–activity relationships in aquatic toxicology. *Pestic. Sci.*, **17**, 287–96.

Hermens, J. L. M. (1989) Quantitative structure–activity relationships of environmental pollutants, in *Handbook of Environmental Chemistry*, Vol. 2E (ed. O. Hutzinger), Springer Verlag, Berlin.

Hermens, J. L. M. (1990) Electrophiles and acute toxicity to fish. *Environ. Health Persp.*, **87**, 218–25.

Hermens, J. and Opperhuizen, A. (eds) (1991) QSAR in environmental toxicology. *Sci. Total Environ.*, **109/110**.

Hermens, J., Leeuwangh, P. and Musch, A. (1984) Quantitative structure–activity relationships and mixture toxicity studies of chloro- and alkylanilines at an acute lethal toxicity level to the guppy (*Poecilia reticulata*). *Ecotoxicol. Environ. Saf.*, **8**, 388–94.

Hermens, J., Canton, H., Janssen, P. and Jong, R. de (1984) Quantitative structure–activity relationships and toxicity studies of mixtures of chemicals with anaesthetic potency: acute lethal and sublethal toxicity to *Daphnia magna. Aquat. Toxicol.*, **5**, 143–54.

Hermens, J., Broekhuyzen, E., Canton, J. and Wegman, R. (1985a) Quantitative structure–activity relationships and mixture toxicity studies of alcohols and chlorohydrocarbons: effects on growth of *Daphnia magna. Aquat. Toxicol.*, **6**, 209–17.

Hermens, J., Busser, F., Leeuwangh, P. and Musch, A. (1985b) Quantitative structure–activity relationships and mixture toxicity of organic chemicals in *Photobacterium phosphoreum*: the Microtox test. *Ecotoxicol. Environ. Saf.*, **9**, 17–25.

Hermens, J., Könemann, H., Leeuwangh, P. and Musch, A. (1985c) Quantitative structure–activity relationships in aquatic toxicity studies of chemicals and complex mixtures of chemicals. *Environ. Toxicol. Chem.*, **4**, 273–9.

Hermens, J., Bruijn, J. de, Pauly, J. and Seinen, W. (1987) QSAR studies for fish toxicity data of organophosphorus compounds and other classes of reactive organic compounds, in *QSAR in Environmental Toxicology – II* (ed. K. L. E. Kaiser), D. Reidel, Dordrecht.

Hermens, J., Verbruggen, E., Busser, F. and Verhaar, H. (1994) A new parameter for the hydrophobicity of mixtures. Abstracts, 15th SETAC Annual Meeting, Denver, CO.

Hermens, J., Balaz, S., Damborsky, J., Karcher, W., Müller, M., Peijnenburg, W., Sabljic, A. and Sjöström, M. (1995) Assessment of QSARs for predicting fate and effects of chemicals in the environment: an international European project. *SAR QSAR Environ. Res.*, **3**, 223–36.

Hetrick, D. M. and McDowell-Boyer, L. M. (1983) *User's Manual for TOXSCREEN: A Multimedia Screening-level Program for Assessing the Potential Fate of Chemicals Released to the Environment*, Office of Pesticides and Toxic Substances, EPA, Washington, DC.

Hickey, J. P. and Passino-Reader, D. R. (1991) Linear solvation energy relationships: 'rules of thumb' for estimation of variable values. *Environ. Sci. Technol.*, **25**, 1753–60.

Hine, J. and Mookerjee, P. K. (1975) The intrinsic hydrophilic character of organic compounds. Correlations in terms of structural contributions. *J. Org. Chem.*, **40**, 292–8.

Hodson, J. and Williams, N. A. (1988) The estimation of the adsorption coefficient ($K_{oc}$) for soils by high performance liquid chromatography. *Chemosphere*, **17**, 67–77.

Hodson, P. V. (1985) A comparison of the acute toxicity of chemicals to fish, rats and mice. *J. Appl. Toxicol.*, **5**, 220–6.

Holcombe, G. W., Phipps, G. L. and Fiandt, J. T. (1983) Toxicity of selected priority pollutants to various aquatic organisms. *Ecotoxicol. Environ. Saf.*, **7**, 400–9.

Howard, P. H., Boethling, R. S., Stiteler, W. M., Meylan, W. M., Hueber, A. E., Beauman, J. A. and Larosche, M. E. (1992) Predictive model for aerobic biodegradability developed from a file of evaluated biodegradation data. *Environ. Toxicol. Chem.*, **11**, 593–603.

Hulzebos, E. M., Adema, D. M. M., Dirven-van Bremen, E. M., Henzen, L. and Gestel, C. A. M. van (1991) QSARs in phytotoxicity. *Sci. Total Environ.*, **109/110**, 493–7.

Hunn, J. B. and Allen, J. L. (1974) Movement of drugs across the gills of fish. *Ann. Rev. Pharmacol.*, **14**, 47–55.

Hunter, B. (1994) A comparative study of chemical similarity distance functions. Abstracts, 6th QSAR Workshop, Belgirate, Italy.

Ikemoto, Y., Motoba, K., Suzuki, T. and Uchida, M. (1992) Quantitative structure–activity relationships of nonspecific and specific toxicants in several organism species. *Environ. Toxicol. Chem.*, **11**, 931–9.

Isensee, A. R. and Jones, G. E. (1975) Distribution of 2,3,7,8-tetrachlorodibenzo-*p*-dioxin (TCDD) in an aquatic model ecosystem. *Environ. Sci. Technol.*, **9**, 668–72.

Isnard, P. and Lambert, S. (1989) Aqueous solubility and n-octanol/water partition coefficient correlations. *Chemosphere*, **18**, 1837–53.

IUCT (1992) *Handbuch zum SAR-Programm Version 3.0*, Fraunhofer-Institut für Umweltchemie und Ökotoxikologie, Schmallenberg.

Iwamura, H. and Fujita, T. (1982) QSAR studies in pesticide research in Japan. *J. Pestic. Sci.*, **7**, 289–99.

Jäckel, H. and Klein, W. (1991) Prediction of mammalian toxicity by quantitative structure–activity relationships: aliphatic amines and anilines. *Quant. Struct.–Act. Relat.*, **10**, 198–204.

Jäckel, H. and Nendza, M. (1994) Reactive substructures in the prediction of aquatic toxicity data. *Aquat. Toxicol.*, **29**, 305–14.

Janardan, S. K., Olson, C. S. and Schaeffer, D. J. (1984) Quantitative comparisons of acute toxicity of organic chemicals to rat and fish. *Ecotoxicol. Environ. Saf.*, **8**, 531–9.

Jaworska, J. S. and Schultz, T. W. (1993) Quantitative relationships of structure–activity and volume fraction for selected nonpolar and polar narcotic chemicals. *SAR QSAR Environ. Res.*, **1**, 3–19.

Johnson, H., Kenley, R. A., Rynard, C. and Golub, M. A. (1985) QSAR for cholinesterase inhibition by organophosphorus esters and CNDO/2 calculations for organophosphorus ester hydrolysis. *Quant. Struct.–Act. Relat.*, **4**, 172–80.

Johnson, R. A. and Wichern, D. W. (1988) *Applied Multivariate Statistical Analysis*. Prentice Hall, Englewood Cliffs, NJ.

Jolliffe, J. T. (1986) *Principal Component Analysis*, Springer Verlag, New York, NY.

Jurs, P. C., Chon, J. T. and Yuan, M. (1979) Computer-assisted structure–activity studies of chemical carcinogens: a heterogeneous data set. *J. Med. Chem.*, **22**, 476–83.

Kaiser, K. L. E., Dixon, D. G. and Hodson, P. V. (1984) QSAR studies on chlorophenols, chlorobenzenes and *para*-substituted phenols, in *QSAR in Environmental Toxicology* (ed. K. L. E. Kaiser), D. Reidel, Dordrecht.

Kamlet, M. J., Doherty, R. M., Veith, G. D. and Taft, R. W., Abraham, M. H. (1986) Solubility properties in polymers and biological media. 7. An analysis of toxicant properties that influence inhibition of bioluminescence in *Photobacterium phosphoreum* (the Microtox test). *Environ. Sci. Technol.*, **20**, 690–5.

Kamlet, M. J., Doherty, R. M., Abraham, M. H., Marcus, Y. and Taft, R. W. (1988a) Linear solvation energy relationships. 46. An improved equation for correlation and prediction of octanol/water partition coefficients of organic nonelectrolytes (including strong hydrogen bond donor solutes). *J. Phys. Chem.*, **92**, 5244–55.

Kamlet, M. J., Doherty, R. M., Abraham, M. H. and Taft, R. W. (1988b) Solubility properties in biological media. 12. Regarding the mechanism of nonspecific toxicity or narcosis by organic nonelectrolytes. *Quant. Struct.–Act. Relat.*, **7**, 71–8.

Kamlet, M. J., Doherty, R. M., Carr, P. W., Mackay, D., Abraham, M. H. and Taft, R. W. (1988c) Linear solvation energy relationships. 44. Parameter estimation rules that allow accurate prediction of octanol/water partition coefficients and other solubility and toxicity properties of polychlorinated biphenyls and polycyclic aromatic hydrocarbons. *Environ. Sci. Technol.*, **22**, 503–9.

Kanazawa, J. (1989) Relationship between the soil sorption constants for pesticides and their physicochemical properties. *Environ. Toxicol. Chem.*, **8**, 477–84.

Kanazawa, J., Isensee, A. R. and Kearney, T. C. (1975) Distribution of carbaryl and 3,5-xylyl methylcarbamate in an aquatic model ecosystem. *J. Agric. Food Chem.*, **23**, 760–3.

Karcher, W. and Devillers, J. (eds) (1990) *Practical Applications of Quantitative Structure–Activity Relationships (QSAR) in Environmental Chemistry and Toxicology*, Kluwer, Dordrecht.

Karickhoff, S. W. (1981) Semi-empirical estimation of sorption of hydrophobic pollutants on natural sediments and soils. *Chemosphere*, **10**, 833–46.

Karickhoff, S. W., Brown, D. S. and Scott, T. A. (1979) Sorption of hydrophobic pollutants on natural sediments and soil. *Water Res.*, **13**, 241–8.

Kawasaki, M. (1980) Experience with the test scheme under the chemical control law of Japan: an approach to structure–activity correlations. *Ecotoxicol. Environ. Saf.*, **4**, 444–54.

Kenaga, E. E. (1980) Correlation of bioconcentration factors of chemicals in aquatic and terrestrial organism with their physical and chemical properties. *Environ. Sci. Technol.*, **14**, 553–6.

Kenaga, E. E. and Goring, C. A. (1980) Relationship between water solubility, soil sorption, octanol–water partitioning and bioconcentration of chemicals in biota, in *Aquatic Toxicology*, (eds J. G. Eaton, P. R. Parrish and A. C. Hendricks), STP 707 ASTM, Philadelphia, PA.

Kier, L. B. and Hall, L. H. (1976) *Molecular Connectivity in Chemistry and Drug Research*, Academic Press, New York, NY.

Kier, L. B., Simons, R. J. and Hall, L. H. (1978) Structure–activity studies on mutagenicity of nitrosamines using molecular connectivity. *J. Pharm. Sci.*, **67**, 725–6.

Klamt, A. (1993) Estimation of gas-phase hydroxyl radical rate constants of organic compounds from molecular orbital calculations. *Chemosphere*, **26**, 1273–89.

Klein, W., Klein, A. W. and Lange, A. W. (eds) (1988) Advances in environmental hazard and risk assessment. *Chemosphere*, **17**, v–viii, 1381–644.

Klopman, G. (1984) Artificial intelligence approach to structure–activity studies. Computer automated structure evaluation of biological activity of organic molecules. *J. Am. Chem. Soc.*, **106**, 7315–21.

Klopman, G., Frierson, M. R. and Rosenkranz, H. S. (1990). The structural basis of the mutagenicity of chemicals in *Salmonella typhimurium*: the gene-tox data basc. *Mutat. Res.*, **228**, 1–50.

Kobayashi, K. (1981) Safety examination of existing chemicals – selection, testing, evaluation and regulation in Japan, in *Proceedings of the Workshop on the Control of Existing Chemicals under the Patronage of the Organisation for Economic Co-operation and Development, June 10–12, 1981*, Umweltbundesamt, Berlin.

Koch, R. (1983) Molecular connectivity index for assessing ecotoxicological behaviour of organic compounds. *Toxicol. Environ. Chem.*, **6**, 87–96.

Koch, R. and Nagel, M. (1988) Quantitative structure–activity relationships in soil ecotoxicology. *Sci. Total Environ.*, **77**, 269–76.

Koch, R. Strobel, K. and Nagel, M. (1985) Mutagene Aktivität von Umweltschadstoffen: Ein Struktur-Wirkungs-Analogie-Modell. *Z. Ges. Hyg.*, **31**, 524–6.

Kondo, M., Nichihara, T., Shimamoto, T., Watabe, K. and Fujii, M. (1988) Screening test method for degradation of chemicals in water. A simple and rapid method for biodegradation test. *Eisei Kagaku*, **34**, 115–22.

Könemann, H. (1981a) Fish toxicity tests with mixtures of more than two chemicals: a proposal for a quantitative approach and experimental results. *Toxicology*, **19**, 229–38.

Könemann, H. (1981b) Quantitative structure–activity relationships in fish toxicity studies. Part I: relationship for 50 industrial pollutants. *Toxicology*, **19**, 209–21.

Könemann, H. and Leeuwen, C. van (1980) Toxicokinetics in fish: accumulation and elimination of six chlorobenzenes in guppies. *Chemosphere*, **9**, 3–19.

Kooijmann, S. A. L. M. (1987) A safety factor for $LC_{50}$ values allowing for differences in sensitivity among species. *Water Res.*, **21**, 269–76.

Kopperman, H. L., Carlson, R. M. and Caple, R. (1974). Aqueous chlorination and ozonation Studies I. Structure–toxicity correlations of phenolic compounds to *Daphnia magna. Chem.–Biol. Interactions*, **9**, 245–51.

Kördel, W., Stutte, J. and Kotthoff, G. (1993) HPLC-screening method for the determination of the adsorption-coefficient on soil – comparison of different stationary phases. *Chemosphere*, **27**, 2341–52.

Kramer, C. R. (1989) Quantitative Struktur-Aktivitätsbeziehungen für die Wachstumshemmung von autotrophen *Chlorella vulgaris*-Suspensionen durch substituierte Aniline und Anilinhydrochloride. *Biochem. Physiol. Pflanzen*, **184**, 461–9.

Kramer, C. R. and Trümper, L. (1986) Quantitative Struktur-Wirkungs-Beziehungen für die Wachstumshemmung von autotrophen *Chlorella vulgaris*-Suspensionen durch monosubstituierte Benzene, Toluene, Halogenbenzene und Methoxybenzene. *Biochem. Physiol. Pflanzen*, **181**, 645–57.

Kramer, C. R., Borns, E. and Böhm, H. (1979) Ergebnisse von Struktur-Wirkungs-Analysen der biologischen Aktivität von Carbonsäurehydraziden auf *Chlorella vulgaris*, in *Wirkungsmechanismen von Herbiziden und synthetischen Wachstumsregulatoren* (ed. H. R. Schütte), Fischer, Jena.

Kramer, C. R., Schelenz, T., Arndt, H., Domaschka, G. and Trümper, L. (1983) Quantitative Struktur-Aktivitäts-Beziehungen für die Wachstumshemmung von *Chlorella vulgaris*-Suspensionen durch 3-substituierte 1-Aminoguanidine und *o*-Alkyl-Carbamate. *Biochem. Physiol. Pflanzen*, **178**, 469–77.

Kubinyi, H. (1977) Quantitative structure–activity relationships. 7. The bilinear model, a new model for nonlinear dependence of biological activity on hydrophobic character. *J. Med. Chem.*, **20**, 625–9.

Kubinyi, H. (1993) *QSAR: Hansch Analysis and Related Approaches*, VCH, Weinheim.

Kuenemann, P. and Vasseur, P. (1988) *Etude Bibliographique des Relations Structure/Biodegradabilité*, Rapport Final du Groupe Biodegradation de la SEFA, Aout 1988, Metz, France.

Kuenemann, P., Vasseur, P. and Devillers, J. (1990) Structure–biodegradability relationships, in *Practical Applications of Quantitative Structure–Activity Relationships (QSAR) in Environmental Chemistry and Toxicology* (eds W. Karcher and J. Devillers), Kluwer, Dordrecht.

Lambert, S. M. (1968) Omega, a useful index of soil sorption equilibria. *J. Agric. Food Chem.*, **16**, 340–3.

Lambert, S. M., Porter, P. E. and Schieferstein, H. (1965) Movement and sorption of chemicals applied to soils. *Weeds*, **13**, 185–90.

Landis, W. G., Matthews, G. B., Matthews, R. A. and Sergeant, A. (1994) Application of multivariate techniques to endpoint determination, selection and evaluation in ecological risk assessment. *Environ. Toxicol. Chem.*, **13**, 1917–27.

Lange, A. W. and Vormann, K. (1995) Experiences with the application of QSAR in the routine of the notification procedure. *SAR QSAR Environ. Res.*, **3**, 171–7.

Lassiter, R. R. and Hallam, T. G. (1990) Survival of the fattest: implications for acute effects of lipophilic chemicals on aquatic populations. *Environ. Toxicol. Chem.*, **9**, 585-95.

Leahy, D. E., Taylor, P. J. and Wait, A. R. (1989) Model solvent systems for QSAR. Part 1. Propylene glycol dipelargonate (PGDP). A new standard solvent for use in partition coefficient determination. *Quant. Struct.–Act. Relat.*, **8**, 17–31.

Leahy, D. E., Morris, J. J., Taylor, P. J., and Wait, A. R. (1992a) Model solvent systems for QSAR. Part 3. An LSER analysis of the 'critical quartet'. New light on hydrogen bond strength and directionality. *J. Chem. Soc. Perkin Trans.*, **2**, 705–22.

Leahy, D. E., Morris, J. J., Taylor, P. J. and Wait, A. R. (1992b) Model solvent systems for QSAR. Part 2. Fragment values (*f*-values) for the 'critical quartet'. *J. Chem. Soc. Perkin Trans.*, **2**, 723–31.

LeBlanc, G. A. (1984) Interspecies relationships in acute toxicity of chemicals to aquatic organisms. *Environ. Toxicol. Chem.*, **3**, 47–60.

Leeuwen, C. J. van and Hermens, J. L. M. (eds) (1995) *Risk Assessment of Chemicals*, Kluwer, Dordrecht.

Leeuwen, C. J. van, Adema, D. M. M. and Hermens, J. (1990) Quantitative structure–activity relationship for fish early life stage toxicity. *Aquat. Toxicol.*, **13**, 321–34.

Leeuwen, C. J. van, Zandt, P. T. J. van der, Aldenberg, T., Verhaar, H. J. M. and Hermens, J. L. M. (1991) The application of QSARs, extrapolation and equilibrium partitioning in aquatic effects assessment for narcotic pollutants. *Sci. Total Environ.*, **109/110**, 681–91.

Leo, A. (1985) Parameter and structure–activity data bases: management for maximum utility. *Environ. Health Persp.*, **61**, 275–85.

Leo, A. J. and Hansch, C. (1971) Linear free-energy relationships between partitioning solvent systems. *J. Org. Chem.*, **36**, 1539–44.

Lewis, D. F. V. (1989) The calculation of molar polarizabilities by the CNDO/2 method: correlation with the hydrophobic parameter, log *P. J. Comput. Chem.*, **10**, 145–51.

Licastro, S. A. de, Villar, M. I. P. de, Wood, E. J., Melgar, F. J. and Zerba, E. N. (1986) Toxicity and acetylcholinesterase activity of organophosphorus compounds in *Triatoma infestans. Comp. Biochem. Physiol.*, **84C**, 131–7.

Lien, E. J., Hansch, C. and Anderson, S. M. (1968) Structure–activity correlations for antibacterial agents on Gram-positive and Gram-negative cells. *J. Med. Chem.*, **11**, 430–41.

Lindgren, F., Hansen, B., Karcher, W., Sjöström, M. and Eriksson, L. (1995) Model validation by permutation tests. Applications to variable selection. Report 'QSAR for prediction of fate and effects of chemicals in the environment', Environmental Technologies RTD Programme (DG XII/D-1) of the European Commission under Contract Number EV5V-CT92-0211.

Lipnick, R. L. (1985a) Editorial: A perspective on quantitative structure–activity relationships in ecotoxicology. *Environ. Toxicol. Chem.*, **4**, 255–7.

Lipnick, R. L. (1985b) Research needs in developing structure activity relationships, in *Aquatic Toxicology and Hazard Assessment* (eds R. C. Bahner and D. J. Hansen), STP 891, ASTM, Philadelphia, PA.

Lipnick, R. L. (1985c) Validation and extension of fish toxicity QSARs and interspecies comparisons for certain classes of organic chemicals, in *QSAR in Toxicology and Xenobiochemistry* (ed. M. Tichy), Elsevier, Amsterdam.

Lipnick, R. L. (1986) Charles Ernest Overton: narcosis studies and a contribution to general pharmacology. *Trends Pharmacol. Sci.* **7**, 161–4.

Lipnick, R. L. (1989a) A quantitative structure–activity relationship study of Overton's data on the narcosis and toxicity of organic compounds to the tadpole, *Rana temporaria*, in *Aquatic Toxicology and Environmental Fate* (eds G. W. Suter II and M. A. Lewis), STP 1007, ASTM, Philadelphia, PA.

Lipnick, R. L. (1989b) Hans Horst Meyer and the lipoid theory of narcosis. *Trends Pharmacol. Sci.*, **10**, 265–9.

Lipnick, R. L. (1989c) Narcosis, electrophile and proelectrophile toxicity mechanisms: application of SAR and QSAR. *Environ. Toxicol. Chem.*, **8**, 1–12.

Lipnick, R. L. (1991) Outliers: their origin and use in the classification of molecular mechanisms of toxicity. *Sci. Total Environ.*, **109/110**, 131–53.

Lipnick, R. L. (1995) Structure–activity relationships, in *Fundamentals of Aquatic Toxicology* (ed. G. R. Rand), Taylor & Francis, London.

Lipnick, R. L. and Hood, M. T. (1986) Correlation of chemical structure and toxicity of industrial organic compounds to *Daphnia*, algae, bacteria, and protozoa. Abstracts, 7th SETAC Annual Meeting, Alexandria, VA.

Lipnick, R. L., Pritzker, C. S. and Bentley, D. L. (1985) A QSAR study of the rat $LD_{50}$ for alcohols, in *QSAR and Strategies in the Design of Bioactive Compounds* (ed. J. K. Seydel), VCH, Weinheim.

Lipnick, R. L., Pritzker, C. S. and Bentley, D. L. (1987) Application of QSAR to model the toxicology of industrial organic chemicals to aquatic organisms and mammals, in *QSAR in Drug Design and Toxicology* (eds D. Hadzi and B. Jerman-Blazic), Elsevier, Amsterdam.

Lipnick, R. L., Watson, K. R. and Strausz, A. K. (1987) A QSAR study of the acute toxicity of some industrial organic chemicals to goldfish. Narcosis, electrophile and proelectrophile mechanisms. *Xenobiotica*, **17**, 1011–25.

Lipnick, R. L., Johnson, D. E., Gilford, J. H., Bickings, C. K. and Newsome, L. D. (1985) Comparison of fish toxicity screening data for 55 alcohols with the quantitative structure–activity relationship predictions of minimum toxicity for nonelectrolyte organic compounds. *Environ. Toxicol. Chem.*, **4**, 281–96.

Lipnick, R. L., Bickings, C. K., Johnson, D. E. and Eastmond, D. A. (1986) Comparison of QSAR predictions with fish toxicity screening data for 110 phenols, in *Aquatic Toxicology and Hazard Assessments* (eds R. C. Bahner and D. J. Hansen), STP 891, ASTM, Philadelphia, PA.

Liu, D., Thomson, K. and Kaiser, K. L. E. (1982) Quantitative structure–toxicity relationship of halogenated phenols on bacteria. *Bull. Environ. Contam. Toxicol.*, **29**, 130–6.

Loewe, S. and Muischnek, H. (1926) Über Kombinationswirkungen. *Arch. Exp. Pathol. Pharmakol.*, **114**, 313–26.

Logan, D. T. and Wilson, H. T. (1995) An ecological risk assessment method for species exposed to contaminant mixtures. *Environ. Toxicol. Chem.*, **14**, 351–9.

LOGKOW (1993), Syracuse Research Corporation, Syracuse, NY.

Lyman, W. J. (1985) Estimation of physical properties, in *Environmental Exposure from Chemicals* (eds W. B. Neely and G. E. Blau), CRC Press, Boca Raton, FL.

Lyman, W. J. (1990a) Adsorption coefficient for soils and sediments, in *Handbook of Chemical Property Estimation Methods* (eds W. J. Lyman, W. F. Reehl and D. H. Rosenblatt), American Chemical Society, Washington, DC.

Lyman, W. J. (1990b) Octanol/water partition coefficient, in *Handbook of Chemical Property Estimation Methods* (eds W. J. Lyman, W. F. Reehl and D. H. Rosenblatt), American Chemical Society, Washington, DC.

Lyman, W. J. (1990c) Solubility in water, in *Handbook of Chemical Property Estimation Methods* (eds W. J. Lyman, W. F. Reehl and D. H. Rosenblatt), American Chemical Society, Washington, DC.

Lyman, W. J., Reehl, W. F. and Rosenblatt, D. H. (1990) *Handbook of Chemical Property Estimation Methods*, American Chemical Society, Washington, DC.

Lynch, D. G., Tirado, N. F., Boethling, R. S., Huse, G. R. and Thom, G. C. (1991) Performance on on-line chemical property estimation methods with TSCA premanufacture notice chemicals. *Sci. Total Environ.*, **109/110**, 643–8.

Macek, K. J., Petrocelli, S. R. and Sleight, B. H. (1979) Considerations in assessing the potential for and significance of biomagnification of chemical residues in aquatic food chains, in *Aquatic Toxicology* (eds L. L. Marking and R. A. Kimerle), STP 667, ASTM, Philadelphia, PA.

Mackay, D. (1979) Finding fugacity feasible. *Environ. Sci. Technol.*, **13**, 1218–23.

Mackay, D. (1982) Correlation of bioconcentration factors. *Environ. Sci. Technol.*, **16**, 274–6.

Mackay, D. and Paterson, S. (1981) Calculating fugacity. *Environ. Sci. Technol.*, **15**, 1006–14.

Mackay, D. and Paterson, S. (1982) Fugacity revisited. *Environ. Sci. Technol.*, **16**, 654–60.

Mackay, D. and Paterson, S. (1990) Fugacity models, in *Practical Applications of Quantitative Structure–Activity Relationships (QSAR) in Environmental Chemistry and Toxicology* (eds W. Karcher and J. Devillers), Kluwer, Dordrecht.

Mackay, D. and Paterson, S. (1991) Evaluating the multimedia fate of organic chemicals: a level III fugacity model. *Environ. Sci. Technol.*, **25**, 427–36.

Mackay, D., Paterson, S. and Shiu, W. Y. (1992) Generic models for evaluating the regional fate of chemicals. *Chemosphere*, **24**, 695–717.

Mackay, D., Bobra, A., Chan, D. W. and Shiu, W. Y. (1982) Vapour-pressure correlations for low volatility environmental chemicals. *Environ. Sci. Technol.*, **16**, 645–9.

Magee, P. S., Henry, D. R. and Block, J. H. (1989) *Probing Bioactive Mechanisms*, American Chemical Society, Washington, DC.

Martens, H. and Naes, T. (1989) *Multivariate Calibration*, Wiley, New York, NY.

Martens, M., Mosselmans, G., Fumero, S., Jacobs, G. and Lafontaine, A. (1984) Some thoughts on a possible regulatory approach at EEC level to the classification and labeling of dangerous preparations. *Reg. Toxicol. Pharmacol.*, **4**, 145–56.

Martin, Y. C. (1978) *Quantitative Drug Design*, Marcel Dekker, New York, NY.

Martin, Y. C. (1983) Studies of relationships between structural properties and biological activity by Hansch analysis, in *Structure-Activity Correlation as a Predictive Tool in Toxicology* (ed. L. Goldberg), McGraw Hill, London.

Matthies, M. and Trenkle, R. (1988) Variability of exposure estimations for hazardous chemicals released to the environment. *Chemosphere*, **17**, 1471– 86.

Matthiessen, P., Thain, J. E., Law, R. J. and Fileman, T. W. (1993) Attempts to assess the environmental hazard posed by complex mixtures of organic chemicals in UK estuaries. *Mar. Pollut. Bull.*, **26**, 90–5.

McCarty, L. S. (1987) Relationship between toxicity and bioconcentration for some organic chemicals. I. Examination of the relationship, in *QSAR in Environmental Toxicology – II* (ed. K. L. E. Kaiser), D. Reidel, Dordrecht.

McCarty, L. S. and Mackay, D. (1993) Enhancing ecotoxicological modeling and assessment. *Environ. Sci. Technol.*, **27**, 1719–28.

McCarty, L. S., Mackay, D., Smith, A. D., Ozburn, G. W. and Dixon, D. G. (1991) Interpreting aquatic toxicity QSARs: the significance of toxicant body residues at the pharmacologic endpoint. *Sci. Total Environ.*, **109/110**, 515–25.

McCarty, L. S., Ozburn, G. W., Smith, A. D. and Dixon, D. G. (1992) Toxicokinetic modeling of mixtures of organic chemicals. *Environ. Toxicol. Chem.*, **11**, 1037–47.

McCutcheon, P., Norager, O., Karcher, W. and Devillers, J. (1990) Sources of data for risk evaluation and QSAR studies, in *Practical Applications of Quantitative Structure–Activity Relationships (QSAR) in Environmental Chemistry and Toxicology* (eds W. Karcher and J. Devillers), Kluwer, Dordrecht.

McFarland, J. W. and Gans, D. J. (1990) Linear discriminant analysis and cluster significance analysis, in *Quantitative Drug Design* (ed. C. A. Ramsden), Pergamon Press, Oxford.

McKim, J. M. and Schmieder, P. K. (1991) Bioaccumulation: does it reflect toxicity?, in *Bioaccumulation in Aquatic Systems* (eds R. Nagel and R. Loskill), VCH, Weinheim.

McKim, J. M., Bradbury, S. P. and Niemi, G. J. (1987) Fish acute toxicity syndromes and their use in the QSAR approach to hazard assessment. *Environ. Health Persp.*, **71**, 171–86.

McKim, J., Schmieder, P. and Veith, G. (1985) Adsorption dynamics of organic chemical transport across trout gills as related to octanol–water partition coefficient. *Toxicol. Appl. Pharmacol.*, **77**, 1–10.

McKinney, J. D. (1985) The molecular basis of chemical toxicity. *Environ. Health Persp.*, **61**, 5–10.

McLeese, D. W., Zitko, V. and Peterson, M. R. (1979) Structure–lethality relationships for phenols, anilines and other aromatic compounds in shrimp and clams. *Chemosphere*, **8**, 53–7.

MedChem (1989) *MedChem Software Version 3.54*, Daylight Chemical Information Systems, Inc., Claremont, CA.

Medven, Z., Güsten, H. and Sabljic, A. (1996) Comparative QSAR study on hydroxyl radical reactivity with unsaturated hydrocarbons: PLS vs. MLR. *J. Chemometr.*, **10**, 135–47.

Meissner, H. P. (1949) Critical constants from parachor and molar refraction. *Chem. Eng. Prog.*, **45**, 149–53.

Mekenyan, O. G. and Veith, G. D. (1994) The electronic factor in QSAR: MO-parameters, competing interactions, reactivity and toxicity. *SAR QSAR Environ. Res.*, **2**, 129–43.

Metcalf, R. L., Kapoor, I. P., Lu, P. Y., Schuth, C. S. and Sherman, P. (1973). Model ecosystem studies of the environmental fate of six organochlorine pesticides. *Environ. Health Persp.*, **35**, 44.

Metcalf, R. L., Sanborn, J. R., Lu, P. Y. and Nye, D. (1975) Laboratory model ecosystem studies of the degradation and fate of radiolabelled tri- tetra- and pentachlorobiphenyl compared with DDE. *Arch. Environ. Contam. Toxicol.*, **3**, 151–65.

Meyer, H. (1899) Lipoidtheorie der Narkose. *Arch. Exp. Pathol. Pharm.*, **42**, 109–18.

Meylan, W. M. and Howard, P. H. (1991) Bond contribution method for estimating Henry's law constants. *Environ. Toxicol. Chem.*, **10**, 1283–93.

Meylan, W., Howard, P. H. and Boethling, R. S. (1992) Molecular topology/fragment contribution method for predicting soil sorption coefficients. *Environ. Sci. Technol.*, **26**, 1560–7.

Meylan, W. M., Howard, P. H. and Boethling, R. S. (1996) Improved method for estimating water solubility from octanol/water partition coefficient. *Environ. Toxicol. Chem.*, **15**, 100–6.

Mill, T. (1982) Hydrolysis and oxidation processes in the environment. *Environ. Toxicol. Chem.*, **1**, 135–41.

Mishra, D. S. and Yalkowsky, S. H. (1991) Estimation of vapor pressure of some organic compounds. *Ind. Eng. Chem. Res.*, **30**, 1609–12.

Mitsutake, K., Iwamura, H., Shimizu, R. and Fujita, T. (1986) Quantitative structure–activity relationship of photosystem II inhibitors in chloroplasts and its link to herbicidal action. *J. Agric. Food Chem.*, **34**, 725–32.

Molinengo, J. (1979) The curve dose vs survival time in the evaluation of acute toxicity. *J. Pharm. Pharmacol.*, **31**, 343–4.

Morgan, M. G. (1985) Risk assessment and risk management decision-making for chemical exposure, in *Environmental Exposure from Chemicals*, Vol. II (eds W. B. Neely and G. E. Blau), CRC Press, Boca Raton, FL.

Moser, P. and Anliker, R. (1991) BCF and P: limitations of the determination methods and interpretation of data in the case of organic colorants, in *Bioaccumulation in Aquatic Systems* (eds R. Nagel and R. Loskill), VCH, Weinheim.

Motais, R., Sola, F. and Cousin, J. L. (1978) Uncouplers of oxidative phosphorylation. A structure–activity study of their inhibitory effect on passive chloride permeability. *Biochim. Biophys. Acta*, **510**, 201–7.

Moulton, M. P. and Schultz, T. W. (1986) Comparisons of several structure–toxicity relationships for chlorophenols. *Aquat. Toxicol.*, **8**, 121–8.

Mudder, T. I. (1981) Development of empirical structure–biodegradability relationships and testing protocol for slightly soluble and volatile priority pollutants. Thesis, University of Iowa, IA (University Microfilms International, Ann Arbor, MI).

Muir, D. C. G., Norstrom, R. J. and Simon, M. (1988) Organochlorine contaminants in arctic marine food chains: accumulation of specific polychlorinated biphenyls and chlordane-related compounds. *Environ. Sci. Technol.*, **22**, 1071–9.

Müller, M. and Klein, W. (1991) Estimating atmospheric degradation processes by SARs. *Sci. Total Environ.*, **109/110**, 261–73.

Müller, M. and Klein, W. (1992) Comparative evaluation of methods predicting water solubility for organic compounds. *Chemosphere*, **25**, 769–82.

Müller, M. and Klein, W. (1993) Fugacity calculations using estimated physico-chemical properties. *SAR QSAR Environ. Res.*, **1**, 245–62.

Müller, M. and Kördel, W. (1996) Comparison of screening methods for the estimation of adsorption coefficients on soil. *Chemosphere*, **32**, 2493–504.

Myrdal, P. and Yalkowsky, S. H. (1994) A simple scheme for calculating aqueous solubility, vapor pressure and Henry's law constant: application to the chlorobenzenes. *SAR QSAR Environ. Res.*, **2**, 17–28.

Nagel, R. and Loskill, R. (eds) (1991) *Bioaccumulation in Aquatic Systems*, VCH, Weinheim.

Neely, W. B. and Blau, G. E. (eds) (1985) *Environmental Exposure from Chemicals*, CRC Press, Boca Raton, FL.

Neely, W. B., Branson, D. R. and Blau, G. E. (1974) Partition coefficients to measure bioconcentration potential of organic chemicals in fish. *Environ. Sci. Technol.*, **8**, 1113–15.

Nendza, M. (1991a) Predictive QSAR models estimating ecotoxic hazard of phenylureas: mammalian toxicity. *Chemosphere*, **22**, 613–23.

Nendza, M. (1991b) QSARs of bioconcentration: validity assessment of log $P_{ow}$/log BCF correlations, in *Bioaccumulation in Aquatic Systems* (eds R. Nagel and R. Loskill), VCH, Weinheim.

Nendza, M. and Hermens, J. (1995) Properties of chemicals and estimation methodologies, in *Risk Assessment of Chemicals* (eds C. J. van Leeuwen and J. L. M. Hermens), Kluwer, Dordrecht.

Nendza, M. and Klein, W. (1990) Comparative QSAR study on freshwater and estuarine toxicity. *Aquat. Toxicol.*, **17**, 63–74.

Nendza, M. and Russom, C. L. (1991) QSAR modeling of the ERL-D fathead minnow acute toxicity database. *Xenobiotica*, **21**, 147–70.

Nendza, M. and Seydel, J. K. (1988a) Multivariate data analysis of various biological test systems used for the quantification of ecotoxic compounds. *Quant. Struct.–Act. Relat.*, **7**, 165–74.

Nendza, M. and Seydel, J. K. (1988b) Quantitative structure–toxicity relationships for ecotoxicologically relevant biotestsystems and chemicals. *Chemosphere*, **8**, 1585–602.

Nendza, M. and Seydel, J. K. (1990) Application of bacterial growth kinetics to *in vitro* toxicity assessment of substituted phenols and anilines. *Ecotoxicol. Environ. Saf.*, **19**, 228–41.

Nendza, M. and Wenzel, A. (1993) Statistical approach to chemicals classification. *Sci. Total Environ.*, Suppl., **1993**, 1459–70.

Nendza, M., Volmer, J. and Klein, W. (1990) Risk assessment based on QSAR estimates, in *Practical Applications of Quantitative Structure–Activity Relationships (QSAR) in Environmental Chemistry and Toxicology* (eds W. Karcher and J. Devillers), Kluwer, Dordrecht.

Nendza, M., Wenzel, A. and Wienen, G. (1995) Classification of contaminants by mode of action based on *in-vitro* assays. *SAR QSAR Environ. Res.*, **4**, 39–50.

Nendza, M., Jäckel, H., Müller, M., Gies-Reuschel, A. and Klein, W. (1993) Estimation of exposure and ecotoxicity related parameters by computer based structure–property and structure–activity relationships. *Toxicol. Environ. Chem.*, **40**, 57–69.

Nendza, M., Herbst, T., Kussatz, C. and Gies, A. (1997). Potential for secondary poisoning and biomagnification in marine organisms. *Chemosphere* (in press).

Neumann, H. G. (1986) Toxication mechanisms in drug metabolism. *Adv. Drug. Res.*, **15**, 1–28.

Niemi, G. J., Veith, G. D., Regal, R. R. and Vaishnav, D. D. (1987) Structural features associated with degradable and persistent chemicals. *Environ. Toxicol. Chem.*, **6**, 515–27.

Nirmalakhandan, N. N. and Speece, R. E. (1988a) Prediction of aqueous solubility of organic chemicals based on molecular structure. *Environ. Sci. Technol.*, **22**, 328–38.

Nirmalakhandan, N. N. and Speece, R. E. (1988b) QSAR model for predicting Henry's law constant. *Environ. Sci. Technol.*, **22**, 1349–57.

Nirmalakhandan, N. N. and Speece, R. E. (1989) Prediction of aqueous solubility of organic chemicals based on molecular structure. 2. Application to PNAs, PCBs, PCDDs, etc. *Environ. Sci. Technol.*, **23**, 708–13.

Nusch, E. A. (1991) Ökotoxikologische Testverfahren. *UWSF-Z. Umweltchem. Ökotox.*, **3**, 12–15.

Nys, G. G. and Rekker, R. F. (1973) Statistical analysis of a series of partition coefficients with special reference to the predictability of folding drug molecules. The introduction of hydrophobic fragmental constant (f values). *Chim. Ther.*, **5**, 521–35.

OECD (1981a) *Guide-line for Testing of Chemicals: Bioaccumulation*, OECD, Paris.

OECD (1981b) *Guide-line for Testing of Chemicals: Partition Coefficient (n-Octanol/Water)*, OECD, Paris.

OECD (1981c) *Guide-line for Testing of Chemicals: Water Solubility*, OECD, Paris.

OECD (1983) *Guide-line for Testing of Chemicals: Adsorption/Desorption*, OECD, Paris.

OECD (1984a) *Guide-line for Testing of Chemicals: Degradation*, OECD, Paris.

OECD (1984b) *Guide-line for Testing of Chemicals: Earthworm Acute Toxicity Test*, OECD, Paris.

OECD (1984c) *Guide-line for Testing of Chemicals: Daphnia sp. Acute Immobilisation Test and Reproduction Test*, OECD, Paris.

OECD (1984d) *Guide-line for Testing of Chemicals: Alga Growth Inhibition Test*, OECD, Paris.

OECD (1989a) *Compendium of Environmental Exposure Assessment Methods for Chemicals*, Environment Monograph No. 27, OECD, Paris.

OECD (1989b) *Guide-line for Testing of Chemicals: Degradation*, OECD, Paris.

OECD (1989c) *Guide-line for Testing of Chemicals: Fish Acute Toxicity Test*, OECD, Paris.

OECD (1992a) *Report of the OECD Workshop on Quantitative Structure Activity Relationships (QSARs) in Aquatic Effects Assessment*, Environment Monograph No. 58, OECD, Paris.

OECD (1992b) *Report of the OECD Workshop on the Extrapolation of Laboratory Aquatic Toxicity Data to the Real Environment*, Environment Monograph No. 59, OECD, Paris.

OECD (1992c) *Report of the OECD Workshop on Effects Assessment of Chemicals in Sediment*, Environment Monograph No. 60, OECD, Paris.

OECD (1993a) *Application of Structure–Activity Relationships to the Estimation of Properties Important in Exposure Assessment*, Environment Monograph No. 67, OECD, Paris.

OECD (1993b) *Structure–Activity Relationships for Biodegradation*, Environment Monograph No. 68, OECD, Paris.

OECD (1993c) *Report of the OECD Workshop on the Application of Simple Models for Environmental Exposure Assessment*, Environment Monograph No. 69, OECD, Paris.

OECD (1994) *US EPA/EC Joint Project on the Evaluation of (Quantitative) Structure Activity Relationships*, Environment Monograph No. 88, OECD, Paris.

OECD (1995) *Guidance Document for Aquatic Effects Assessment*, Environment Monograph No. 92, OECD, Paris.

Oepen, B. von (1990) *Sorption organischer Chemikalien an Böden.* Dissertation, Universität Duisburg, Duisburg (Wissenschaftsverlag Maraun, Frankfurt).

Oepen, B. von, Kördel, W. and Klein, W. (1991) Sorption of nonpolar and polar compounds to soil: processes, measurements and experience with the modified OECD guide-line 106. *Chemosphere*, **22**, 285–304.

Oepen, B. von, Kördel, W., Klein, W. and Schüürmann, G. (1990) Predictive QSAR models for estimating soil sorption coefficients: potential and limitations based on dominating processes. *Sci. Total Environ.*, **109/110**, 343–54.

Ogata, M., Fujisawa, K., Ogino, Y. and Mano, E. (1984) Partition coefficients as a measure of bioconcentration potential of crude oil compounds in fish and shellfish. *Bull. Environ. Contam. Toxicol.*, **33**, 561–7.

Oliver, B. G. (1984) The relationship between bioconcentration factor in rainbow trout and physical-chemical properties for some halogenated compounds, in *QSAR in Environmental Toxicology* (ed. K. L. E. Kaiser), D. Reidel, Dordrecht.

Oliver, B. G. and Niimi, A. (1983) Bioconcentration of chlorobenzenes from water to rainbow trout: correlation with partition coefficients and environmental residues. *Environ. Sci. Technol.*, **17**, 287–91.

Opperhuizen, A., Serné, P. and Steen, J. M. D. van der (1988) Thermodynamics of fish/water and octan-1-ol/water partitioning of some chlorinated benzenes. *Environ. Sci. Technol.*, **22**, 286–92.

Opperhuizen, A., Velde, E. W. van der, Gobas, F. A. P. C., Lem, D. A. K., Steen, J. M. D. van der and Hutzinger, O. (1985) Relationship between bioconcentration in fish and steric factors of hydrophobic chemicals. *Chemosphere*, **14**, 1871–96.

Ou, X. C., Ouyang, Y. and Lien, E. J. (1986) Examination of quantitative relations of partition coefficient (log *P*) and molecular weight, dipole moment and hydrogen bond capability of miscellaneous compounds. *J. Molec. Sci.*, **4**, 89.

Overton, E. (1897) Über die osmotischen Eigenschaften der Zelle in ihrer Bedeutung für die Toxikologie und Pharmakologie. *Z. Phys. Chem.*, **22**, 189–209.

Paris, D. F., Wolfe, N. L. and Steen, W. C. (1982) Structure–activity relationships in microbial transformation of phenols. *Appl. Environ. Microbiol.*, **44**, 153–8.

Paris, D. F., Wolfe, N. L., Steen, W. C. and Baughman, G. L. (1983) Effect of phenol molecular structure on bacterial transformation rate constants in ponds and river samples. *Appl. Environ. Microbiol.*, **45**, 1153–5.

Parsons, J. R. and Govers, H. A. J. (1990) Quantitative–structure activity relationships (QSARs) for biodegradation. *Ecotoxicol. Environ. Saf.*, **19**, 212–27.

Passino, D. B. M., Hickey, J. P. and Frank, A. M. (1988) Linear solvation energy relationships for toxicity of selected organic chemicals to *Daphnia magna* and *Daphnia pulex*, in *Proceedings of the Third International Workshop on Quantitative Structure–Activity Relationships in Environmental Toxicology* (eds J. E. Turner, M. W. England, T. W. Schultz and N. J. Kwaak), NTIS, Springfield, VA.

Pedersen, F., Kristensen, P., Damborg, A. and Christensen, H. W. (1994) *Ecotoxicological Evaluation of Industrial Wastewater*, Ministry of the Environment, Copenhagen.

Peijnenburg, W. J. G. M. (1991) The use of quantitative structure–activity relationships for predicting rates of environmental hydrolysis processes. *Pure Appl. Chem.*, **63**, 1667–76.

Peijnenburg, W. (1994) Structure–activity relationships for biodegradation: a critical review. *Pure Appl. Chem.*, **66**, 1931–41.

Peijnenburg, W. J. G. M. and Karcher, W. (eds) (1995) *Proceedings of the Workshop 'Quantitative Structure Activity Relationships for Biodegradation'* [September 1994, Belgirate, Italy], Report No. 719101021, National Institute of Public Health and Environmental Protection, Bilthoven.

Peijnenburg, W. J. G. M., Hart, M. J. 't, Hollander, H. A. den, Meent, D. van de, Verboom, H. and Wolfe, N. L. (1991) QSARs for predicting biotic and abiotic reductive transformation rate constants of halogenated hydrocarbons in anoxic sediment systems. *Sci. Total Environ.*, **109/110**, 283–300.

Peijnenburg, W. J. G. M., Beer, K. G. M. de, Hollander, H. A. den, Stegeman, M. H. L. and Verboom, H. (1993) Kinetics, products, mechanisms and QSARs for the hydrolytic transformation of aromatic nitriles in anaerobic sediment slurries. *Environ. Toxicol. Chem.*, **12**, 1149–61.

Perrin, D. D., Dempsey, B. and Serjeant, E. P. (1981) *$pK_a$ Prediction for Organic Acids and Bases*, Chapman & Hall, London.

Pfeifer, S., Pflegel, P. and Borchert, H. H. (1984) *Grundlagen der Biopharmazie*, Verlag Chemie, Weinheim.

Pitter, P. (1976) Determination of biological degradability of organic substances. *Water Res.*, **10**, 231–5.

Pitter, P. (1985) Correlation of microbial degradation with the chemical structure. *Acta Hydrochim. Hydrobiol.*, **13**, 453–60.

Pollard, J. E. and Hern, S. C. (1985) A field test of the EXAMS model in the Monogahela river. *Environ. Toxicol. Chem.*, **4**, 361–9.

ProLogP (1992) *An Expert System for the Calculation of log P*, CompuDrug Chemistry Ltd, Budapest.

Purchase, I. F. H. (1983) Carcinogenicity, in *Animals and Alternatives in Toxicity Testing* (eds M. Balls, R. J. Riddell and A. N. Worden), Academic Press, London.

QCPE (1986) *Quantum Chemistry Program Exchange*, AMPAC, Program No. 506, Department of Chemistry, Indiana University, Bloomington, IN.

QCPE (1990) *Quantum Chemistry Program Exchange*, MOPAC, Department of Chemistry, Indiana University, Bloomington, IN.

Raevsky, O. A., Grigor'ev, V. Y. and Dolmatova, L. (1994) Hydrogen bonding in QSAR. Abstracts, 6th QSAR Workshop, Belgirate, Italy.

Raevsky, O. A., Grigor'ev, V. Y., Kireev, D. B. and Zefirov, N. S. (1992a) Complete thermodynamic description of H-bonding in the framework of multiplicative approach. *Quant. Struct.–Act. Relat.* **11**, 49–63.

Raevsky, O. A., Grigor'ev, V. Y., Kireev, D. B. and Zefirov, N. S. (1992b) Correlation analysis and H-bond ability in framework of QSAR. *J. Chim. Phys.*, **89**, 1747–53.

Randic, M. (1975) On characterization of molecular branching. *J. Am. Chem. Soc.*, **97**, 6609–15.

Rechsteiner, C. E. (1990) Boiling point, in *Handbook of Chemical Property Estimation Methods* (eds W. J. Lyman, W. F. Reehl and D. H. Rosenblatt), American Chemical Society, Washington, DC.

Rekker, R. F. (1977) *The Hydrophobic Fragmental Constant*, Elsevier Scientific, New York, NY.

Rekker, R. F. (1980) $LD_{50}$ values: are they about to become predictable? *Trends Pharmacol. Sci.*, **10**, 383–4.

Ribo, J. M. and Kaiser, K. L. E. (1983) Effects of selected chemicals to photoluminiscent bacteria and their correlation with acute and sublethal effects on other organisms. *Chemosphere*, **12**, 1421–42.

Ribo, J. M. and Kaiser, K. L. E. (1984) Toxicities of chloroanilines to *Photobacterium phosphoreum* and their correlations with effects on other organisms and structural parameters, in *QSAR in Environmental Toxicology* (ed. K. L. E. Kaiser), D. Reidel, Dordrecht.

Ripley, B. D. (1993) Statistical Aspects of Neural Networks, Proceedings Sem Stat (Séminaire Européen de Statistique), Sandbjerg, Denmark, 1992, Chapman & Hall, London.

Roberts, D. W. (1989) Acute lethal toxicity quantitative structure–activity relationships for electrophiles and pro-electrophiles: mechanistic and toxicokinctic principles, in *Aquatic Toxicology and Environmental Fate* (eds G. W. Suter II and M. A. Lewis), STP 1007, ASTM, Philadelphia, PA.

Roberts, D. W. (1991) QSAR issues in aquatic toxicity of surfactants. *Sci. Total Environ.*, **109/110**, 557–68.

Rojas, R. (1993) *Theorie der neuronalen Netze*, Springer Verlag, Berlin.

Rorije, E., Wezel, M. C. van and Peijnenburg, W. J. G. M. (1995) On the use of back-propagation neural networks in modelling environmental degradation. *SAR QSAR Environ. Res.* **4**, 219–35.

Rorije, E., Langenberg, J. H., Richter, J. and Peijnenburg, W. J. G. M. (1995) Modeling reductive dehalogenation with quantum-chemically derived descriptors. *SAR QSAR Environ. Res.*, **4**, 237–52.

Rubino, J. T. and Yalkowsky, S. H. (1987) Cosolvency and cosolvent polarity. *Pharm. Res.*, **4**, 220–30.

Russom, C. L., Bradbury, S. P., Broderius, S. J., Hammermeister, D. E. and Drummond, R. A. (1997). Predicting modes of toxic action from chemical structure: acute toxicity in the fathead minnow (*Pimephales promelas*). *Environ. Toxicol, Chem.*, **16**, 948–67.

Saarikoski, J. and Viluksela, M. (1982) Relation between physicochemical properties of phenols and their toxicity and accumulation in fish. *Ecotoxicol. Environ. Saf.*, **6**, 501–12.

Sabljic, A. (1983) Quantitative structure–toxicity relationships of chlorinated compounds: a molecular connectivity investigation. *Bull. Environ. Contam. Toxicol.*, **30**, 80–3.

Sabljic, A. (1987a) Nonempirical modeling of environmental distribution and toxicity of major organic pollutants, in *QSAR in Environmental Toxicology – II* (ed. K. L. E. Kaiser), D. Reidel, Dordrecht.

Sabljic, A. (1987b) On the prediction of soil sorption coefficients of organic pollutants from molecular structure: application of molecular topological model. *Environ. Sci. Technol.*, **21**, 358–66.

Sabljic, A. (1987c). The prediction of fish bioconcentration factors of organic pollutants from the molecular connectivity model. *Z. Ges. Hyg.*, **33**, 493–6.

Sabljic, A. (1990) Topological indices and environmental chemistry, in *Practical Applications of Quantitative Structure–Activity Relationships (QSAR) in Environmental Chemistry and Toxicology* (eds W. Karcher and J. Devillers), Kluwer, Dordrecht.

Sabljic, A. and Güsten, H. (1990) Predicting the night-time $NO_3$ radical reactivity in the troposphere. *Atmosph. Environ.*, **24A**, 73–8.

Sabljic, A. and Protic, M. (1982) Relationship between molecular connectivity indices and soil sorption coefficients of polycyclic aromatic hydrocarbons. *Bull. Environ. Contam. Toxicol.*, **28**, 162–5.

Sabljic, A., Güsten, H., Schönherr, J. and Riederer, M. (1990) Modelling plant uptake of airborne organic chemicals. I. Plant cuticle/water partitioning and molecular connectivity. *Environ. Sci. Technol.*, **24**, 1321–6.

Sabljic, A., Güsten, H., Verhaar, H. and Hermens, J. (1995) QSAR modelling of soil sorption. Improvements and systematics of log $K_{oc}$ vs. log $K_{ow}$ correlations. *Chemosphere*, **31**, 4489–514.

Schaper, K. J. (1983) Rational selection of test series for QSAR analysis. *Quant. Struct.–Act. Relat.*, **2**, 111–20.

Schaper, K. J. and Kaliszan, R. (1987) Applications of statistical methods to drug design, in *Trends in Medicinal Chemistry* (eds E. Mutschler and E. Winterfeld), VCH, Weinheim.

Schaper, K. J. and Seydel, J. K. (1985) Multivariate methods in quantitative structure–pharmacokinetics relationship analysis, in *QSAR and Strategies in the Design of Bioactive Compounds* (ed. J. K. Seydel), VCH, Weinheim.

Schelenz, T. and Kramer, C. R. (1985) Quantitative Struktur-Aktivitäts-Beziehungen für die Hemmung des autotrophen Wachstums synchroner *Chlorella vulgaris* — Kulturen durch substituierte Thioharnstoffe. *Biochem. Physiol. Pflanzen*, **130**, 353–69.

Scherrer, R. A. (1984) The treatment of ionizable compounds in quantitative structure–activity studies with special consideration to ion partitioning, in *Pesticide Synthesis Through Rational Approaches* (eds P. S. Magee, G. K. Kohn and J. J. Menn), American Chemical Society, Washington, DC.

Schild, R., Donkin, P., Cotsifis, P. A. and Donkin, M. E. (1993) A quantitative structure–activity relationship for the effects of alcohols on neutral red retention by the marine macroalga *Enteromorpha intestinalis*. *Chemosphere*, **27**, 1777–88.

Schmieder, P., Lothenbach, D., Tietge, J., Erickson, R. and Johnson, R. (1995) [³H]-2,3,7,8-TCDD uptake and elimination kinetics of medaka (*Oryzias latipes*). *Environ. Toxicol. Chem.*, **14**, 1735–43.

Schrap, S. M. and Opperhuizen, A. (1990) Relationship between bioavailability and hydrophobicity: reduction of the uptake of organic chemicals by fish due to the sorption on particles. *Environ. Toxicol. Chem.*, **9**, 715–24.

Schultz, T. W. (1987) The use of the ionization constant ($pK_a$) in selecting models of toxicity in phenols. *Ecotoxicol. Environ. Saf.*, **14**, 178–83.

Schultz, T. W. and Tichy, M. (1993) Structure–toxicity relationships for unsaturated alcohols to *Tetrahymena pyriformis*: $C_5$ and $C_6$ analogs and primary propargylic alcohols. *Bull. Environ. Contam. Toxicol.*, **51**, 681–8.

Schultz, T. W., Holcombe, G. W. and Phipps, G. L. (1986) Relationship of quantitative structure–activity to comparative toxicity of selected phenols in the *Pimephales promelas* and *Tetrahymena pyriformis* test systems. *Ecotoxicol. Environ. Saf.*, **12**, 146–53.

Schultz, T. W., Kier, L. B. and Hall, L. H. (1982) Structure–toxicity relationships of selected nitrogenous heterocyclic compounds. III. Relations using molecular connectivity. *Bull. Environ. Contam. Toxicol.*, **28**, 373–8.

Schultz, T. W., Lin, D. T., Wilke, T. S. and Arnold, L. M. (1990) Quantitative structure–activity relationships for the *Tetrahymena pyriformis* population growth endpoint: a mechanisms of action approach, in *Practical Applications of Quantitative Structure–Activity Relationships (QSAR) in Environmental Chemistry and Toxicology* (eds W. Karcher and J. Devillers), Kluwer, Dordrecht.

Schüürmann, G. (1990a) QSAR analysis of the acute fish toxicity of organic phosphorothionates using theoretically derived molecular descriptors. *Environ. Toxicol. Chem.*, **9**, 417–28.

Schüürmann, G. (1990b) Quantitative structure–property relationships for the polarizability, solvatochromic parameters and lipophilicity. *Quant. Struct.–Act. Relat.*, **9**, 326–33.

Schüürmann, G. (1995) Quantum chemical approach to estimate physicochemical compound properties: application to substituted benzenes. *Environ. Toxicol. Chem.*, **14**, 2067–76.

Schüürmann, G. and Klein, W. (1988) Advances in bioconcentration prediction. *Chemosphere*, **17**, 1551–74.

Schüürmann, G. and Müller, E. (1994) Back-propagation neural networks – recognition vs. prediction capability. *Environ. Toxicol. Chem.*, **13**, 743–7.

Schwarzenbach, R. P., Gschwend P. M. and Imboden, D. M. (1993) *Environmental Organic Chemistry*, Wiley, New York, NY.

Scow, K. M. (1990) Rate of biodegradation, in *Handbook of Chemical Property Estimation Methods* (eds W. J. Lyman, W. F. Reehl and D. H. Rosenblatt), American Chemical Society, Washington, DC.

Scribner, J. D. (1985) Chemical carcinogenesis, in *Environmental Pathology* (ed. N. K. Mottet), Oxford University Press, New York, NY.

Seydel, J. K. (1971) Physicochemical approaches to the rational development of new drugs. *Drug Design*, **1**, 343–79.

Seydel, J. K. and Schaper, K. J. (1979) *Chemische Struktur und Biologische Wirkung*, Verlag Chemie, Weinheim.

Seydel, J. K., Coats, E. A., Cordes, H. P. and Wiese, M. (1994) Drug membrane interaction and the importance for drug transport, distribution, accumulation, efficacy and resistance. *Arch. Pharm.*, **327**, 601–10.

Shaw, G. R. and Connell, D. W. (1982) Factors influencing concentrations of polychlorinated biphenyls in organisms from an estuarine ecosystem. *Aust. J. Mar. Freshw. Res.*, **33**, 1057–70.

Sheehan, P., Korte, F., Klein, W. and Bourdeau, P. (1985) *Appraisal of Tests to Predict the Environmental Behaviour of Chemicals*, Wiley, New York, NY.

Shigeoka, T., Sato, Y., Takeda, Y., Yoshida, K. and Yamauchi, F. (1988) Acute toxicity of chlorophenols to green algae, *Selenastrum capricornutum* and *Chlorella vulgaris*, and quantitative structure–activity relationships. *Environ. Toxicol. Chem.*, **7**, 847–54.

Shiu, W. Y., Doucette, W., Gobas, F. A. P. C., Andren, A. and Mackay, D. (1988) Physical-chemical properties of chlorinated dibenzo-*p*-dioxins. *Environ. Sci. Technol.*, **22**, 651–8; cited by Müller, M. and Klein, W. (1992) Comparative evaluation of methods predicting water solubility for organic compounds. *Chemosphere*, **25**, 769–82.

Sijm, D. T. H. M., Schipper, M. and Opperhuizen, A. (1993) Toxicokinetics of halogenated benzenes in fish: lethal body burden as a toxicological endpoint. *Environ. Toxicol. Chem.*, **12**, 1117–27.

Silber, J. R. and Loeb, L. A. (1985) The molecular basis of environmental mutagenesis, in *Environmental Pathology* (ed. N. K. Mottet), Oxford University Press, New York, NY.

Singer, G. M., Taylor, H. W. and Lijinsky, W. (1977) Liposolubility as an aspect of nitrosamine carcinogenicity: quantitative correlations and qualitative observations. *Chem. Biol. Interact.*, **19**, 133–42.

Slooff, W., Canton, J. H. and Hermens, J. L. M. (1983) Comparison of the suscepti-bility of 22 freshwater species to 15 chemical compounds. I. (Sub)acute toxicity tests. *Aquat. Toxicol.*, **4**, 113–28.

Slooff, W., Oers, J. A. M. van and Zwart, D. de (1986) Margins of uncertainty in eco-toxicological hazard assessment. *Environ. Toxicol. Chem.*, **5**, 841–52.

Smissaert, H. R. and Jansen, A. A. M. (1984) On the variation of toxic effects over species, its cause, and analysis by 'structure–selectivity relations'. *Ecotoxicol. Environ. Saf.*, **8**, 294–302.

Smith, A. D., Bharath, A., Mallard, C., Orr, D., McCarty, L. S. and Ozburn, G. W. (1990) Bioconcentration kinetics of some chlorinated benzenes and chlorinated phenols in american flagfish, *Jordanella floridae* (Goode and Bean). *Chemosphere*, **20**, 379–86.

Spacie, A. and Hamelink, J. L. (1982) Alternative models for describing the biocon-centration of organics in fish. *Environ. Toxicol. Chem.*, **1**, 309–20.

Spacie, A., Landrum, P. F. and Leversee, G. J. (1983) Uptake, depuration, and bio-transformation of anthracene and benzo($a$)pyrene in bluegill sunfish. *Ecotoxicol. Environ. Saf.*, **7**, 330–41.

Speece, R. E. (1990) Reply to 'Comment on predictions of aqueous solubility of organic chemicals based on molecular structure. 2. Application to PNAs, PCBs, PCDDs, etc.'. *Environ. Sci. Technol.*, **24**, 929–30.

Sprague, J. B. (1970) Measurement of pollutant toxicity to fish. II. Utilizing and applying bioassay results. *Water Res.*, **4**, 3–32.

Stephan, C. E. (1977) Methods for calculating an $LC_{50}$, in *Aquatic Toxicology and Hazard Evaluation* (eds F. L. Mayer and J. L. Hamelink), STP 634, ASTM, Philadelphia, PA.

Stewart, J. J. P. (1989) Optimization of parameters for semiempirical methods. I. Method. *J. Comp. Chem.*, **10**, 209–20.

Suedel, B. C., Boraczek, J. A., Peddicord, R. K., Clifford, P. A. and Dillon, T. M. (1994) Trophic transfer and biomagnification potential of contaminants in aquatic ecosystems. *Rev. Environ. Contam. Toxicol.*, **136**, 21–89.

Suter II, G. W. and Rosen, A. E. (1986) *Comparative Toxicology of Marine Fishes and Crustaceans*, NOAA Technical Memorandum NOS OMA 30, National Oceanic and Atmospheric Administration, Rockville, MD.

Suter II, G. W., Barnthouse, G. W. and O'Neill, R. V. (1987) Treatment of risk in environmental impact assessment. *Environ. Management*, **11**, 295–303.

Suter II, G. W., Vaughan, D. S. and Gardner, R. H. (1983) Risk assessment by analy-sis of extrapolation error, a demonstration for effects of pollutants on fish. *Environ. Toxicol. Chem.*, **2**, 369–78.

Sykes, P. (1982) *Reaktionsmechanismen der Organischen Chemie*, Verlag Chemie, Weinheim. German translation from: Sykes, P. (1981) *Guidebook to Mechanisms in Organic Chemistry*, Longman, London.

Szydlo, R. M., Ford, M. G., Greenwood, R. and Salt, D. W. (1985) The use of multi-variate data sets in the study of structure–activity relationships of synthetic pyrethroid insecticides: Part II. The relationships between pharmacokinetics and toxicity, in *QSAR and Strategies in the Design of Bioactive Compounds* (ed. J. K. Seydel), VCH, Weinheim.

Taft, R. W. (1956) Separation of polar, steric, and resonance effects in reactivity, in *Steric Effects in Organic Chemistry* (ed. M. S. Newman), Wiley, New York, NY.

Taft, R. W., Abraham, M. H., Doherty, R. M. and Kamlet, M. J. (1985) The molecular properties governing solubilities of organic nonelectrolytes in water. *Nature*, **313**, 384–6.

Takemoto, J., Yoshida, R., Sumida, S. and Kamoshita, K. (1985) Quantitative structure–activity relationships of herbicidal $N'$-substituted phenyl-$N$-methoxy-$N$-methyl-ureas. *Pestic. Biochem. Physiol.*, **23**, 341–8.

Tanii, H., Tsuji, H. and Hashimoto, K. (1986) Structure–toxicity relationship of monoketones. *Toxicol. Lett.*, **30**, 13–17.

Taylor, P. J. (1990) Hydrophobic properties of drugs, in *Quantitative Drug Design* (ed. C. A. Ramsden), Pergamon Press, Oxford.

Terada, H. (1990) Uncouplers of oxidative phosphorylation. *Environ. Health Persp.*, **87**, 213–18.

Tewari, Y. B., Miller, M. M. and Wasik, S. P. (1982). Calculation of aqueous solubility of organic compounds. *J. Res. NBS*, **87**, 155–8.

Thielemans, A. and Massart, D. L. (1985) The use of principal component analysis as a display method in the interpretation of analytical chemical, biochemical, environmental, and epidemiological data. *Chimia*, **39**, 236–42.

Thurston, R. V., Gilfoil, T. A., Meyn, E. L., Zajdel, R. K., Aoki, T. I. and Veith, G. D. (1985) Comparative toxicity of ten organic chemicals to ten common aquatic species. *Water Res.*, **19**, 1145–55.

Tichy, M. (1987) QSAR study on a distribution of xenobiotics in a body – biosolubility, in *QSAR in Drug Design and Toxicology* (eds D. Hadzi and B. Jerman-Blazic), Elsevier, Amsterdam.

Tichy, M. (1991) QSAR approach to estimation of the distribution of xenobiotics and the target organ in the body. *Drug Metab. Drug Interact.*, **9**, 191–200.

Tichy, M., Cikrt, M. and Roth, Z. (1994) QSAR analysis in mixture toxicity assessment. Abstracts, 6th QSAR Workshop, Belgirate, Italy.

Topliss, J. G. and Edwards, R. P. (1979) Chance factors in studies of quantitative structure–activity relationships. *J. Med. Chem.*, **22**, 1238–44.

Tosato, M. L., Vigano, L., Skagerberg, B. and Clementi, S. (1991) A new strategy for ranking chemical hazards. Framework and application. *Environ. Sci. Technol.*, **25**, 695–702.

Travis, C. C., Baes III, C. F., Barnthouse, L. W., Etnier, E. L., Holton, G. A., Murphy, B. D., Thompson, G. P., Suter II, G. W. and Watson, A. P. (1983) *Exposure Assessment Methodology and Reference Environments for Synfuel Analysis*, ORNL/TM-8672, Oak Ridge National Laboratory, Oak Ridge, TN.

Tulp, M. T. M. and Hutzinger, O. (1978) Some thoughts on aqueous solubilities and partition coefficients of PCB, and the mathematical correlation between bioaccumulation and physico-chemical properties. *Chemosphere*, **7**, 849–60.

Turner, L., Choplin, F., Dugard, P., Hermens, J., Jaeckh, R., Marsmann, M. and Roberts, D. (1987) Structure–activity relationships in toxicology and ecotoxicology: an assessment. *Toxic. in Vitro*, **1**, 143–71.

Umweltbundesamt (1990) *Grundzüge der Bewertung von neuen Stoffen nach dem ChemG*, Umweltbundesamt, Berlin.

Unger, S. H. and Chiang, G. H. (1991) Octanol-physiological buffer distribution coefficients of lipophilic amines by reversed-phase HPLC and their correlation with biological activity. *J. Med. Chem.*, **24**, 262–70.

Unger, S. H. and Feuerman, T. F. (1979) Octanol–aqueous partition, distribution and ionization coefficients of lipophilic acids and their anions by rp-HPLC. *J. Chromatogr.*, **176**, 426–9.

Unger, S. H., Cook, J. R. and Hollenberg, J. S. (1978) Simple procedure for determining octanol–aqueous partition, distribution and ionization coefficients by rp-HPLC. *J. Pharm. Sci.*, **67**, 1364–7.

Urano, K. and Kato, Z. (1986) A method to classify biodegradability of organic compounds. *J. Hazard. Materials*, **13**, 135–45.

Urban, D. J. and Cook, N. J. (1986) *Hazard Evaluation Division Standard Evaluation Procedure Ecological Risk Assessment*, EPA 540/9-85-001, Environmental Protection Agency, Washington, DC.

Vaishnav, D. D., Boethling, R. S. and Babeu, L. (1987) Quantitative structure–biodegradability relationships for alcohols, ketones and alicyclic compounds. *Chemosphere*, **16**, 695–703.

Valvani, S. C. and Yalkowski, S. H. (1980) Solubility and partitioning in drug design, in *Physical Chemical Properties of Drugs* (eds S. H. Yalkowski, A. A. Sinkula and S. C. Valvani), Marcel Dekker, New York, NY.

Valvani, S. C., Yalkowsky, S. H. and Roseman, T. J. (1981) Solubility and partitioning IV: aqueous solubility and octanol–water partition coefficients of liquid non-electrolytes. *J. Pharm. Sci.*, **70**, 502–7.

Veith, G. D. and Broderius, S. J. (1987) Structure–toxicity relationships for industrial chemicals causing type (II) narcosis syndrome, in *QSAR in Environmental Toxicology – II* (ed. K. L. E. Kaiser), D. Reidel, Dordrecht.

Veith, G. D. and Broderius, S. J. (1990) Rules for distinguishing toxicants that cause type I and type II narcosis syndromes. *Environ. Health Persp.*, **87**, 207–11.

Veith, G. D. and Kosian, P. (1983) Estimating bioconcentration potential from otanol/water partition coefficients, in *Physical Behaviour of PCBs in the Great Lakes* (eds D. Mackay, S. Paterson, S. J. Eisenreich and M. S. Simmons), Ann Arbor Science, Ann Arbor, MI.

Veith, G. D. and Mekenyan, O. G. (1993) A QSAR approach for estimating the aquatic toxicity of soft electrophiles. *Quant. Struct.–Act. Relat.*, **12**, 349–56.

Veith, G. D., Call, D. J. and Brooke, L. T. (1983) Structure–toxicity relationships for the fathead minnow, *Pimephales promelas*: narcotic industrial chemicals. *Can. J. Fish. Aquat. Sci.*, **40**, 743–8.

Veith, G. D., Defoe, D. L. and Bergsted, B. V. (1979) Measuring and estimating the bioconcentration factor of chemicals in fish. *J. Fish. Res. Board Can.*, **36**, 1040–8.

Veith, G. D., Lipnick, R. L. and Russom, C. L. (1989) The toxicity of acetylenic alcohols to the fathead minnow, *Pimephales promelas*: narcosis and proelectrophile activation. *Xenobiotica*, **11**, 555–65.

Veith, G. D., Macek, K. J., Petrocelli, S. R. and Carrol, J. (1980) An evaluation of using partition coefficients and water solubility to estimate bioconcentration factors for organic chemicals in fish, in *Aquatic Toxicology* (eds J. G. Eaton, P. R. Parrish and A. C. Hendricks), STP 707, ASTM, Philadelphia, PA.

Veith, G. D., Greenwood, B., Hunter, R. S., Niemi, G. J. and Regal, R. R. (1988) On the intrinsic dimensionality of chemical structure space. *Chemosphere*, **17**, 1617–30.

Verbruggen, E., Loon, W. van and Hermens, J. (1994). A Parameter for the hydrophobicity of mixtures. Abstracts, 6th QSAR Workshop, Belgirate, Italy.

Verhaar, H. J. M., Busser, F. J. M. and Hermens, J. L. M. (1995) A surrogate parameter for the baseline toxicity content of contaminated water. Simulating the bioconcentration of unknown mixtures and counting their molecules. *Environ. Sci. Technol.*, **29**, 726–34.

Verhaar, H. J. M., Leeuwen, C. J. van and Hermens, J. L. M. (1992) Classifying environmental pollutants. 1: Structure–activity relationships for prediction of aquatic toxicity. *Chemosphere*, **25**, 471–91.

Verhaar, H. J. M., Ramos, E. U. and Hermens, J. L. M. (1995) Classifying environmental pollutants 2. Separation of class 1 (baseline toxicity) and class 2 (polar narcosis) type compounds based on chemical descriptors, in *Predictive Methods in Aquatic Toxicology* (by H. J. M. Verhaar). Thesis, Utrecht University, Utrecht.

Verhaar, H. J. M., Leeuwen, C. J. van, Bol, J. and Hermens, J. L. M. (1994) Application of QSARs in risk management of existing chemicals. *SAR QSAR Environ. Res.*, **2**, 39–58.

Verloop, A., Hoogenstraaten, W. and Tipker, J. (1976) Development and application of new steric substituent parameters in drug design. *Drug Res.*, **7**, 165–207.

Walker, C. H. (1987) Kinetic models for predicting bioaccumulation of pollutants in ecosystems. *Environ. Pollut.*, **44**, 227–40.

Wallace, K. B. and Niemi, G. J. (1988) Structure–activity relationships of species-selectivity in acute chemical toxicity between fish and rodents. *Environ. Toxicol. Chem.*, **7**, 201–12.

Waterbeemd, H. van de and Kansy, M. (1992) Hydrogen-bonding capacity and brain penetration. *Chimia*, **46**, 299–303.

Weber, E. J. and Wolfe, N. L. (1987) Kinetic studies of the reduction of aromatic azo compounds in anaerobic sediment/water systems. *Environ. Toxicol. Chem.*, **6**, 911–19.

Weinbach, E. C. and Garbus, J. (1969) Mechanism of action of reagents that uncouple oxidative phosphorylation. *Nature*, **221**, 1016–18.

Weisberg, S. (1985) *Applied Linear Regression*, Wiley, New York, NY.

Wenzel, A., Nendza, M., Hartmann, P. and Kanne, R. (1997) Test battery for the assessment of aquatic toxicity. *Chemosphere* (in press).

Wezel, A. P. van, Vries, D. A. M. de, Kostense, S., Sijm, D. T. H. M. and Opperhuizen, A. (1995) Intraspecies variation in lethal body burdens of narcotic compounds. *Aquat. Toxicol.*, **33**, 325–42.

Wilson, L. Y. and Famini, G. R. (1991) Using theoretical descriptors in quantitative structure–activity relationships: some toxicological indices. *J. Med. Chem.*, **34**, 1668–74.

Wishnok, J. S., Archer, M. C., Edelmann, A. S. and Rand, W. M. (1978) Nitrosamine carcinogenicity: a quantitative Hansch–Taft structure–activity relationship. *Chem. Biol. Interact.*, **20**, 43–54.

Wold, S., Albano, C., Dunn III, W. J., Esbensen, K., Hellberg, S., Johansson, E. and Sjöström, M. (1983) Pattern recognition: finding and using regularities in multivariate data, in *Food Research and Data Analysis* (eds H. Martens and H. Russwurm), Applied Science, London.

Wold, S., Albano, C., Dunn III, W. J., Esbensen, K., Hellberg, S., Johansson, E., Lindberg, W. and Sjöström, M. (1984) Modelling data tables by principal components and PLS: class patterns and quantitative predictive patterns. *Analusis*, **12**, 477–85.

Wolf, W. de, Canton, J. H., Deneer, J. W., Wegman, R. C. C. and Hermens, J. L. M. (1988) Quantitative structure–activity relationships and mixture toxicity studies of alcohols and chlorohydrocarbons: reproducibility of effects on growth and reproduction of *Daphnia magna*. *Aquat. Toxicol.*, **12**, 39–49.

Wolf, W. de, Bruijn, J. de, Seinen, W. and Hermens, J. L. M. (1992a) Influence of biotransformation on the relationship between bioconcentration factors and octanol–water partition coefficients. *Environ. Sci. Technol.*, **26**, 1197–201.

Wolf, W. de, Seinen, W., Opperhuizen, A. and Hermens, J. L. M. (1992b) Bioconcentration and lethal body burden of 2,3,4,5-tetrachloroaniline in guppy, *Poecilia reticulata*. *Chemosphere*, **25**, 853–63.

Wolfe, N. L., Steen, W. C. and Burns, L. A. (1980) Phthalate ester hydrolysis: linear free energy relationships. *Chemosphere*, **9**, 403–8.

Wolfe, N. L., Zepp, R. G. and Paris, D. F. (1978) Use of structure–reactivity relationships to estimate hydrolytic persistence of carbamate pesticides. *Water Res.*, **12**, 561–3.

Wong, P. T. S., Chau, Y. K., Rhamey, J. S. and Docker, M. (1984) Relationship between water solubility of chlorobenzenes and their effects on a freshwater green alga. *Chemosphere*, **13**, 991–6.

Yalkowsky, S. H. and Mishra, D. S. (1991) Vapor pressure estimations for organic compounds. *Sci. Total Environ.*, **109/110**, 243–50.

Yalkowsky, S. H., Valvani, S. C. and Mackay, D. (1983) Estimation of aqueous solubility of some aromatic compounds. *Resid. Rev.*, **85**, 43–55.

Yalkowsky, S. H., Carpenter, O. S., Flynn, G. L. and Slunick, T. G. (1973) Drug absorption kinetics in goldfish. *J. Pharm. Sci.*, **62**, 1949–54.

Yoshioka, Y., Ose, Y. and Sato, T. (1986) Correlation of the five test methods to assess chemical toxicity and relation to physical properties. *Ecotoxicol. Environ. Saf.*, **12**, 15–21.

Yuan, M. and Jurs, P. C. (1980) Computer-assisted structure–activity studies of chemical carcinogens: a polycyclic aromatic hydrocarbon data set. *Appl. Pharmacol.*, **52**, 294–312.

Zandt, P. T. J. van der and Leeuwen, C. J. van (1992) *A Proposal for Priority Setting of Existing Chemical Substances*, Directorate General for Environmental Protection, The Hague.

Zaroogian, G., Heltshe, J. F. and Johnson, M. (1985) Estimation of toxicity to marine species with structure–activity models developed to estimate toxicity to freshwater fish. *Aquat. Toxicol.*, **6**, 251–70.

Zeeman, M., Auer, C. M., Clements, R. G., Nabholz, J. V. and Boethling, R. S. (1995) U.S. EPA regulatory perspectives on the use of QSAR for new and existing chemical evaluations. *SAR QSAR Environ. Res.*, **3**, 179–201.

Zitko, V. (1980) Metabolism and distribution by aquatic animals, in *Handbook of Environmental Chemistry* (ed. O. Hutzinger), Springer, Berlin.

# *Index*

Page numbers in *italics* refer to tables; page numbers in **bold** refer to illustrations.

GPSR Compliance
The European Union's (EU) General Product Safety Regulation (GPSR) is a set
of rules that requires consumer products to be safe and our obligations to
ensure this.

If you have any concerns about our products, you can contact us on

ProductSafety@springernature.com

In case Publisher is established outside the EU, the EU authorized
representative is:

Springer Nature Customer Service Center GmbH
Europaplatz 3
69115 Heidelberg, Germany


www.ingramcontent.com/pod-product-compliance
Ingram Content Group UK Ltd.
Pitfield, Milton Keynes, MK11 3LW, UK
UKHW020918130726
13719UKWH00012B/170